METHODS IN MOLECULAR BIOLOGY™

Prenatal Diagnosis

Edited by

Sinuhe Hahn

Laboratory for Prenatal Medicine and Gynecological Oncology,
Department of Biomedicine,
University Womens's Hospital, Basel, Switzerland

and

Laird G. Jackson

Department of Obstetrics and Gynecology,
Drexel University College of Medicine, Philadelphia, Pennsylvania

Editors
Sinuhe Hahn
Laboratory for Prenatal Medicine
and Gynecological Oncology,
Department of Biomedicine
University Womens's Hospital
Basel, Switzerland
shahn@uhbs.ch

Laird G. Jackson
Department of Obstetrics and Gynecology
Drexel University College of Medicine
Philadelphia, PA
laird.jackson@drexelmed.edu

Series Editor
John M. Walker
School of Life Sciences
University of Hertfordshire
Hatfield, Hertfordshire, UK

ISBN: 978-1-58829-803-4 e-ISBN: 978-1-59745-066-9
ISSN: 1064-3745 e-ISSN: 1940-6029

Library of Congress Control Number: 2007940754

Cover illustration: Figure 1, Chapter 3, "Rapid Prenatal Aneuploidy Screening by Fluorescene In Situ Hybridization (FISH)," by Anja Weise and Thomas Liehr.

Printed on acid-free paper

9 8 7 6 5 4 3 2 1

springer.com

Prenatal Diagnosis

METHODS IN MOLECULAR BIOLOGY™

***John M. Walker**, Series Editor*

465. **Mycobacteria**, *Second Edition*, edited by *Tanya Parish and Amanda Claire Brown, 2008*
460. **Essential Concepts in Toxicogenomics**, edited by *Donna L. Mendrick and William B. Mattes, 2008*
459. **Prion Protein Protocols**, edited by *Andrew F. Hill, 2008*
458. **Artificial Neural Networks:** *Methods and Applications*, edited by *David S. Livingstone, 2008*
457. **Membrane Trafficking**, edited by *Ales Vancura, 2008*
456. **Adipose Tissue Protocols**, *Second Edition*, edited by *Kaiping Yang, 2008*
455. **Osteoporosis**, edited by *Jennifer J. Westendorf, 2008*
454. **SARS- and Other Coronaviruses:** *Laboratory Protocols*, edited by *Dave Cavanagh, 2008*
453. **Bioinformatics, Volume 2:** *Structure, Function, and Applications*, edited by *Jonathan M. Keith, 2008*
452. **Bioinformatics, Volume 1:** *Data, Sequence Analysis, and Evolution*, edited by *Jonathan M. Keith, 2008*
451. **Plant Virology Protocols:** *From Viral Sequence to Protein Function*, edited by *Gary Foster, Elisabeth Johansen, Yiguo Hong, and Peter Nagy, 2008*
450. **Germline Stem Cells**, edited by *Steven X. Hou and Shree Ram Singh, 2008*
449. **Mesenchymal Stem Cells:** *Methods and Protocols*, edited by *Darwin J. Prockop, Douglas G. Phinney, and Bruce A. Brunnell, 2008*
448. **Pharmacogenomics in Drug Discovery and Development**, edited by *Qing Yan, 2008*
447. **Alcohol:** *Methods and Protocols*, edited by *Laura E. Nagy, 2008*
446. **Post-translational Modification of Proteins:** *Tools for Functional Proteomics, Second Edition*, edited by *Christoph Kannicht, 2008*
445. **Autophagosome and Phagosome**, edited by *Vojo Deretic, 2008*
444. **Prenatal Diagnosis**, edited by *Sinuhe Hahn and Laird G. Jackson, 2008*
443. **Molecular Modeling of Proteins**, edited by *Andreas Kukol, 2008*
439. **Genomics Protocols:** *Second Edition*, edited by *Mike Starkey and Ramnanth Elaswarapu, 2008*
438. **Neural Stem Cells:** *Methods and Protocols, Second Edition*, edited by *Leslie P. Weiner, 2008*
437. **Drug Delivery Systems**, edited by *Kewal K. Jain, 2008*
436. **Avian Influenza Virus**, edited by *Erica Spackman, 2008*
435. **Chromosomal Mutagenesis**, edited by *Greg Davis and Kevin J. Kayser, 2008*
434. **Gene Therapy Protocols: Volume 2:** *Design and Characterization of Gene Transfer Vectors*, edited by *Joseph M. LeDoux, 2008*
433. **Gene Therapy Protocols: Volume 1:** *Production and In Vivo Applications of Gene Transfer Vectors*, edited by *Joseph M. LeDoux, 2007*
432. **Organelle Proteomics**, edited by *Delphine Pflieger and Jean Rossier, 2008*
431. **Bacterial Pathogenesis:** *Methods and Protocols*, edited by *Frank DeLeo and Michael Otto, 2008*
430. **Hematopoietic Stem Cell Protocols**, edited by *Kevin D. Bunting, 2008*
429. **Molecular Beacons:** *Signalling Nucleic Acid Probes, Methods and Protocols*, edited by *Andreas Marx and Oliver Seitz, 2008*
428. **Clinical Proteomics:** *Methods and Protocols*, edited by *Antonio Vlahou, 2008*
427. **Plant Embryogenesis**, edited by *Maria Fernanda Suarez and Peter Bozhkov, 2008*
426. **Structural Proteomics:** *High-Throughput Methods*, edited by *Bostjan Kobe, Mitchell Guss, and Huber Thomas, 2008*
425. **2D PAGE: Volume 2:** *Applications and Protocols*, edited by *Anton Posch, 2008*
424. **2D PAGE: Volume 1:** *Sample Preparation and Pre-Fractionation*, edited by *Anton Posch, 2008*
423. **Electroporation Protocols**, edited by *Shulin Li, 2008*
422. **Phylogenomics**, edited by *William J. Murphy, 2008*
421. **Affinity Chromatography:** *Methods and Protocols, Second Edition*, edited by *Michael Zachariou, 2008*
420. **Drosophila:** *Methods and Protocols*, edited by *Christian Dahmann, 2008*
419. **Post-Transcriptional Gene Regulation**, edited by *Jeffrey Wilusz, 2008*
418. **Avidin-Biotin Interactions:** *Methods and Applications*, edited by *Robert J. McMahon, 2008*
417. **Tissue Engineering**, *Second Edition*, edited by *Hannsjörg Hauser and Martin Fussenegger, 2007*
416. **Gene Essentiality:** *Protocols and Bioinformatics*, edited by *Svetlana Gerdes and Andrei L. Osterman, 2008*
415. **Innate Immunity**, edited by *Jonathan Ewbank and Eric Vivier, 2007*
414. **Apoptosis in Cancer:** *Methods and Protocols*, edited by *Gil Mor and Ayesha Alvero, 2008*
413. **Protein Structure Prediction**, *Second Edition*, edited by *Mohammed Zaki and Chris Bystroff, 2008*
412. **Neutrophil Methods and Protocols**, edited by *Mark T. Quinn, Frank R. DeLeo, and Gary M. Bokoch, 2007*
411. **Reporter Genes for Mammalian Systems**, edited by *Don Anson, 2007*
410. **Environmental Genomics**, edited by *Cristofre C. Martin, 2007*
409. **Immunoinformatics:** *Predicting Immunogenicity In Silico*, edited by *Darren R. Flower, 2007*
408. **Gene Function Analysis**, edited by *Michael Ochs, 2007*
407. **Stem Cell Assays**, edited by *Vemuri C. Mohan, 2007*
406. **Plant Bioinformatics:** *Methods and Protocols*, edited by *David Edwards, 2007*
405. **Telomerase Inhibition:** *Strategies and Protocols*, edited by *Lucy Andrews and Trygve O. Tollefsbol, 2007*

Preface

Molecular biology has transformed prenatal diagnosis because it permits an accurate diagnosis to be made from very small quantities of fetal material, even single cells. Although the latter permits the analysis of preimplantation embryos and, perhaps in the future, the analysis of fetal cells enriched from maternal blood, a major current focus is to facilitate rapid, cost-effective diagnoses. In this manner, the use of fluorescence in situ hybridization (FISH) or polymerase chain reaction (PCR)-based approaches on uncultured amniocytes or chorionic villi already permits a rapid assessment of the most common fetal aneuploidies (chromosomes 13, 18, 21, X, and Y) to be obtained within 24 h, thereby obviating the need for a 2-week culture period previously required for a traditional karyotype. Although the accuracy of karyotypic analysis is greatly enhanced by methodologies such as spectral karyotyping (SKY) or comparative genomic hybridization (CGH), the advent of high-density nucleotide arrays (chips) facilitates rapid assessment of the fetal genotype for a large number of mutations of frequent Mendelian disorders, e.g., cystic fibrosis, the hemoglobinopathies, and Tay-Sachs syndrome.

A further focus is the development of noninvasive, and hence, risk-free alternatives for prenatal diagnosis that no longer rely on invasive procedures such as amniocentesis or chorionic villus sampling. The most successful of these approaches is the analysis of placentally derived cell-free DNA in maternal blood. This approach has been found to be very reliable for the assessment of fetal loci absent from the maternal genome, and it is consequently already being used clinically for the determination of the fetal Rhesus D status and sex in pregnancies at risk for X-linked disorders.

As such, cutting-edge applications for the rapid assessment of fetal aneuploidies and Mendelian disorders on fetal material gained by invasive approaches and procedures being validated for routine clinical analysis of cell-free fetal DNA are covered in *Prenatal Diagnosis*. In addition, several chapters address developments that are on the brink of clinical applications.

We thank all the authors for their excellent contributions and trust that *Prenatal Diagnosis* will be useful to molecular biologists interested in clinical applications and those involved in basic research in prenatal medicine.

Sinuhe Hahn
Basel, Switzerland
and
Laird G. Jackson
Philadelphia, Pennsylvania

Contents

Contributors

NEIL D. AVENT • *Centre for Research in Biomedicine, University of the West of England, Bristol, UK*
DIANA W. BIANCHI • *Division of Genetics, Departments of Pediatrics, Obstetrics, and Gynecology, Tufts-New England Medical Center and Floating Hospital for Children, Tufts University School of Medicine, Boston, MA*
MARINUS A. BLANKENSTEIN • *Department of Clinical Chemistry, VU University Medical Center, Amsterdam, The Netherlands*
LAURA CREMONESI • *Genomic Unit for Diagnosis of Human Pathologies, San Raffaele Scientific Institute, Milan, Italy; and Diagnostica e Ricerca San Raffaele S.p.A, Milan, Italy*
CHUNMING DING • *Centre for Emerging Infectious Diseases, Chinese University of Hong Kong, Hong Kong, China*
RÉGEN DROUIN • *Science of Genetics, Department of Pediatrics, Faculty of Medicine and Health Sciences, Université de Sherbrooke, Quebec, Canada*
LECH DUDAREWICZ • *Department of Genetics, Polish Mother's Memorial Hospital, Lodz, Poland*
PAUL EIJK • *Department of Pathology, VU University Medical Center, Amsterdam, The Netherlands*
MAURIZIO FERRARI • *Genomic Unit for Diagnosis of Human Pathologies, San Raffaele Scientific Institute, Milan, Italy; Diagnostica e Ricerca San Raffaele S.p.A.; and Università Vita-Salute San Raffaele, Milan, Italy*
BARBARA FOGLIENI • *Genomic Unit for Diagnosis of Human Pathologies, San Raffaele Scientific Institute, Milan, Italy*
MACOURA GADJI • *Science of Genetics, Department of Pediatrics, Faculty of Medicine and Health Sciences, Université de Sherbrooke, Quebec, Canada*
SILVIA GALBIATI • *Genomic Unit for Diagnosis of Human Pathologies, San Raffaele Scientific Institute, Milan, Italy*
ROBERT-JAN GALJAARD • *Erasmus University of Rotterdam, Rotterdam, The Netherlands*

ATTIE T. J. J. GO • *Department of Obstetrics/Gynecology, VU University Medical Center, Amsterdam, The Netherlands*
ESTHER GUETTA • *Danek Gertner Institute of Human Genetics, Chaim Sheba Medical Center, Tel-Hashomer, Israel*
SINUHE HAHN • *Laboratory of Prenatal Medicine and Gynecological Oncology, Department of Biomedicine, University Women's Hospital, Basel, Switzerland*
KLAUS-HINRICH HEERMANN • *Department of Virology, University of Göttingen, Göttingen, Germany*
WOLFGANG HOLZGREVE • *Laboratory of Prenatal Medicine, University Women's Hospital/Department of Research, Basel, Switzerland*
DOROTHY J. HUANG • *Laboratory of Prenatal Medicine, University Women's Hospital/Department of Research, Basel, Switzerland*
LAIRD G. JACKSON • *Department of Obstetrics and Gynecology, Drexel University College of Medicine, Philadelphia, PA*
KIRBY L. JOHNSON • *Division of Genetics, Departments of Pediatrics, Obstetrics, and Gynecology, Tufts-New England Medical Center and Floating Hospital for Children, Tufts University School of Medicine, Boston, MA*
EMMANUEL KANAVAKIS • *Department of Medical Genetics, National and Kapodistiran University of Athens, St. Sophia's Children's Hospital, Athens, Greece*
CATHERINE D. KASHORK • *Signature Genomics Laboratories, LLC, Spokane, WA*
KADA KRABCHI • *Science of Genetics, Department of Pediatrics, Faculty of Medicine and Health Sciences, Université de Sherbrooke, Quebec, Canada*
OLAV LAPAIRE • *Division of Genetics, Departments of Pediatrics, Obstetrics, and Gynecology, Tufts-New England Medical Center and Floating Hospital for Children, Tufts University School of Medicine, Boston, MA; University Women's Hospital/Department of Obstetrics and Gynecology, Basel, Switzerland*
TOBIAS J. LEGLER • *Department of Transfusion Medicine, University of Göttingen, Göttingen, Germany*
YING LI • *Laboratory of Prenatal Medicine, University Women's Hospital/Department of Research, Basel, Switzerland*
THOMAS LIEHR • *Institute of Human Genetics and Anthropology, Friedrich-Schiller-University Jena, Jena, Germany*
ZHONG LIU • *Hefei Red Cross Blood Center, Hefei, People's Republic of China*

Y. M. DENNIS LO • *Department of Chemical Pathology, The Chinese University of Hong Kong, Prince of Wales Hospital, Hong Kong, China*
DEBORAH G. MADDOCKS • *Centre for Research in Biomedicine, University of the West of England, Bristol, UK*
KATHY MANN • *Cytogenetics Department, Guy's Hospital, London, UK*
SUSANNE MERGENTHALER-GATFIELD • *Laboratory of Prenatal Medicine, University Women's Hospital/Department of Research, Basel, Switzerland*
ANDRES METSPALU • *Department of Biotechnology, Institute of Molecular and Cell Biology, University of Tartu, Tartu, Estonia; The Estonian Biocentre, MDC of United Laboratories of the Tartu University Hospital, Tartu, Estonia; Estonian Genome Project Foundation, Tartu, Estonia*
MONIQUE A. M. MULDERS • *Department of Clinical Chemistry, VU University Medical Center, Amsterdam, The Netherlands*
MATTHEW R. NELSON • *Sequenom, Inc., San Diego, CA*
TAKASHI OKAI • *Department of Obestetrics and Gynecology, Showa University School of Medicine, Tokyo, Japan*
CEES B. M. OUDEJANS • *Department of Clinical Chemistry, VU University Medical Center, Amsterdam, The Netherlands*
LEAH PELEG • *Danek Gertner Institute of Human Genetics, Chaim Sheba Medical Center, Tel-Hashomer, Israel*
ERWIN PETEK • *Department of Human Biology and Medical Genetics, Medical University of Graz, Graz, Austria*
BARBARA PERTL • *Department of Obstetrics and Gynecology, Medical University of Graz, Graz, Austria*
JANNE PULLAT • *Department of Biotechnology, Institute of Molecular and Cell Biology, University of Tartu, Tartu, Estonia; The Estonian Biocentre, Tartu, Estonia*
YUDITIYA PURWOSUNU • *Department of Obstetrics and Gynecology, Showa University School of Medicine, Tokyo, Japan*
CHINNAPAPAGARI SATHEESH KUMAR REDDY • *Laboratory of Prenatal Medicine, University Women's Hospital/Department of Research, Basel, Switzerland*
JAN SCHOUTEN • *MRC-Holland bv, Amsterdam, The Netherlands*
AKIHIKO SEKIZAWA • *Department of Obstetrics and Gynecology, Showa University School of Medicine, Tokyo, Japan*
LISA G. SHAFFER • *Signature Genomics Laboratories, LLC, Spokane, WA; Health Research and Education Center, Washington State University, Spokane, WA*
AICHA AIT SOUSSAN • *Sanquin Research, Amsterdam, The Netherlands*
AARON THEISEN • *Signature Genomics Laboratories, LLC, Spokane, WA*

JOANNE TRAEGER-SYNODINOS • *Department of Medical Genetics, National and Kapodistiran University of Athens, St. Sophia's Children's Hospital, Athens, Greece*
NANCY B. Y. TSUI • *Department of Chemical Pathology, The Chinese University of Hong Kong, Prince of Wales Hospital, Hong Kong, China*
C. ELLEN VAN DER SCHOOT • *Sanquin Research, Amsterdam, The Netherlands*
MARIE VAN DIJK • *Department of Clinical Chemistry, VU University Medical Center, Amsterdam, The Netherlands*
JOHN M. G. VAN VUGT • *Department of Obstetrics/Gynecology, VU University Medical Center, Amsterdam, The Netherlands*
ALLERDIEN VISSER • *Department of Clinical Chemistry, VU University Medical Center, Amsterdam, The Netherlands*
CHRISTINA VRETTOU • *Department of Medical Genetics, National and Kapodistiran University of Athens, St. Sophia's Children's Hospital, Athens, Greece; Athens University Research Institute for Prevention and Treatment of Genetic and Malignat Diseases of Childhood, St. Sophia's Children's Hospital, Athens, Greece*
ANJA WEISE • *Institute of Human Genetics and Anthropology, Friedrich-Schiller-University Jena, Jena, Germany*
JU YAN • *Science of Genetics, Department of Paediatrics, Faculty of Medicine and Health Sciences, Université de Sherbrooke, Quebec, Canada*
BAUKE YLSTRA • *Department of Pathology, VU University Medical Center, Amsterdam, The Netherlands*
BERNHARD G. ZIMMERMANN • *Centre for Research in Biomedicine, University of the West of England, Bristol, UK; DRI (Dental Research Institute), Center for Health Sciences, University of California, Los Angeles, CA*
XIAO YAN ZHONG • *Laboratory of Prenatal Medicine, University Women's Hospital/Department of Research, Basel, Switzerland*

I

Invasive Approaches

1

Spectral Karyotyping (SKY): *Applications in Prenatal Diagnostics*

Susanne Mergenthaler-Gatfield, Wolfgang Holzgreve, and Sinuhe Hahn

Summary

The method of spectral karyotyping (SKY) is based on a combination of the technologies of charge-coupled device imaging and spectrometry. The engineering feasibility has been realized in the SpectraCube system from Applied Spectral Imaging Inc., and it allows the simultaneous identification of all 24 human chromosomes. This is performed by characterizing the spectral signature of every image pixel in relation to a fluorochrome combinatorial library translating the image and spectral information into chromosome classification. Applications for SKY include pre- and postnatal characterization of certain numerical and structural rearrangements and complex karyotypes and highly informative analysis of sample materials with only single or few cells available for investigation.

Key Words: Spectral karyotyping (SKY); multicolor fluorescence in situ hybridization (M-FISH); spectroscopy; Fourier transformation; spectral signature; single cell analysis; pre- and postnatal diagnosis.

1. Introduction

For cytogenetical analysis in prenatal diagnosis, the routine methodological spectrum includes Giemsa-banding (G-banding) and fluorescence in situ hybridization (FISH). G-banding produces a black-white banding pattern that is individual to every chromosome, and FISH takes advantage of fluorochrome-labeled probes to identify known sequences of interest. Giemsa staining as the basic method allows the detection of numeric aberrations and also definite structural rearrangements such as deletions, insertions, duplications, inversions, and translocations.

From: *Methods in Molecular Biology, vol. 444: Prenatal Diagnosis*
Edited by: S. Hahn and L. G. Jackson © Humana Press, Totowa, NJ

However, it depends on the resolution of the chromosomes and on the experience of the operator which type and size of aberrations can be detected. Application of FISH allows refining of the karyotype analysis, in that it can be applied to identify cryptic rearrangements such as microdeletions and insertions, and it can confirm translocation breakpoints. FISH probe categories include chromosome enumeration probes (CEP) hybridizing to centromeres, satellite DNA, or telomeric DNA; single locus probes complementary to unique genome sequences (LSI), and whole chromosome painting probes staining whole chromosomes (WCP). These painting probes were generated for the first time in the late 1980s and early 1990s for individual chromosomes based on unique DNA sequences ***(1,2)***. Pinkel et al. ***(3)*** established a chromosome 4-specific painting probe by combining in situ hybridization of 120 fluorescent DNA probes along the entire length of this chromosome.

In 1996, it became possible to paint all 24 chromosomes in the human genome simultaneously in unique and distinct fluorescent colors per chromosome ***(4,5)***. This technique was named spectral karyotyping (SKY).

For many questions in prenatal diagnostics, G-banding or FISH using distinct probes are sufficient. However, a range of certain cytogenetic aberrations cannot be sufficiently characterized by Giemsa or by FISH. In addition sample materials with only single or few cells available for analysis require multitasking investigational methods. Then, the simultaneous interrogation of many or all chromosomes is necessary, which is directly correlated to the need of at least 24 distinct different fluorochrome dyes.

Applied Spectral Imaging designed such a set of FISH-probes representing all 24 chromosomes (SKY Paint) together with the appropriate image capturing and analysis tools (SpectraCube, Spectral Imaging, and SKYView).

1.1. Technical Details

Although a broad spectral range of different fluorochrome dyes is available, the human eye is not capable to precisely resolve between small wavelength differences. Therefore, a technical translation system has been constructed by Applied Spectral Imaging Inc. that is the SD200SpectraCube Spectral Imaging System (ASI Ltd., Migdal Haemek, Israel).

The SpectraCube system is based on spectral imaging that combines the technologies of charge-coupled device (CCD)-imaging and spectrometry ***(6)***. CCD imaging produces a finely detailed monochrome image of an object, whereas spectrometry measures the spectrum of each pixel in the object and then displays each spectrum as a separate graph. Thus, the image itself contains the spectral information for each point in the image. The SpectraCube system is attached to a normal fluorescence microscope and consists in an optical head (Sagnac interferometer) coupled to a CCD-camera (Hammamatsu, Bridgewater, NJ). The image data acquired by the system are stored in a computer and analyzed by the system software.

A xenon lamp is used as light source because it contains many different wavelengths; thus, it is able to excite all fluorochrome dyes present in a sample. However, before the light impinges on the sample is passes a multiple band pass filter set custom-designed (SKY-1, Chroma Technology Corp., Brattleboro, VT) for the excitation of the SKY kit-fluorochromes, which provides broad emission bands as well (a fractional spectral measuring of approx. 450 to 850 nm). As the light is reflected from the selected area in the sample, now with a longer wavelength than before impinging on the sample, it is split into two beams by a beam splitter in the optical head. A set of mirrors directs these two beams down two paths of different lengths. At the end of the paths, the two beams merge and they are thus superimposed. The intensity of the merged beam is then focused to and measured by the CCD-camera. Such a measurement is called a frame. The intensity is not measured only once, but 80–130 of these interferometric frames are acquired. The reason is that this total intensity of the two superimposed beams is taken as a function of the difference in the distances between the two paths the split beams pass through. As the beam-splitter-reflector unit is rotating during the image acquisition, many different pathlength combinations of the two split beams are traversed, and for each of these positions or optical path differences (OPD), respectively, representing a specific path difference with changing light intensity, such a frame is captured. For each pixel in the image, this process is performed simultaneously. The stacks of the interferogramic frames are summarized in an interferogram, which is a representation of the light intensity as it changes with each changing OPD. This information is then Fourier transformed, and the data are sorted in the software and translated into the spectral information of each pixel (*see* **Fig. 1**). After the spectrum for each image pixel has been measured, the image is analyzed by mapping each spectrum to a unique red green blue (RGB)-color. As dyes, five spectrally distinct fluorochromes within the green-to-red spectrum (human SKY Paint kit from ASI Ltd. uses Rhodamine, Texas-Red, cyanine [Cy]5, fluorescein isothiocyanate [FITC], and Cy5.5, with excitation between 495 and 675 nm and emission between 525 and 694 nm) are used either alone or in combination with each other to create 24 color mixtures with distinct proportions of each dye, as defined in a color library. Thus, each chromosome is characterized by a specific spectral signature, which is based on the specific combination of one or more of the five basic dyes. The reference library is build by analyzing a metaphase plate with known chromosome identities. The spectra of the pure dyes (no combinations!) are stored as reference spectra. Corresponding to certain combinations, pseudocolors are assigned. For preparation of the DNA probes, chromosomes

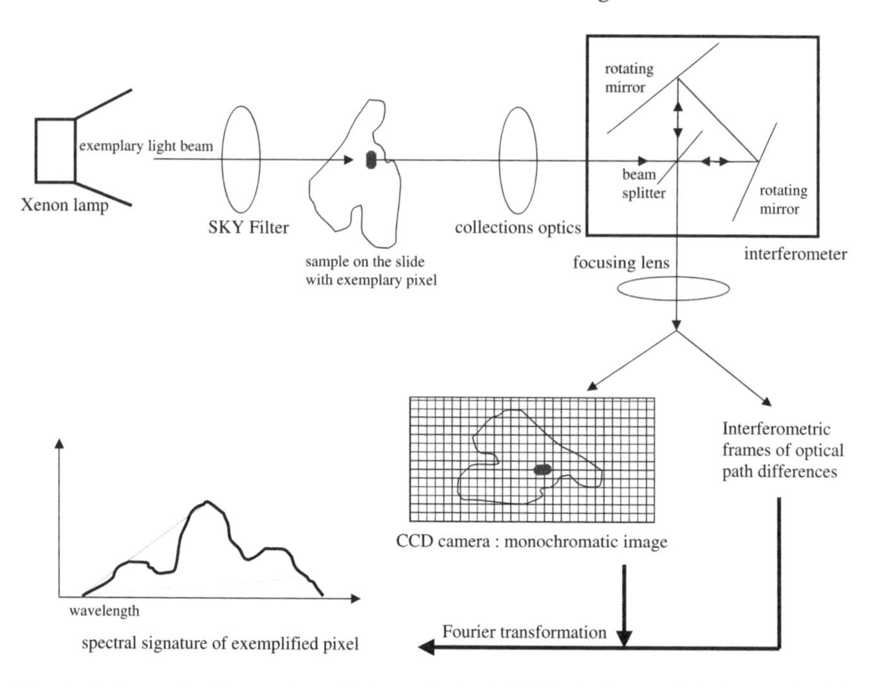

Fig. 1. Schematic illustration of the technical SKY-platform with the optical head adapted from ASI Ltd.

are flow sorted and labeled with color mixtures according to the library file. This is achieved by degenerate oligonucleotide-primed polymerase chain reaction (DOP-PCR) during which either directly labeled molecules or haptens are incorporated. The latter need to be detected with secondary- or tertiary-labeled molecules. The chromosomal regions rich in repetitive DNA sequences as found in constitutive heterochromatin and satellite regions are not labeled by this method but suppressed before the hybridization process to avoid cross-reactions. These labeled probes are hybridized to the metaphase sample, with both DNAs denatured before, detected, and washed, and chromosomes are counterstained with 4,6-diamidino-2-phenylindole (DAPI). After image capturing with the spectral and DAPI filter sets, the software compares the captured spectral image to the combinatorial library and accordingly classifies the chromosomes. Hence, the classified chromosomes are displayed in pseudocolors for the operator.

1.2. Applications for SKY

Cells that are still cell cycle active or in which mitosis can be induced to generate metaphases are inevitable. Although chromosomes in interphase of the cell cycle do engage specific nucleus domains, they are not appropriate

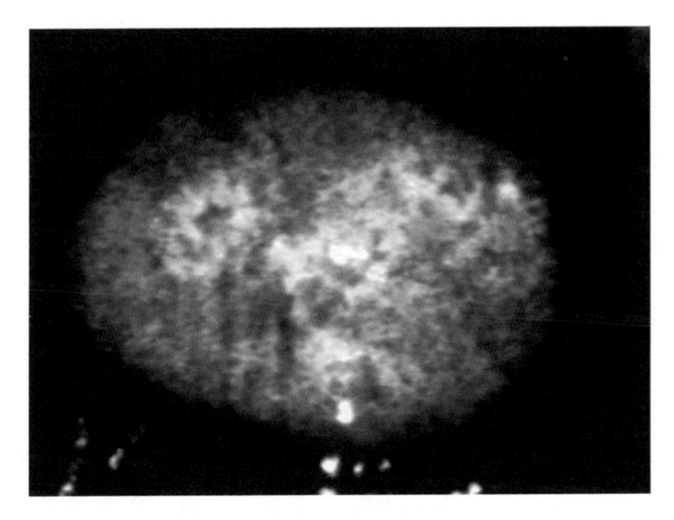

Fig. 2. Hybridization of SKY Paint to a lymphocyte interphase nucleus does not allow to extract precise genomic information.

because the chromosomal chromatin is decondensed and the areas overlap to a significant extent (*see* **Fig. 2**).

Two main categories of investigations profit from the information provided by SKY: identification of certain numerical and structural aberrations ***(7,8)*** and situations where only one or few cells are available for interrogation or where metaphases are difficult to obtain.

It has been estimated that the prenatal frequency of de novo supernumerary marker chromosomes is approximately 1/2500 ***(9)***. Because the origin of the marker chromosomes is normally unknown, SKY analysis can help to identify the chromosomal material of these additional chromosomes, as exemplified in reports for either postnatal analysis and prenatal investigation or pre- and postnatal identification ***(10–12)*** (*see* **Fig. 3**).

Daniel et al. ***(13)*** reported that 1/600 individuals is a carrier of a balanced translocation, and this implies a risk for an unbalanced aberration in the progeny of such individuals varying between <1–50%. Robertsonian translocations are estimated to be the most frequent aberration type in human population, with incidence of 1.2/1,000. Although the balanced carriers are phenotypically normal, they hold an increased risk for offspring with unbalanced chromosomes combinations and uniparental disomy ***(14)***. Using spectral karyotyping, the chromosomes involved in even complex and cryptic translocations can be identified. Furthermore, SKY is useful to resolve complex karyotypes (involving more than two chromosomes with three or more breakpoints), which are often present in hematologic and cancer disorders ***(15,16)***. In addition,

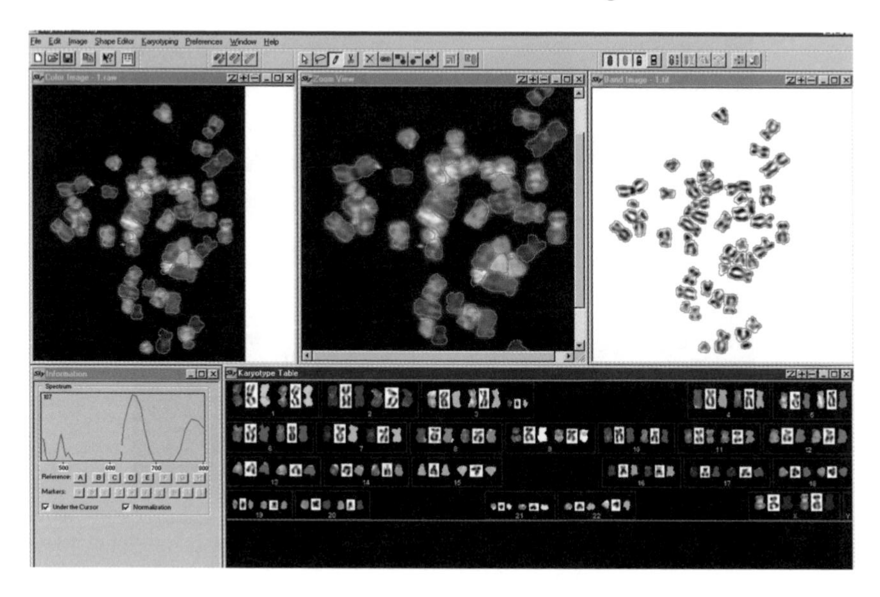

Fig. 3. Screen shot of the SKYView software operator interface. (*Left*) metaphase chromosomes in spectral staining where assignment to pseudocolors has already been performed. (*Right*) DAPI-counterstain presenting a G-like banding pattern on the same metaphase. The karyogram below shows the sorted chromosomes with their spectral, classified, and DAPI-counterstained features. By means of SKY, the additional marker chromosome in this pathological karyotype 46,XX+mar, was proved to originate from chromosome 3 material. (Courtesy of Friedel Wenzel, University Children's Hospital, Basel, Switzerland).

the chromosome morphology in these samples can be rather poor; consequently, further karyotype information apart from G-banding might be essential ***(17)***. Chen et al. ***(18)*** reviewed the aspects of de novo complex chromosomal rearrangements in prenatal diagnosis.

Situations where only single or few cells are available for investigation include preimplantation diagnosis (PGD) on blastomeres, or alternatively, the analysis of polar bodies as the genetic complement of female oocytes. Also, amniocentesis for rapid FISH on uncultured amniocytes and prenatal diagnosis on isolated fetal cells from maternal blood have to deal with this problem (*see* **Note 1**). To provide a proper basis for the SKY-application in PGD, Willadsen et al. ***(19)*** tested a procedure to enable human blastomeres to reenter the cell cycle and thus generate good-quality metaphases. Marquez et al. ***(20)*** demonstrated the successful feasibility of applying SKY to unfertilized human oocytes and fresh polar bodies, with a mean hybridization efficiency for oocytes of 95.2%. Another study combines the two application categories for SKY to an unfertilized oocyte and its first polar body from a Robertsonian translocation carrier undergoing in vitro fertilization and PGD for translocations and aneuploidy screening were assessed by SKY and comparative genomic hybridization *(21)*. SKY also was used to investigate as many chromosomes simultaneously as possible to determine the mechanisms generating aneuploidy in female meiosis I in oocytes *(22)*. Application of the commercial SKY system was even suggested to be useful as an additional diagnostic tool for the differentiation of chronic lymphocytic leukemia cells versus normal small lymphocytes on the spectral level *(23)*.

1.3. Limitations of SKY

The SKY application has its limitations in cases of intrachromosomal aberrations and rearrangements involving pericentric regions, false insertions at the interface of translocated segments, and coamplifications of material from nonhomologous chromosomes, where misleading results by SKY analysis were obtained on various tumor and leukemia samples *(24)*. Their collective analysis suggested that most errors in multicolor karyotyping are based on similar mechanisms, which are mainly represented by "flaring," whereby structural rearrangements juxtapose nonhomologous chromosomal material, which results in overlapping fluorescence at the interface of the translocated sequences and consequently disturbs the recognition pattern of fluorescence used by the multicolor karyotyping systems *(24)*. For interchromosomal rearrangements, a detection limit with the resolution of approximately 2 megabases of DNA was published *(25)*. This nevertheless allows to identify subtle translocations that are difficult to ascertain by conventional banding analysis. Yaron et al. *(26)* tested the application of SKY in the characterization of de novo extra structurally abnormal chromosomes (ESACs) in prenatal diagnosis. Whereas the evaluation of ring chromosomes and nonsatellited ESACs benefited greatly from the additional analysis, the characterization of uni- and especially bisatellited ESACs did not gain further information by SKY.

In addition, the results of multicolor karyotyping often need to be confirmed or refined by the use of then better-chosen FISH probes for detection of subtle anomalies. Therefore, Heng et al. *(27)* proposed a strategy effective in monetary, time, and technology-availability aspects combining G-banding, SKY, and FISH for pre- and postnatal chromosomal analysis.

1.4. SKY and Multicolor-FISH (M-FISH)

During 1996 Multicolor-FISH (M-FISH) was established alternatively to SKY. This technique also uses the combinatorial labeling of five basic dyes (spectral interval approx 350–770 nm), but instead of each pixel spectrum, the intensities of the five fluorochromes are detected by a system of filters with defined emission spectra. Each image is captured once in all five filters, one after the other, and for the final evaluation the single filter information layers are summarized. Pseudocolors are assigned on the basis of gray-scale intensities. A technical advantage of the SKY system is the measurement of the spectrum without using restrictive filters, which results in more signal-photons being available for detection on the CCD-camera. This finally provides a better signal-to-noise ratio. A further disadvantage of M-FISH is the use of intensity as a measurement parameter, because variation in intensity is highly frequent in painting probes, especially toward the chromosome ends.

Spectral karyotyping is normally performed using the commercial SKY kit and the SpetraCube system. In contrast, M-FISH is often performed on the basis of user-fabricated probes; thus, it is more flexible in the number of probes used. Whereas 24 painting probes applied on an interphase nucleus provide no informative content (*see* **Fig. 2**), M-FISH with 8 or 10 chromosome enumeration probes can be used with good impact, e.g., on single lymphocytes, amniocytes, and blastomeres ***(28–30)***.

These applications of M-FISH even exceed the commercially available M-FISH kits for the interphase detection of five chromosomes by using filter-based fluorescence microscopy (Aneuvysion assay kit for chromosomes X, Y, 13, 18, and 21, Vysis and MultiVysion PB Multicolor probe panel for chromosomes 13, 16, 18, 21, and 22, Vysis). Sequential M-FISH could be applied as well to score additional information from a single cell. However, the quality of the cells during sequential FISH is deteriorating, and cells might even be lost from the slide during processing ***(31)***. Furthermore, availability of FISH probes that can be flexibly combined with each other might be limited. Nietzel et al. ***(32)*** finally reported the generation of centromer-specific multicolor FISH characterization of all 24 chromosomes simultaneously. In contrast to the painting probes used otherwise for M-FISH and SKY, these probes hybridize against the repetitive sequences at the centromeric regions of the chromosomes. Because these probes do not paint whole chromosomes, they could be better used in situations with only single cells available where as much information as possible has to be extracted from this material.

1.5. Protocol Details

1.5.1. Sample Preparation, Slide Pretreatment, and Denaturation

Because the application of SKY focuses on the staining of all chromosomes within a cell, it is essential that good-quality metaphase chromosomes are available on the slide (*see* **Note 2**). Thus, only tissues with mitotically active cells or cells that can be activated to reenter the cell cycle can be used for this analysis. Slide preparation includes short-term culture of the cells terminated by arresting the cells in the metaphase stadium by means of mitotic spindle toxic agents such as colcemide. Depending on the tissue, time and concentration of the colcemide incubation have to be optimized (*see* **Note 3**). In the following example, the cells are treated hypotonically to flow-separate the chromosomes within the cells, and fixative addition maintains this state. Dropping the cells on the slides should be performed under standardized and reproducible temperature and humidity conditions, because they are crucial for the slide quality in terms of metaphase quality and slide background. Further fixation of cells on the slide is necessary followed by aging of the slides before they can be used for hybridization.

Precedent to SKY, the slides should be quality assessed in phase contrast microscopy to ensure the quantity and quality of metaphases present; this monitoring step also determines the further pretreatment of the slide (*see* **Note 4**). Remaining cellular and cytoplasmic debris should be removed by means of protease treatment, otherwise this debris might contribute to unspecific background later on. After providing a maximally clean slide, the metaphase DNA needs to be denatured to allow access for the probe DNA during hybridization. This is achieved by supplementing heat treatment with formamide addition, because this lowers the actually necessary temperatures to separate the DNA double strands of the target DNA (*see* **Note 5**). To stabilize the two single strands apart from each other and prevent reannealing, the slides have to be immediately incubated in an increasing ethanol series. This also helps to remove any residual moisture in the sample (*see* **Note 6**). The denaturation step is a crucial part in the slide pretreatment, because too much denaturation destroys the DNA, whereas too little denaturation denies the accessibility of the probe to the target DNA. Both outcomes result in poor hybridization efficiency. It might be necessary to optimize denaturation time and temperature for different sample cell types.

1.5.2. Probe Preparation

The painting probes of SKY are generated by flow sorting human chromosomes and after DNA amplification by using degenerate oligonucleotide primed PCR (DOP-PCR) *(33)*. During the PCR-amplification, the probe fragments are labeled either directly with a fluorochrome or indirectly with a hapten.

The SKY probe containing labeled sequences complementary to all 24 human chromosomes has to be denatured before the hybridization as well. In the commercial kit, this probe is provided in a hybridization cocktail of formamide and dextran sulfate, helping to reduce the necessary denaturing temperature and facilitating hybridization conditions. It also includes unlabeled COT1-DNA to block repetitive sequences that are widely distributed throughout the whole genome from hybridization. Therefore, the probe cocktail is first denatured at 80°C and then it is incubated at 37°C for 1 h to allow preannealing between the repetitive probe sequences and the COT1-DNA. Because the sequences blocked by COT1-DNA belong to the category of the DNA annealing with maximal speed, no danger of reannealing of the labeled DNA properties is expected.

Alternatively to separate denaturation, the labeled probe DNA and the target DNA might be codenatured as well. In this case, 10 μl of the probe should be placed on the slide, covered with a coverslip, denatured together for 1.5 min at 75°C, and followed by immediate incubation at 37°C.

1.5.3. Hybridization

After separate denaturation of both DNA sequences, the probe is applied to the slide, and hybridization 37°C for at least 48 h is allowed to achieve maximal annealing between the target DNA on the slide and the labeled probe DNA, with a simultaneous minimum of unspecific pairing. Precedent cleaning of the slides and complete denaturation of both DNAs thereby ensures optimal conditions for the probe DNA to access the chromosomal DNA and produce a specific strong signal.

1.5.4. Posthybridization: Washes and Sandwich-Detection

To remove excess of unbound probe DNA and reopen unspecific annealing between the DNAs, stringent posthybridization washes are performed (*see* **Note 7**). This includes a first wash in high-temperature formamide followed by washes in heated standard saline citrate (SSC) as detergent (*see* **Note 8**). A blocking step is inserted to reduce any background or noise during the subsequent immunological detections steps (*see* **Note 9**). Some of the probe DNA is directly fluorochrome labeled, whereas other parts are coupled to a hapten. Complement molecules or antibodies directed against these haptens are either fluorochrome-labeled themselves or target of a second labeled antibody. Thus, three layers are topped onto the sample DNA. Intermediate washings steps with detergents ensure that unbound antibodies are removed and thus prevent background noise. Finally, the chromosomes are counterstained with the blue dye DAPI, which is combined with a antifade agent to avoid fast fluorescence bleaching. Slides can be stored at –20°C for several weeks with signals remaining stable.

1.5.5. Image Acquisition

Although the spectral image capturing by means of the SpectraCube system is the central aspect of the analysis, not much can be done by the user to optimize these steps. Care has to be taken that the fluorescence integrity of the kit is guaranteed, because this is the basis for the spectral signature of each image pixel. SpectraCube SD200 combines spectroscopy and imaging and it enables measurement of the visible spectrum of each pixel in the observed image. The RGB image is initially displayed in the software by assigning the red, green, and blue display color ranges to specific emission ranges (red = near infrared, green = red emission, and blue = green emission). This RGB display color image is not a true 24-color image and it does not display all the spectral information as measured by the SKY system. The ambiguity inherent in the RGB display colors is overcome by performing a spectrally based mathematical pixel classification such that every chromosome is classified in a different user-selectable pseudocolor. Classification means computer-based separation of spectra. Before capturing the image, it has to be ensured that the image is in sharp focus on the monitor; this will guarantee the best segmentation and sharpest bands possible.

1.5.6. SKY in Combination with Giemsa or FISH

Combining Giemsa or FISH with SKY might be of interest to verify inconclusive chromosomal aberrations on a specific metaphase or for the identification of an unknown gene locus. First, Giemsa (*see* **Note 10**) or FISH (*see* **Note 11**) is performed according to standard procedures, the position of the metaphase in question is marked, SKY is performed according to protocol, and the same metaphase relocated and evaluated. However, obstacles might include increased background and mixtures of fluorochromes impairing the SKY identification. The storage conditions concerning humidity and temperature in between the processes greatly influences the outcome. Slides should not be older than 2 weeks after G-banding. Furthermore, the DNA in the G-banded chromosomes already has endured harsh treatment, such as trypsin digestion for cytoplasmic debris removal, stripping away DNA-associated proteins for the banding effect, and heat treatment to enhance the contrast of the binding. In addition, the immersion oil from previous slide analysis or the mounting medium under the coverslip might degrade the DNA. Thus, gentle treatment of the slides is recommended to prevent further DNA damage, e.g., by omitting any digestive treatment and adjusting the denaturation time.

1.5.7. Troubleshooting

Many parameters influence a good experimental outcome in spectral karyotyping. Thus, in case of dissatisfying results, it can be hard to decide where to start for changing the influencing factors. The following describes some supportive measurements, which might help when administered singularly or in combination to achieve better results. Poor morphology of the metaphases can be caused by overdenaturation of the chromatin; therefore, it is important to correlate especially the denaturation time to the sample type and to the age of the slide preparations. It is also essential to check the denaturation temperature in the solution in the coplin jar, and eventually to increase the temperature of the water bath. Furthermore, the temperature should be monitored and eventually adjusted during or before the slide incubation as well, as per incubated slide the heated solution looses approximately 1°C. Generally, never more than four slides should be handled in parallel.

Background fluorescence can be due to cytoplasmic debris, and removal by RNase digestion and stronger proteolytic treatment with pepsin or proteinase K is recommended. Utilizing the SKY Paint kit a carefully balanced combination of probes is hybridized to the slide. However, with home-made probes, the concentration or volumes of probes applied might be excessive. Enhancing the stringency of posthybridization washes by increasing the temperature or decreasing the salt concentration also might help to reduce the background noise. Gentle agitation of the slides in the washing solutions might result in higher washing efficiency as well. In addition, microorganisms such as yeast or bacteria contaminating the sample during tissue culturing might cause background noise.

Cytoplasmic debris around the metaphases also might be a reason for weak probe signals, because they imprison the probe; thus, it is not capable of reaching and staining the target DNA. Using probes stored at incorrect conditions or after the expiration date can contribute to background noise and weak signals as well. In addition, weak probe signals can lie either in a combination of the slide age and the slide denaturation or in the hybridization or detection processes. In the first scenario, slides too young can easily be overdenatured, whereas on slides that are too old, the chromatin might be harder to denature or contain time-degraded DNA already. Too little moisture in the incubation environment, either caused by incomplete sealing of the coverslip, which allows dilution of the probe on the sample, or by the slides drying out between the blocking and antibody detection steps, can cause weak signals and unspecific background (*see* **Note 12**). Slides that were evaluated with immersion oil need to be cleaned in xylene to remove the oil, which prevents probe hybridization. Using antifade solution past the expiration date might result in quickly fading signals and the development of a colored veil across the slide, even though the slides are stored at proper conditions. Cross-hybridizations resulting in reduced probe specificity, as seen in ambiguous chromosomes classification according to the combinatorial table during SKYView evaluation, might be reduced by spiking an increased amount of COT1-DNA to the probe cocktail.

2. Materials

2.1. Sample Preparation, Slide Pretreatment, and Denaturation

1. Sterile tissue culture flasks.
2. 15-ml conical tubes (e.g., Falcon; BD Biosciences Discovery Labware, Bedford, MA).
3. SuperFrost glass slides (e.g., Menzel Gläser, Braunschweig, Germany).
4. Slide storage box.
5. Glass coplin jars.
6. Phase contrast microscope.
7. Waterbath at 37°C.
8. Waterbath at 75°C.
9. 5% CO_2 incubator at 37°C.
10. Sample tissue, e.g., peripheral whole blood with heparin as anticoagulants.
11. Tissue culture media for short-term culturing (e.g., for peripheral whole blood, Sigma PB-Medium or RPMI 1640 medium supplemented with penicillin/streptomycin (final concentration 100 U/ml/100 µg/ml), phytohemagglutinin (final concentration 5 µg/ml) and fetal calf serum. Provide 10 ml/sample.
12. 20× SSC: 3 M NaCl/0.3 M Na-citrate. Prepare with double distilled water, adjust to pH 7.0, autoclave, and store at room temperature.
13. Deionized, distilled high-grade formamide.
14. 1× phosphate-buffered saline (PBS). Store at room temperature (3 × 50 ml in a coplin jar).
15. 1× PBS, 50 mM $MgCl_2$ at room temperature (50 ml in a coplin jar). It is recommended to filter the solution (22-µm pore size) for sterilization. Check for contamination before use. Store at room temperature.
16. 1% formaldehyde/1× PBS/50 mM $MgCl_2$. Store at room temperature (50 ml in a coplin jar). Can be stored for 3–4 applications maximum 1 week at 4°C.
17. 0.075 M KCl prewarmed to 37°C (10 ml/sample).
18. 0.01 MHCl prewarmed to 37°C (50 ml in a coplin jar).
19. 3:1 methanol-to-glacial acetic acid (freshly prepared each time, approx 35 ml/sample on day 1 and approx 5 ml per sample on day 2), store at –20°C until use.
20. Ethanol series at room temperature and –20°C: 70, 80, 100% in coplin jars.
21. 70% formamide, 2× SSC preheated to 72°C (50 ml in a coplin jar).
22. Colcemid (10 µg/ml) (e.g., KaryoMax colcemid, Invitrogen, Carlsbad, CA).
23. 10% (w/v) pepsin.

2.2. Probe Preparation

1. SKY Paint kit (ASI Ltd.): Vial 1 human spectral karyotyping reagent. Stored at 4°C. Protect from light at any time.
2. Water bath or PCR-machine at 75°C.
3. Water bath or PCR-machine at 37°C.

2.3. Hybridization

1. Denatured metaphase slide (*see* **Subheadings 2.1.** and **3.1.**).
2. Denatured labeled probe (*see* **Subheading 2.2** and **3.2.**).
3. Coverslips (12 × 12 mm^2).
4. Rubber cement (e.g., FixoGum, Marabu, Tamm, Germany).
5. Hybridization system (e.g., Vysis HYBrite).

2.4. Posthybridization: Washes and Sandwich-Detection

1. Waterbath at 45°C.
2. Three coplin jars containing 50% formamide, 2× SSC at 45°C (50 ml each).
3. Three coplin jars containing 1× SSC at 45°C (50 ml each).
4. Seven coplin jars containing 0.01% Tween 20, 4× SSC at 45°C (50 ml each).
5. 1 coplin jar containing 2× SSC at room temperature (50 ml).
6. SKY Paint kit (ASI Ltd.): vial 2, blocking reagent; vial 5, DAPI/antifade.
7. Concentrated antibodies detection (CAD) kit (ASI Ltd.): vial 3, detection reagent antidogoxigenin/Cy5 streptavidin; vial 4, detection reagent Cy5.5 sheep anti-mouse. Protect regents from light at any time.
8. Coverslips 24 × 60 mm^2.
9. Storage box for glass slides at –20°C.

2.5. Image Acquisition

1. Fluorescence microscope equipped with DAPI filter and SpetraCube filter.
2. Optical SpectraCube-Head (ASI Ltd.) attached to the fluorescent microscope.
3. Spectral Imaging acquisition software (ASI Ltd.).
4. SKY View analysis software (ASI Ltd.).

2.6. SKY in Combination with Giemsa or FISH

1. 2 coplin jars with xylene (50 ml each).
2. 2 coplin jars with methanol (50 ml each).
3. Ethanol series: 70, 80 95%; ethanol stored at –20°C.

3. Methods

3.1. Sample Preparation, Slide Pretreatment and Denaturation

Protocol for metaphase chromosome preparation from peripheral whole blood

1. Prewarm 0.075 M KCl to 37°C.
2. Prepare fresh fixative (3:1 methanol-to-glacial acetic acid) and store at –20°C until use.
3. Prewarm 10 ml of medium to 37°C.
4. To the prewarmed medium, add 500 μl of whole peripheral anticoagulated blood and mix well.

5. Transfer mixture in a culture flask and incubate at 37°C under 5% CO_2 for 72 h with intermediate gentle shaking.
6. Add 100 µl of colcemid (final concentration 0.1µg/ml) to the flask and mix well.
7. Incubate at 37°C under 5%v CO_2 for 25 min with intermediate gentle shaking.
8. Transfer the content of the flask in a conical 15-ml tube.
9. Centrifuge at 240*g* for 10 min.
10. Discard supernatant until 0.5 ml and resuspend cells.
11. Slowly add 1 ml of the prewarmed 0.075 M KCl in a dropwise manner and carefully mix during and after adding each drop (*see* **Note 13**).
12. Slowly add the remaining 9 ml of KCl under continuous mixing and finally gently invert tube several times.
13. Incubate in a water bath at 37°C for 20 min.
14. Slowly add 1 ml of fresh cooled fixative (3:1 methanol-to-glacial acetic acid) in a dropwise manner and carefully mix during and after adding each single drop; finally gently invert the tube several times (*see* **Note 14**).
15. Incubate at room temperature for 5 min.
16. Centrifuge at 240*g* for 10 min.
17. Aspirate the supernatant until 1 ml is leftover above the pellet, and resuspend the pellet.
18. Add 2 ml of fresh cooled fixative in a dropwise manner under continuous mixing.
19. Add the remaining 8 ml of fixative slowly under continuous mixing.
20. Gently invert the tube several times.
21. Incubate at +4°C for 30 min.
22. Centrifuge at 240*g* for 10 min.
23. Aspirate the supernatant until 1 ml is leftover above the pellet, and resuspend the pellet.
24. Repeat 2 times **steps 17–21** without **step 20**.
25. The cell suspension may be stored for several months as pellet under fixative at –20°C.
26. To prepare cytogenetic slides, aspirate the fixative until 1 ml is leftover above the pellet, and resuspend the pellet.
27. Add 2 ml of fresh cooled fixative and resuspend the cells.
28. Incubate at –20°C overnight.
29. Centrifuge at 240*g* for 10 min.
30. Discard the supernatant completely.
31. Slowly add sufficient fresh cooled fixative until the solution is slightly translucent.
32. Clean a glass slide with 70% ethanol and wipe dry using a Kimwipe. Dip the slide in A.dest and coat the upper slide surface evenly with a water film.
33. Immediately drop 100 µl of the cell suspension along the tilted (angle of approximately 45°) clean glass slides at room temperature under approximately 45–50% humidity.
34. Airdry the slides for 5–10 min.
35. Allow the slides to age at room temperature for at least 2–3 days or place the slides overnight in an incubator at 37°C or for 3 h in an oven at 60°C.

36. Prewarm 50 ml of 70% formamide/2× SSC in a coplin jar to 75°C.
37. Prewarm 50 ml of 0.01 M HCl in a coplin jar to 37°C.
38. Dehydrate slides in an increasing ethanol series (70, 80, 100% ethanol) for 5 min each.
39. Airdry the slides for 5–10 min.
40. Add 4–10 µl of 10% pepsin stock to prewarmed 0.01 M HCl in the coplin jar.
41. Incubate the slides for 5 min in the Pepsin/HCl solution.
42. Incubate the slides two times for 5 min in 1× PBS at room temperature.
43. Incubate slides for 5 min in 1× PBS/50 mM$MgCl_2$ at room temperature.
44. Incubate slides in 1% formaldehyde/1× PBS/50 mM $MgCl_2$ for 10 min at room temperature (*see* **Note 15**).
45. Incubate slides for 5 min in 1× PBS at room temperature.
46. Pass slides through the increasing ethanol series (70, 80, and 100%) with 5 min incubation each.
47. Air dry the slides for 5–10 min.
48. Check that formamide solution achieved at 72°C.
50. Process slides through increasing ethanol series (70, 80, and 100%) for 2 min each at –20°C.
51. Air dry slides for 5–10 min.

3.2. Probe Preparation

CAVE: Work fast but precise and provide light protection in each handling step!

1. Briefly centrifuge probe cocktail (Human SKY reagent vial 1 from SKY Paint kit, ASI Ltd.) to collect all probe material in the tube bottom.
2. Take 10 µl of labeled probe per slide and denature at 80°C for 7 min either in a water bath or a PCR-machine.
3. Immediately after the denaturation, preanneal the probe at 37°C for 1 h either in a water bath or a PCR-machine.

3.3. Hybridization

CAVE: Wok fast but precise and provide light protection in each handling step!

1. Carefully aspirate the denatured probe and place 10 µl on a coverslip (12 × 12 mm^2).
2. To combine the denatured probe DNA and the denatured target DNA on the slide for hybridization lower the slide with the denatured metaphases face downwards down on the coverslip with the probe; avoid air bubbles.
3. Carefully seal the brim of the cover slip with rubber cement and transfer the slide in a humid hybridization chamber at 37°C (*see* **Note 16**).
4. Coincubate labeled probe and sample DNA for hybridization at constant 37°C for 48 h in the humid chamber.

3.4. Posthybridization: Washes and Sandwich-Detection

1. Prewarm all washing-solutions to 45°C in a water bath. CAVE: Be careful to protect slides from light during all handling steps as much as possible!
2. Remove the slides from the hybridization system and carefully detach the rubber cement around the cover slip; take care to gently remove the cover slip from the slide; if it does not come off easily place the slides in the first washing solution, gently shake for several seconds and gently shift the cover slip towards the edge of the slide that it can finally be taken away.
3. Wash the slides three times for 5 min each in 50% formamide/2× SSC at 45°C.
4. Wash slides two times for 5 min each in 1× SSC at 45°C.
5. Dip slides shortly in 4× SSC/0.01 %Tween 20 for 2 min at 45°C.
6. Optional!
   Apply 80 µl of blocking agent (SKY Paint kit vial 2) on the slide, cover with a coverslip (24 × 60 mm^2) and incubate for 30 min at 37°C in a humid chamber.
7. Prepare working solution for the first detection layer: To 1 ml of 4× SSC, add 10 µl of vial 3 of the CAD kit containing antidogoxigenin and Cy5 streptavidin; prepare this solution always fresh, store at 4°C until use, and discard at the end of the day.
8. Carefully remove coverslip and place slide in a tilted position; allow fluid to drain. Apply 80 µl of antidigoxigenin/Cy5 working solution, cover with a coverslip (24 × 60 mm^2) and incubate for 45 min at 37°C in a humid chamber.
9. Carefully remove the cover slip and wash slides three times for 3 min each in 4× SSC/0.01%Tween 20 at 45°C.
10. Prepare working solution for second detection layer: to 1 mL of 4× SSC; add 5 µl of vial 4 of the CAD kit containing Cy5.5 sheep anti-mouse; prepare this solution always fresh, store at 4°C until use, and discard at the end of the day.
11. Apply 80 µl of Cy5.5 working solution, cover with a coverslip (24 × 60 mm^2), and incubate for 45 min at 37°C in a humid chamber.
12. Carefully remove the coverslip and wash slides three times for 3 min each in 4× SSC/0.01%Tween 20 at 45°C.
13. Wash slides briefly in A.dest and air dry the slides for 5–10 min.
14. Add 20 µl of DAPI/antifade reagent (kit vial 5) on the slide, cover with a coverslip (24 × 60 mm^2); be careful to avoid air bubbles.
15. The slides are now ready for evaluation and should be stored at –20°C.

3.5. Image Acquisition

1. Image capturing and analysis should be performed as soon as possible. Although storage at –20°C and avoiding light irradiation should maintain a stable fluorescence for several months, the signal intensity will diminish over time.
2. Slides are screened in the DAPI filter for good-quality metaphases with long and nonoverlapping chromosomes (*see* **Notes 17** and **18**).
3. Images are firstly captured in the SpectraCube filter followed by the counterstain image capturing using the DAPI filter.

4. Both image files need to be stored and imported into the SKYView software for further analysis (*see* **Notes 19** and **20**).
5. Training for the software use has to be carried out when the system is installed in the laboratory.

3.6. SKY in Combination with Giemsa or FISH

1. In case of residual oil on the slides, wash slides two times for 5 minutes each in xylene at room temperature.
2. Destain the slides in methanol for 10–15 min followed by a degrading series of ethanol (100, 80, and 70%) 5 min each.
3. Air dry the slides for 5–10 min.
4. Fix slides in 1% formaldehyde/1× PBS/50 mM $MgCl_2$ for 2 min at room temperature.
5. Place slides in 1× PBS for 5 min and dehydrate in 70, 80, and 100% ethanol for 2 minutes each.
6. Air dry the slides for 5–10 min.
7. Denature slides according to protocol, but only 30 s.
8. Continue with the regular protocol.

4. Notes

1. Although amniocytes for rapid FISH and fetal cells isolated noninvasively from maternal blood might be good candidates for the application of SKY on the first glance, critical considerations do not confirm this approach. The benefit of using uncultured amniocytes is the short experimental time necessary until the result is available. Generating metaphases for SKY needs culturing time, which ruins the aim of providing rapid first karyotype impressions. Fetal erythroblasts isolated from maternal blood do not represent a pure fetal cell population, thereby requiring differentiation between fetal and maternal cells. This poses less problems in cases of male pregnancies, but female fetal karyotypes need further confirmation regarding their origin, which cannot be provided by SKY. Furthermore, generation of metaphases of these fetal cells might provide difficulties according to culture experiments *(34)*. In addition, fetal erythroblasts were shown to be impervious to FISH *(35)*. Thus, these cell types are not suitable for SKY analysis.
2. Preparation of metaphase slides is performed differently in every lab. However, common critical aspects are to control the general temperature and humidity conditions that might be changed by placing the slides either on a hot-plate for better drying or providing steam by a boiling kettle or a humidifier. Regardless, slides should be quality-assessed by phase contrast microscopy. No debris should be visible, and chromosomes should look dark gray/black with the banding pattern slightly recognizable. Good-quality metaphases include chromosomes not too short or too long without any overlapping chromosomes. Slides can be

stored at room temperature but should then be used within two weeks and as early as the next day. Further aging will deteriorate the chromatin structure and necessitates further optimized denaturation conditions.

3. Metaphase quality directly influences the quality of the spectral images and the quality of the metaphases depends largely on the effects mediated by colcemid; because different tissues and cell types react differently to concentration and incubation time of colcemid, it might be worthwhile to precheck the optimal conditions before analyzing the sample of interest. Also, the chromosomes size and condensation grade are very suspectible to the influence of the colcemid.
4. Even with slides with good phase contrast, monitoring all pretreatment steps should be performed because the quality of slides is the basis for a successful SKY-hybridization.
5. Formamide reduces the melting temperature necessary to split a DNA double strand into single strands. However, depending on the age and quality of the slide, time and temperature of the formamide incubation have to be carefully estimated. Younger slides might not have had enough time to age properly; thus, the DNA may be more delicate, so shorter incubation times are sufficient for an effective denaturation. If there is cytoplasmic debris around the metaphases, longer incubation might achieve better probe access to the target DNA. Too much or too little denaturation can be monitored in the chromosome appearance, although in the normal procedure the slides are not checked between denaturation and hybridization anymore. However, because this step provides the basic conditions for efficient probe hybridization and DAPI staining, it is essential to consider this step in the slide evaluation for possible further hybridization of slides of the same batch. Like in all incubation steps that need specific temperatures above room temperature, it is absolutely necessary to ensure that not the water in the water bath shows the target temperature, but the fluid in the coplin jar; in this respect glass coplin jars transfer the temperature more efficiently to the containing fluid than do plastic coplin jars. Also, the number of incubated slides influences the temperature of the fluid, so that, on average, approximately 1°C per slide should be calculated and the temperature of the water bath thereby adjusted. Nevertheless, in no case more than four slides should be processed simultaneously.
6. After denaturation it is essential to maintain the target DNA in single-strand conformation. Therefore, the slides need to be immersed in 70% ethanol at once after denaturation. The alcohol series can be at room temperature or –20°C. It is recommended to substitute the 70% ethanol each time, whereas the 80% and 100% ethanol can be replenished after the experiment. For storage, tight sealing of the containers prevents the evaporation.
7. Gentle agitation during the washing steps might increase the washing effect. It might help to remove unbound probe or antibodies from the slide to obtain a better background for the microscopic evaluation.
8. Instead of posthybridization washing **steps 2** and **3**, a substitutive single rapid wash can be performed by placing slides in 0.5× SSC for 2–5 min at 72°C.

9. During washing, blocking, and antibody detection steps, the slides should never dry our, otherwise the background might be enormously enhanced.
10. G-banded slides can be used for subsequent spectral karyotyping according to ASI Ltd.; however, the time interval between these two stainings should not be too long; good results are obtainable until approximately 2 weeks.

    a. Remove residual oil form the slides by briefly dipping slides in xylene.
    b. Incubate slides for 5 min in methanol.
    c. Pass slides through a decreasing ethanol series of 100, 80, and 70% for 3 min each.
    d. Fix in 1% formaldehyde/1× PBS/50 mM $MgCl_2$ for 2 min at room temperature.
    e. Wash slides in 1× PBS at room temperature for 5 min.
    f. Dehydrate slides in an increasing ethanol series 70, 80, and 100% for 2 min each.
    g. Air dry slides for 5–10 min.
    h. Denature slides according to protocol but only for 30 s!
    i. Continue with the regular protocol; no further pretreatment is necessary.

11. FISH can be performed following to SKY either to confirm a result or to map a unique probe.

    a. Wash slides 2× for 5 min in 4× SSC/0.01% Tween 20.
    b. Rinse in running water.
    c. Dehydrate in 70, 80, and 100% ethanol for 2 min each.
    d. Air dry slides for 5–10 min.
    e. Continue with the regular protocol but denature the slides only for 30 s.

12. In case of weak Cy5 or Cy5.5 staining ASI Ltd. recommends rehybridization of SKY.

    a. Wash slides 2× for 5 min in 4× SSC/0.01% Tween 20.
    b. Rinse slides in running water.
    c. Dehydrate in 70, 80, and 100% ethanol for 2 min each.
    d. Air dry slides for 5–10 min.
    e. Continue with the regular protocol without any pretreatments and denature the slides only for 30 s.

13. Adding the first milliliter of KCL drop-wise to the cells helps to prevent a sudden rush of water into the cells with the consequential bursting of the cell. Gentle mixing between the single drops ensures that the solution effects slowly on the cell suspension. The 37°C temperature serves to assist the swelling of the cells more rapidly. If the cells do not swell adequately increase the incubation time in 5-min increments.
14. The hypotonic treatment leaves the cells in a very sensitive state; therefore, the slow adding of the first milliliter of fixative helps to harden the cell membrane for subsequent centrifugation and fixation. A side effect of the treatment is that residual red blood cells are lysed. After the centrifugation following the

hypotonic and fixative treatment, the size of the cell pellet should have doubled to indicate that the treatment has been successful.

15. Formaldehyde assists to maintain the chromosomal morphology and thus helps to make the DNA a little bit more resistant to denaturation. Hence, for older slides, the formaldehyde treatment can be reduced in time incubation. On previously G-banded slides that underwent rough treatment to the banding pattern by trypsin digestion and heat application for the aging process, the banding pattern is better maintained by introducing a formaldehyde incubation into the protocol.
16. The HYBrite system from Vysis presents a hybridization chamber with user-determinable incubation temperature and incubation time-programs. The system can be run either dry or provide a humid environment by means of damp paper towels. Alternatively a light-proof box can be used and incubated in prewarmed incubator. However, because in the HYBrite system the slides are in direct contact with the heating plate, the temperature transfer is probably more close and exact than in the box in the incubator. Damp paper towels also can be lined around the wall of the box to provide the humid environment, which seems to enhance the hybridization kinetics.
17. For image capturing, metaphases with cytoplasmic debris around and with overlapping chromosomes should be avoided if enough metaphase plates are available. They might provide unnecessary and intriguing fluorescence and spectral information and also irritate in evaluation for false detection of translocations.
18. The images of the metaphases should represent close sections of the chromosome plate without much space around; this occupies unnecessarily space on the hard disk, reduces acquisition time, and might introduce further background fluorescence in the evaluation.
19. Helpful tools to monitor the true constancy of a spectral signature are the so-called pixel confidence and subset classification in the SKYView software.
20. In case a chromosome is not present in the image analyzed, e.g., the Y chromosome in a female karyotype, then omitting its specific color combination from the subset classification improves the classification for the other chromosomes by sort of fine-tuning.

References

1. Pinkel, D., Straume, T. and Gray, J. W. (1986) Cytogenetic analysis using quantitative high-sensitivity fluorescence in situ hybridization. *Proc. Natl. Acad. Sci. USA* **83,** 2934–2938.
2. Tkachuk, D. C., Pinkel, D., Kuo, W.-L., Weier, H.-U., and Gray, J. W. (1991) Clinical applications of fluorescence in situ hybridization. *Genet. Anal. Tech. Appl.* **8,** 67–74.
3. Pinkel, D., Landegent, J., and Collins, C. (1988) Fluorescence in situ hybridization with human chromosome-specific libraries: detection of trisomy 21 and translocations of chromosome 4. *Proc. Natl. Acad. Sci. USA* **85,** 9138–9142.

4. Garini, Y., Macville, M., du Manoir, S., et al. (1996) Spectral karyotyping. *Bioimaging* **4,** 65–72.
5. Schröck, E., du Manoir, S., Veldman, T., et al. (1996) Multicolor spectral karyotyping of human chromosomes. *Science* **273,** 494–497.
6. Malik, Z., Dishi, M., and Garini, Y. (1996) Fourier transform multipixel spectroscopy and spectral imaging of protoporphyrin in single melanoma cells. *Photochem. Photobiol.* **63,** 608–614.
7. Bayani, J. and Squire, J. A. (2002) Spectral karyotyping. In: Fan, Y.-S. (ed.). Molecular cytogenetics protocols and application. Humana Press, Totowa, NJ, pp 85–104.
8. Bayani, J. and Squire, J. A. (2001) Advances in the detection of chromosomal aberrations using spectral karyotyping. *Clin. Genet.* **59,** 65–73.
9. Warburton, D. (1991) De novo balanced chromosome rearrangements and extra maker chromosomes. Identified at prenatal diagnosis: clinical significance and distribution of breakpoints. *Am. J. Hum. Genet.* **49,** 995–1013.
10. Huang, B., Ning, Y., Lamb, A. N., et al. (1998) Identification of an unusual maker chromosome by spectral karyotyping. *Am. J. Med. Genet.* **80,** 368–372.
11. Ning, Y., Laundon, C. H., Schröck, E., Buchanan, P., and Ried, T. (1999) Prenatal diagnosis of a mosaic extra structurally abnormal chromosome by spectral karyotyping. *Prenat. Diagn.* **19,** 480–482.
12. Guanciali-Franchi, P., Calabrese, G., Morizio, E., et al. (2004) Identification of 14 rare marker chromosomes and derivatives by spectral karyotyping in prenatal and postnatal diagnosis. *Am. J. Med. Genet.* **127,** 144–148.
13. Daniel, A., Hook, E. B., and Wulf, G. (1989) Risk of unbalanced progeny at amniocentesis to carriers of chromosome rearrangements: data from United States and Canadian laboratories. *Am. J. Med. Genet.* **31,** 14–53.
14. Wang, J.-C., Passage, M. B., Yen, P. H., Shapiro, L. J., and Mohandas, T. K. (1991) Uniparental disomy heterodisomy for chromosome 14 in a phenotypically abnormal familial balanced 13/14 Robertsonian translocation carrier. *Am. J. Hum. Genet.* **48,** 1069–1074.
15. Tchinda, J., Volpert, S., McNeil, N., et al. (2003) Multicolor karyotyping in acute myeloid leukemia. *Leuk. Lymphoma* **44,** 1843–1853.
16. McNeil, N. and Ried, T. (2000) Novel molecular cytogenetic techniques for identifying complex chromosomal rearrangements: technology and applications in molecular medicine. *Expert Rev. Mol. Med.* **14,** 1–14.
17. Kolialexi, A., Tsangaris, G. T., Kitsiou, S., Kanavakis, E., and Mavrou, A. (2005) Impact of cytogenetic and molecular cytogenetic studies on hematologic malignancies. *Anticancer Res.* **25,** 2979–2983.
18. Chen, C. P., Chern, S. R., Lee, C. C., et al. (2006) prenatal diagnosis of de novo t(2;18;14)(q33.1;q12.2;q31.2), dup(5)(q34q34), del(7)(p21.1p21.1), and del(10)(q25.3q25.3) and a review of the prenatally ascertained de novo apparently balanced complex and multiple chromosomal rearrangements. *Prenat. Diagn.* **26,** 138–146.

19. Willadsen, S., Levron, J., Munne, S., Schimmel, T., Marquez, C., Scott, R., and Cohen, J. (1999) Rapid visualization of metaphase chromosomes in single human blastomeres after fusion with in vitro matured bovine eggs. *Hum. Reprod.* **2**, 470–475.
20. Marquez, C., Cohen, J., and Munne, S. (1998) Chromosome identification in human oocytes and polar bodies by spectral karyotyping. *Cytogenet. Cell. Genet.* **81,** 254–258.
21. Gutierrrez-Mateo, C., Gadea, L., Benet, J., Wells, D., Munne, S. and Navarro, J. (2005) Aneuploidy in a Robertsonian (13;14) carrier: case report. *Hum. Reprod.* **20**, 1256–1260.
22. Sandalinas, M., Marquez, C., and Munne, S. (2002) Spectral karyoytping of fresh, non-inseminated oocytes. *Mol. Hum. Reprod.* **8,** 80–585.
23. Malik, Z., Rothmann, C., Cycowith, T., Cycowitz, Z. J., and Cohen, A. M. (1998) Spectral morphometric characterization of B-CLL cells versus normal small lymphocytes. *J. Histochem. Cytochem.* **46,** 1113–1118.
24. Lee, C., Gisselsson, D., Jin, C., et al. (2001) Limitations of chromosome classification by multicolor karyotyping. *Am. J. Hum. Genet.* **68,** 1043–1047.
25. Fan, Y. S., Siu, V. M., Jung, J. H., and Xu, J. (2000) Sensitivity of multiple color spectral karyotyping in detecting small interchromosomal rearrangements. *Genet. Test.* **4,** 9–14.
26. Yaron, Y., Carmon, E., Goldstein, M., et al. (2003) The clinical application of spectral karyotyping (SKY™) in the analysis of prenatally diagnosed extra structurally abnormal chromosomes (ESACs). *Prenat. Diagn.* **23**, 74–79.
27. Heng, H. H., Ye, C. J., Yang, F., et al. (2003) Analysis of marker or complex chromosomal rearrangements present in pre- and post-natal karyotypes utilizing a combination of G-banding, spectral karyotyping and fluorescence in situ hybridization. *Clin. Genet.* **63,** 358–367.
28. Fung, J., Hyun, W., Dandekar, P., Pedersen, R. A., and Weier, H.-U. G. (1998) Spectral imaging in preconception/preimplantation diagnosis (PGD) of aneuploidy: multi-colour, multi-chromosome screening of cells. *J. Assist. Reprod. Genet.* **15,** 322–329.
29. Fung, J., Weier, H.-U. G., Pedersen, R. A., and Zitzelsberger, H. F. (2002) Spectral imaging analysis of metaphase and interphase cells. In: Rautenstrauss B. and Liehr T. (eds.). FISH technology, Springer-Verlag, Berlin Heidelberg, New York, pp 363–387.
30. Fung, J, Weier, H.-U. G, Goldberg, J. D., and Pedersen, R. A. (2000) Multilocus genetic analysis of single interphase cells by spectral imaging. *Hum. Genet.* **107,** 615–622.
31. Vollmer, M., Wenzel, F., deGeyter, C., Zhang, H., Holzgreve, W., and Miny, P. (2000) Assessing the chromosome copy number in metaphase II oocytes by sequential fluorescence in situ hybridization. *J. Assist. Reprod. Genet.* **17,** 596–602.
32. Nietzel, A., Rocchi, M., Heller, A., et al. (2001) A new multicolor-FISH approach for the characterization of marker chromosomes: centromer-specific multicolor-FISH (cenM-FISH). *Hum. Genet.* **108,** 199–204.

33. Telenius, H., Pelmear, A. H., Tunnaclife, A., et al. (1992) Cytogenetic analysis by chromosomes painting using DOP-PCR amplified flow-sorted chromosomes. *Genes Chromosomes Cancer* **4,** 257–1.
34. Zimmermann, B., Holzgreve, W., Zhong, X. Y., and Hahn, S. (2002) Inability to clonally expand fetal progenitors from maternal blood. *Fetal Diagn. Ther.* **17,** 97–100.
35. Babochkina, T., Mergenthaler, S., De Napoli, G., et al. (2005) Numerous erythroblasts in maternal blood are impervious to fluorescent in sity hybridization analysis, a feature related to a dense compact nucleus with apoptotic character. *Haematologica* **90,** 740–745.

2

Characterization of Prenatally Assessed De Novo Small Supernumerary Marker Chromosomes by Molecular Cytogenetics

Thomas Liehr

Summary

Small supernumerary marker chromosomes (sSMC) are structurally abnormal chromosomes that cannot be identified or characterized unambiguously by conventional banding cytogenetics alone, and they are generally equal in size or smaller than a chromosome 20 of the same metaphase spread. sSMC are reported in 0.043% of newborn infants and 0.075% of prenatal cases. Molecular cytogenetics is necessary to characterize the origin of an sSMC, and many highly sophisticated approaches are available throughout the literature for their comprehensive description. However, because in a prenatal diagnostic laboratory such techniques are not available, I suggest here a straightforward scheme to characterize at least the sSMC's chromosomal origin as quickly as possible. Based on this scheme, it is possible to compare the actual present case with similar cases from the literature, which are summarized on http://www.med.uni-jena.de/fish/sSMC/00START.htm. For a more wide-ranging sSMC characterization, a specialized laboratory should be contacted, e.g., my laboratory.

Key Words: Cytogenetics; prenatal diagnosis; molecular cytogenetics; fluorescence in situ hybridization (FISH); metaphase FISH; small supernumerary marker chromosome (sSMC).

1. Introduction

Small supernumerary marker chromosomes (sSMC) are still a major problem in clinical cytogenetics, because they are too small to be characterized for their chromosomal origin by traditional banding techniques; molecular cytogenetic techniques are needed for their identification. sSMC are present in ~2.5 million

From: *Methods in Molecular Biology, vol. 444: Prenatal Diagnosis*
Edited by: S. Hahn and L. G. Jackson © Humana Press, Totowa, NJ

people worldwide, and they were recently defined as structurally abnormal chromosomes that cannot be identified or characterized unambiguously by conventional banding cytogenetics alone; they are generally equal in size or smaller than a chromosome 20 of the same metaphase spread; sSMC can either be present additionally in (1) an otherwise normal karyotype, (2) a numerically abnormal karyotype (such as Turner- or Down-syndrome), or (3) a structurally abnormal but balanced karyotype with or without ring chromosome formation *(1)*.

SMC are detected mainly during prenatal diagnosis by banding cytogenetics (*see* **Fig. 1**). Still, the statement of Paoloni-Giacobino et al. *(2)* is valid, i.e., that cases with a de novo sSMC, particularly prenatally ascertained sSMC, are not easy to correlate with a clinical outcome. It is known that 32% of sSMC are derived from chromosome 15; 11% are i(12p) = Pallister-Killian, ∼10% are der(22)-, ∼7% are inv dup (22)-cat-eye-, and ∼6% are i(18p)-syndrome associated sSMC *(1)*. In general, the risk for an abnormal phenotype in prenatally ascertained de novo cases with sSMC is given as ∼13% *(3)*. This risk has been refined to 7% (for sSMC from chromosome #13, #14, #21, or

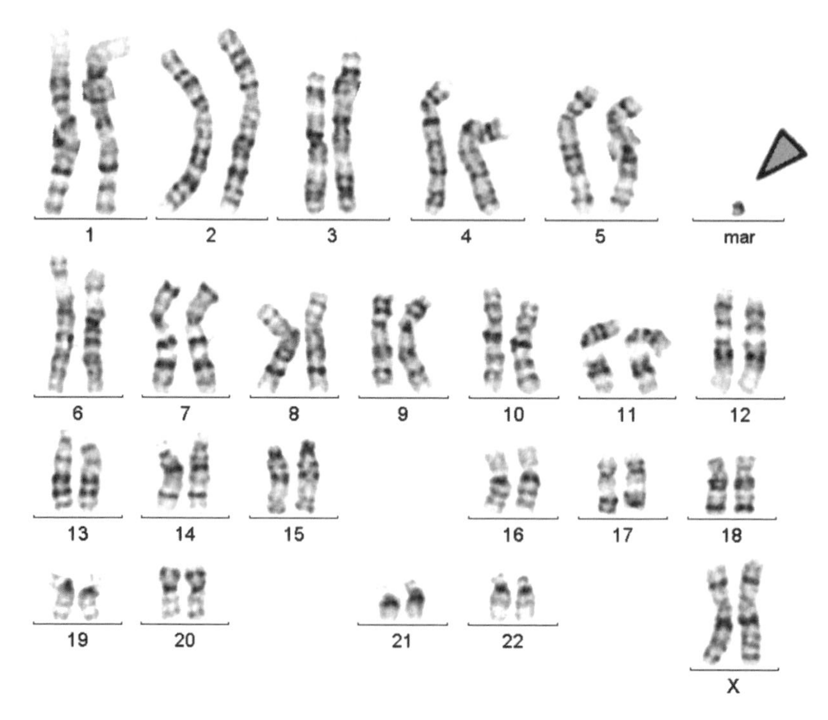

Fig. 1. An sSMC was detected in banding cytogenetics. Apart from a normal female karyotype, a minute sSMC also is present (arrowhead). This sSMC was characterized by molecular cytogenetics (data not shown) later as a min(7)(:p11.2->q11.21:)[63%]/r(7)(::p11.2->q11.21::)[25%]/r(7;7)(::p11.2->q11.21::p11.2->q11.21::)[12%].

#22) and 28% (for all nonacrocentric autosomes) *(4)*. Also, generally, sSMC inherited by a normal sSMC carrier to its children are usually not correlated with clinical problems *(5)*, even though exceptions are described *(6,* case I*)*.

Recently, my laboratory collected all available reported sSMC cases on the regularly updated sSMC homepage *(7)*. Based on these data of ~1,800 cases, **Figure 2** summarizes the frequency of the chromosomal origin in all reported sSMC cases. Even though these data are biased because not all but mainly "the interesting sSMC cases" are published, they provide the only available picture of the chromosomes most frequently involved in sSMC formation. Chromosome #15 (32%) is most frequently participating, followed by chromosomes #14/#22 (29%), #12 (11%), #18 (7%), #13/#21 (4%), #8 (~2%), #1 (~2%), #20 (~1%), X (~1%), 3 (~1%), and all others. Interestingly, a very similar distribution is observed in neocentric sSMC (*see* **Fig. 3**): chromosome #15 (24%), #8 (15%), #13 (15%), #3 (9%), #1 (7%), and #12 (4%), which seems to be in connection with the mechanisms of sSMC formation (summarized in **ref.** *(1)*).

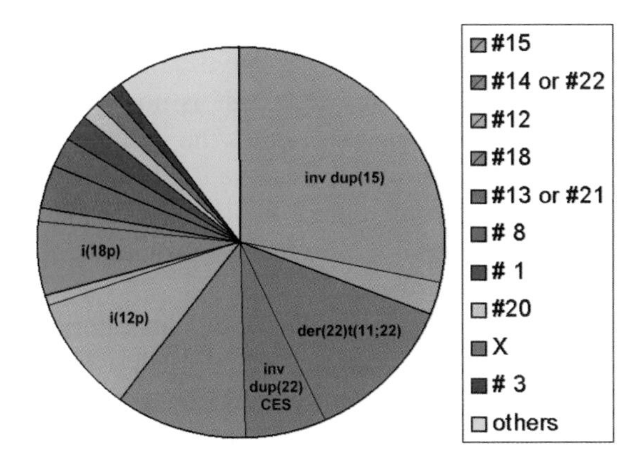

Fig. 2. Chromosomal origin of ~1,800 sSMC cases reviewed from the literature *(7)*. The 10 chromosomes most frequent involved in sSMC formation are specified in different color codes as explained in the figure the remaining 14 chromosomes are summarized under "others" in yellow. Separated from the sSMC(15) are cases with inverted duplication chromosomes 15 {(inv dup(15))}. The same was done for sSMC(14/22) for cases with derivative chromosome 22 syndrome {der(22)t(11;22)} and cat-eye syndrome (CES) with an inv(dup)(22) for the second most frequent sSMC group derived from chromosomes #14 or #22. These two chromosomes cannot be distinguished by application of the centromere-specific probe D14/22Z1. Within the two groups sSMC(12) and sSMC(18), cases with Pallister–Killian syndrome with an isochromosome of the short arm of chromosome 12 {i(12p)} and with i(18p)-syndrome, respectively, are marked in different colors.

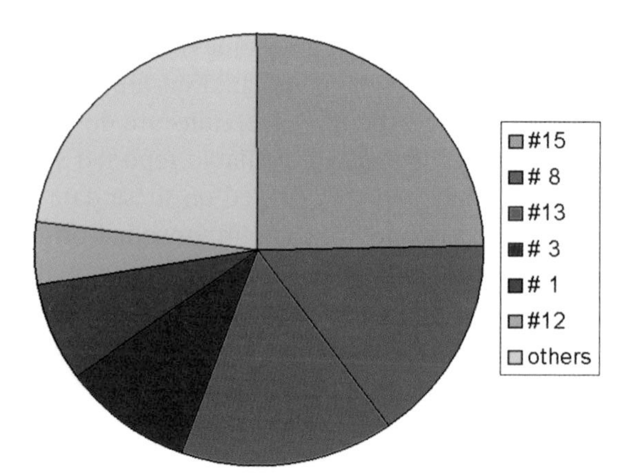

Fig. 3. Similar to **Figure 2** the chromosomal origin of 66 neocentric sSMC cases reviewed from the literature *(7)* are summarized here. The six chromosomes most frequent involved in sSMC formation are specified in different color codes as explained in the figure the remaining 18 are summarized under "others" in yellow.

Although it has to be emphasized that a comprehensive marker chromosome characterization, especially in prenatal de novo sSMC cases, would be the best for the couple and for sSMC research, this is not always possible. Thus, here a scheme is suggested that at least allows the chromosomal origin of the sSMC to be determined. This knowledge can be used to determine a better risk assessment of the clinical outcome based on similar cases summarized on the sSMC homepage *(7)*. Additionally, we recently provided a first attempt of a genotype–phenotype correlation for sSMC *(8)*.

We suggest the following scheme for diagnostic laboratories that do not have sophisticated molecular cytogenetic possibilities, such as microdissection and reverse fluorescence in situ hybridization (FISH) *(9)*, multicolor FISH applying all whole chromosome painting probes (i.e., multiplex-FISH = M-FISH *(10)*, spectral karyotyping *(11)*), or (sub)centromere-specific probes (=cenM-FISH *(12)*, subcenM-FISH *(13)*):

1. Clarify by conventional chromosome banding analysis of parental peripheral blood whether the sSMC is de novo. If the sSMC is inherited go to **step 2**; if it is de novo or its parental origin cannot be determined, go to **step 3**.
2. If the sSMC is inherited from one clinically normal parent, the identification of the sSMC origin may be replaced by genetic counseling and closely monitoring the pregnancy by high-resolution ultrasound controls. However, as mentioned, an inherited sSMC can be connected with clinical abnormalities in exceptional cases *(6,* case I*)*. In cases where the sSMC origin is be clarified nonetheless, go to **step 4**.
3. If the de novo sSMC has almost the size of chromosome 20 of the same metaphase spread, first the presence of a large inverted duplication chromosome 15 {inv

dup(15)}, an isochromosome 18 {i(18p)}, or an i(12p) should be excluded by the corresponding centromeric probe, whole chromosome painting probe, or both. In ~35% of the cases, the sSMC's origin is clarified by use of these probes; thus, go to **step 9**. In cases where the sSMC origin was not determined by the probes, go on to **step 4**.

4. If a clear positive nucleolus organization region (NOR) silver staining result is available ***(14)*** for the sSMC, its origin can be determined by hybridizing commercially available centromeric probes for all acrocentric chromosomes, exclusively, i.e., #13/#21, #14/#22, 15. In ~75% of the cases, the sSMC's origin is now clarified; thus, go to **step 9**. In cases where the sSMC origin was not determined, go to **step 5**.
5. To determine sSMC origin, use commercially available centromeric probes, testing sequentially for #8, #1, #20, X, #18, #3, and #12. Finally, even in sSMC that were NOR-negative, test for #14/#22, #15, and #13/#21, because there are cases reported with sSMC derived from acrocentric chromosomes, but without NOR *(7)*. In ~90% of the cases, the sSMC's origin is clarified; thus, go to **step 9**. In cases where the sSMC origin was not determined, go to **step 6**.
6. To determine whether the studied case is a rare case with a neocentric sSMC, a commercially available pan centromeric probe should be applied. This test has to be performed, because neocentric sSMC nearly always have a clinically adverse prognosis ***(1,7)***. In ~4% of the cases, no α-satellite DNA is present on the sSMC. If the sSMC has α-satellite DNA, go to **step 7**; if the sSMC has no α-satellite DNA, go to **step 8**.
7. Still, the sSMC with α-satellite DNA can derive from 12 different human chromosomes. If there is enough material to continue the analysis, go on in the following sequence (applying centromeric probes if nothing else is mentioned): #19 (whole chromosome painting probe), #9, #16, #17, #7, #6, #2, #4, #5 (whole chromosome painting probe), #11, and Y. Now, in ~100% of the cases, the origin of a centric sSMC is clarified; thus, go to **step 9**.
8. To characterize the origin of a neocentric sSMC according to **Figure 3**, conduct FISH applying the following sequence of whole chromosome painting probes: #15, #8, #13, #3, #1, and #12. In ~75% of the cases, the neocentric sSMC's origin is now clarified; thus, go to **step 9**.
9. In ~10% of sSMC cases, a uniparental disomy (UPD) of the sSMC's sister-chromosome is described ***(1)***. Thus, we and others ***(1,15,16)*** previously recommended that, after identification of the origin of the sSMC, its normal sister-chromosomes should be tested for their parental origin to exclude a possible UPD. UPD can be tested by molecular genetic approaches, such as microsatellite analysis ***(17)*** or methylation-specific polymerase chain reaction (PCR) ***(18)***, and at the present state of research, in our opinion, should be done for every sSMC case in which parental cell material is available. Apart from the chromosomes known to be in connection with imprinting (#6, #7, #14, #15, and #20), other chromosomes also should be tested, because uniparental isodisomy can lead to homozygotization of an otherwise recessive, disease-causing gene (e.g., ***19***).

By applying this scheme, the chromosomal origin of a sSMC can be determined and a possibly disease-causing UPD can be excluded. However, either subcenM-FISH *(13)* or array comparative genomic hybridization *(20)* are the only possibilities to exclude or define a small partial tri- or polysomy of the centromere-near region of the sSMC. After exclusion of a UPD of the sSMC's sister-chromosomes, such an imbalance determines the clinical effect of the marker chromosome *(8,21)*. When such possibilities are not available one can at least find all reported sSMC cases sorted by chromosomal origin on the sSMC homepage *(7)*.

2. Materials

2.1. Slide Pretreatment

1. 20× standard saline citrate (SSC) stock solution.
   a. 3.0 M NaCl and 0.3 M Na-citrate.
   b. Prepare with double-distilled water; adjust to pH 7.0.
   c. Autoclave.
   d. Store at room temperature (RT).
2. Pepsin stock solution 10% (w/v): Dissolve 100 mg of pepsin (Serva, Heidelberg, Germany, cat. no. 31855) in 1 ml of filtered double-distilled water at 37°C. Aliquot and store at –20°C.
3. Pepsin buffer: Add 1 ml of 1M HCl to 99 ml of distilled water and incubate at 37°C for about 20 min. Add 50 µl of the pepsin stock solution 10% (w/v), then leave the coplin jar at 37°C. Make fresh as required.
4. 1× phosphate-buffered saline (PBS), 5% $MgCl_2$ (v/v):
   a. 1 M $MgCl_2$ in 1× PBS.

2.2. FISH and Evaluation

2.2.1. Slide Denaturation

1. Deionized formamide: Add 5 g of ion exchanger Amberlite MB1 (Serva, cat. no. 40701) to 100 ml of formamide (Merck, Darmstadt, Germany, cat. no. 1.09684). Stir for 2 h (at RT), then filter twice through Whatman no. 1 filter paper. Aliquot and store at –20°C.
2. Phosphate buffer: Prepare 0.5 M Na_2HPO_4 and 0.5 M NaH_2PO_4. Mix these two solutions 1:1 (v/v) to get pH 7.0. Aliquot and store at –20°C.
3. Denaturation buffer: 70% (v/v) deionized formamide, 10% (v/v) filtered double-distilled water, 10% (v/v) 20× SSC, 10% (v/v) phosphate buffer. Make fresh as required.

2.2.2. Hybridization

1. Directly labeled centromere-specific FISH probes plus corresponding hybridization buffer (e.g., provided by Vysis, Downers Grove, Illinois, USA or

Q-Biogene, Heidelberg, Germany) for all human chromosomes: #1, #2#, #3, #4, #1/#5/#19, #6, #7, #8, #9, #10, #12, #13/#21, #14/#22, #15, #16, #17, #18, #20, X, and Y.
2. Directly labeled whole chromosome painting FISH probes plus corresponding hybridization buffer (e.g., provided by Vysis or Q-Biogene) for chromosomes, #1, #3, #5, #8, #12, #13, #21, #14/#22, #15, #18, and #19.

2.2.3. Posthybridization Washing

1. 20× SSC (*see* **Subheading 2.1., item 1**).
2. DAPI-solution: Dissolve 5 µl of 4,6-diamidino-2-phenylindol 2HCl (DAPI) stock solution (Serva, cat. no. 18860) in 100 ml of 4× SSC/0.2% Tween. Make fresh as required.

3. Methods

This section describes only the molecular cytogenetic part of the sSMC characterization—the starting material for the analysis is cell suspension in Carnoys fixative or chromosome spread prepared in a slide. Also, the suggested UPD-analysis is not described here.

3.1. Slide Pretreatment

1. Drop cell suspension of the sSMC case delivered in Carnoys fixative on dry and clean slide(s) and air dry for 10 min at RT (*see* **Note 1**).
2. Dehydrate in an ethanol series (70, 90, 100% for 3 min each) and air dry.
3. Slides are incubated in 2 SSC for 1 min at RT (in a 100-ml coplin jar on a shaker).
4. Replace 1× PBS with 100 ml of prewarmed pepsin buffer (37°C) and incubate the slides for 5 min at 37°C without agitation (*see* **Note 2**).
5. Replace fluid with 100 ml of 1× PBS/$MgCl_2$ and incubate at RT for 5 min with gentle agitation. $MgCl_2$ will block the enzymatic activity of pepsin.
6. Postfix nuclei on the slide surface by replacing 1× PBS/$MgCl_2$ with 100 ml of formalin buffer for 10 min at RT (with gentle agitation).
7. Formalin-buffer is replaced by 100 ml of 1× PBS for 2 min at RT (with gentle agitation).
8. Finally, slides are dehydrated by an ethanol series (70, 90, 100% for 3 min each) and air-dried (*see* **Note 3**).

3.2. FISH and Evaluation

The FISH procedure itself is divided in several steps: denaturation (*see* **Subheadings 3.2.1.** and **3.2.2.**), hybridization (*see* **Subheading 3.2.3.**), posthybridization washing (*see* **Subheading 3.2.4.**), and evaluation (*see* **Subheading 3.2.5.**).

3.2.1. Denaturation of Target DNA

1. Add 100 µl of denaturation buffer to the slides and cover with (24 × 50 mm) coverslips.
2. Incubate slides on a heating plate for 3 min at 75°C (*see* **Note 2**).
3. Remove the coverslips immediately by forceps and place slides in a coplin jar filled with 70% ethanol (4°C) to conserve target DNA as single strands.
4. Dehydrate slides in ethanol (70, 90, 100% all at 4°C for 3 min each) and air dry.

3.2.2. Denaturation of Probe DNA

1. For each slide to be hybridized, mix 1 µl (for centromeric probes) or 4 µl (for whole chromosome painting probes) of the probe solution with the corresponding hybridization buffer provided by the company (e.g., Vysis or Q-Biogene). Normally, the probe DNA is diluted in a final volume of 10 µl.
2. The probe DNA is denatured for 5 min at 75°C either in a thermocycler or in a water bath.
3. With a centromeric probe, the reaction tube is put on ice immediately after denaturation; with a whole chromosome painting probe, a prehybridization of 15 min has to be performed.

3.2.3. Hybridization

1. Add 10 µl of the ready probe mix (*see* **Subheading 3.2.2., step 2**) on the dry slide (*see* **Subheading 3.2.1., step 4**) and put 20- × 20-mm coverslip(s) on the drop(s) and seal with rubber cement (Marabu, Tamm, Germany, Fixogum).
2. Incubate slides for one night at 37°C in a humid chamber (*see* **Note 4**).

3.2.4. Posthybridization Washing

1. Take the slides out of 37°C, remove rubber cement with forceps, and remove coverslips by letting them "swim off" in 4× SSC/0.2%Tween at RT (100-ml coplin jar) (*see* **Note 5**).
2. Postwash the slides 1× 2 min in 0.4× SSC (56–62°C) followed by 1× 1 min in 4× SSC/0.2% Tween at RT (100 ml).
3. Counterstain the slides with DAPI solution (100 ml in a coplin jar at RT) for 8 min.
4. Wash slides in water for a few seconds and air dry.
5. Add 15 µL of Vectashield antifade (Vector Laboratories via Camon, Wiesbaden, Germany, cat. no. H1000), cover with coverslips, and examine with a fluorescence microscope.

3.2.5. Evaluation

Look for well-spread metaphase, identify the sSMC, and check whether it has been stained by the corresponding probe(s). In our experience, slides can be used in two to three further rounds of FISH experiments, which might be necessary to characterize the sSMC origin in prenatal cases with low amounts of available material (*see* **Note 6**).

4. Notes

1. If cells were grown in situ, the slides are provided with spread chromosomes and you can start immediately with steps under **Subheading 3.1.2.**
2. Pepsin pretreatment conditions and denaturation time of the target DNA should be tested in each laboratory on a single slide first. Pepsin concentration can be too stringent, resulting in clean slides without any remaining metaphase plates, nuclei, or both.
3. The pretreated slides can be hybridized immediately or stored at RT for up to 3 weeks. If longer storage is necessary, slides are stable at –20°C for several months. However, in this application this possibility will be the exception.
4. Incubation can be stopped, if necessary, after 3 h. However, especially the signals of the whole chromosome painting probes may be too weak for evaluation.
5. During the washing steps, it is important to prevent the slide surface from drying out; otherwise, background problems may arise.
6. The main problem of sSMC evaluation is that in ~40% of the cases, the sSMC is present in mosaic, i.e., the sSMC may be present between 5 and <100% of the cells. This certainly hampers the analysis, and in some cases, also a final result.

Acknowledgments

This work was supported by Dr. Robert Pfleger-Stiftung, the Deutsche Forschungsgemeinschaft (436 RUS 17/109/04, 436 WER 17/5/05).

References

1. Liehr, T., Claussen, U., and Starke, H. (2004) Small supernumerary marker chromosomes (sSMC) in humans. *Cytogenet. Genome Res.* **107,** 55–67.
2. Paoloni-Giacobino, A., Morris, M. A., and Dahoun, S. P. (1998) Prenatal supernumerary r(16) chromosome characterized by multiprobe FISH with normal pregnancy outcome. *Prenat. Diagn.* **18,** 751–752.
3. Warburton, D. (1991) De novo balanced chromosome rearrangements and extra marker chromosomes identified at prenatal diagnosis: clinical significance and distribution of breakpoints. *Am. J. Hum. Genet.* **49,** 995–1013.
4. Crolla, J. A. (1998) FISH and molecular studies of autosomal supernumerary marker chromosomes excluding those derived from chromosome 15: II. Review of the literature. *Am. J. Med. Genet.* **75,** 367–381.

5. Bröndum-Nielsen, K., and Mikkelsen, M. (1995) A 10-year survey, 1980–1990, of prenatally diagnosed small supernumerary marker chromosomes, identified by FISH analysis. Outcome and follow-up of 14 cases diagnosed in a series of 12,699 prenatal samples. *Prenat. Diagn.* **15**, 615–619.
6. Anderlid, B. M., Sahlen, S., Schoumans, J., et al. (2001) Detailed characterization of 12 supernumerary ring chromosomes using micro-FISH and search for uniparental disomy. *Am. J. Med. Genet.* **99,** 223–233.
7. Liehr, T. (2006) sSMC homepage. http://www.med.uni-jena.de/fish/sSMC/00START.htm.
8. Liehr, T., Mrasek, K., Weise, A., et al. (2006) Small supernumerary marker chromosomes–progress towards a genotype-phenotype correlation. *Cytogenet. Genome Res.* **112**(1–2), 23–34.
9. Starke, H., Raida, M., Trifonov, V., et al. (2001) Molecular cytogenetic characterization of an acquired minute supernumerary marker chromosome as the sole abnormality in a case clinically diagnosed as atypical Philadelphia-negative chronic myelogenous leukaemia. *Br. J. Haematol.* **113,** 435–438.
10. Speicher, M. R., Gwyn Ballard, S., and Ward, D. C. (1996) Karyotyping human chromosomes by combinatorial multi-fluor FISH. *Nat. Genet.* **12,** 368–375.
11. Schröck, E., du Manoir, S., Veldman, T., et al. (1996) Multicolor spectral karyotyping of human chromosomes. *Science* **273,** 494–497.
12. Nietzel, A., Rocchi, M., Heller, A., et al. (2001) A new multicolor-FISH approach for the characterization of marker chromosomes: centromere-specific multicolor-FISH (cenM-FISH). *Hum. Genet.* **108,** 199–204.
13. Starke, H., Nietzel, A., Weise, A., et al. (2003) Small supernumerary marker chromosomes (SMCs): genotype-phenotype correlation and classification. *Hum. Genet.* **114,** 51–67.
14. Bloom, S. E., and Goodpasture, C. (1975) An improved technique for selective silver staining of nucleolar organizer regions in human chromosomes. *Hum. Genet.* **34,** 199–206.
15. von Eggeling, F., Hoppe, C., Bartz, U., et al. (2002) Maternal uniparental disomy 12 in a healthy girl with a 47,XX,+der(12)(:p11–>q11:)/46,XX karyotype. *J. Med. Genet.* **39,** 519–521.
16. Kotzot, D. (2004) Advanced parental age in maternal uniparental disomy (UPD): implications for the mechanism of formation. *Eur. J. Hum. Genet.* **12,** 343–346.
17. Salafsky, I. S., MacGregor, S. N., Claussen, U., and von Eggeling, F. (2001) Maternal UPD 20 in an infant from a pregnancy with mosaic trisomy 20. *Prenat. Diagn.* **21,** 860–863.
18. Nietzel, A., Albrecht, B., Starke, H., et al. (2003) Partial hexasomy 15pter–>15q13 including SNRPN and D15S10: first molecular cytogenetically proven case report. *J. Med. Genet.* **40,** e28.
19. Thompson, D. A., McHenry, C. L., Li, Y., et al. (2002) Retinal dystrophy due to paternal isodisomy for chromosome 1 or chromosome 2, with homoallelism for mutations in RPE65 or MERTK, respectively. *Am. J. Hum. Genet.* **70,** 224–229.

20. Shaw, C. J., Stankiewicz, P., Bien-Willner, G., et al. (2004) Small marker chromosomes in two patients with segmental aneusomy for proximal 17p. *Hum. Genet.* **115,** 1–7.
21. Mrasek, K., Starke, H., and Liehr, T. (2005) Another small supernumerary marker chromosome (sSMC) derived from chromosome 2–towards a genotype/phenotype correlation. *J. Histochem. Cytochem.* **53,** 367–370.

3

Rapid Prenatal Aneuploidy Screening by Fluorescence In Situ Hybridization (FISH)

Anja Weise and Thomas Liehr

Summary

The most common aneuploidies in prenatal diagnostics of the second trimenon are trisomies of chromosomes 13, 18, and 21 and gonosomal abnormalities. To detect these trisomies as quickly as possible after amniocentesis, besides using polymerase chain reaction, fluorescence in situ hybridization (FISH), applying corresponding centromeric or locus-specific probes, is the method of choice. Results of a rapid prenatal aneuploidy screening in uncultured amniocytes by using FISH are available within 24 hr or less. However, care has to be taken against possible pitfalls in connection with the commercially available probe sets and thus interpretation of results in general. Here, we explain how rapid prenatal aneuploidy screening is performed using the Food and Drug Administration-approved Aneu Vysion kit (Abbott/Vysis), and a review is given of drawbacks and opportunities of this method.

Key Words: Prenatal diagnosis; molecular cytogenetics; fluorescence in situ hybridization (FISH); interphase FISH; pitfalls.

1. Introduction

In the 1980s, it took ≈3–4 weeks to obtain a foetal karyotype from amniocytic fluid cells, mainly due to the necessary cell cultivation *(1)*. This timeframe was shortened to ≈2 weeks *(2)*, due to progress in cytogenetic techniques, i.e., in cell growth media and also in cell preparation *(3)*. Although this is a relatively short time compared with the 3–4 weeks necessary in the 1980s, it was recognized that such moderately long waiting times cause psychological distress for the pregnant women *(1)*.

From: *Methods in Molecular Biology, vol. 444: Prenatal Diagnosis*
Edited by: S. Hahn and L. G. Jackson © Humana Press, Totowa, NJ

This distress was one of the main causes for the introduction of molecular (cytogenetics) methods for prenatal diagnosis of the most common chromosome disorders in the second trimenon: trisomy 13, 18, and 21; monosomy X; and other gonosomal aberrations, including triploidy *(4)*. Because in such approaches no time-consuming cell culture is required, results can routinely be obtained within 24 h. Quantitative fluorescence polymerase chain reaction (PCR) or fluorescence in situ hybridization (FISH) are the two methods of choice *(1)*.

The only Food and Drug Administration-approved FISH test for rapid aneuploidy screening in uncultivated amniocytic cells is the Aneu Vysion kit (commercially available at Abbott/Vysis, Downer's Grove, IL). This kit consists of two locus-specific probes for 13q14 (LSI 13) and 21q22.13˜22.2 (LSI 21) and three α-satellite DNA-probes for chromosome X, Y, and 18 (cep X, cep Y, and cep 18, respectively) (*see* **Figs. 1** and **2**). The two locus-specific and the three centromeric probes are applied in two different hybridization experiments to the samples.

Be aware that other genetic disorders, such as (un)balanced structural rearrangements, numerical aberrations besides the five tested chromosomes, microdeletions, microduplications, uniparental disomy, or mutations detectable only by molecular genetics, are not excluded after obtaining a "normal" result for the rapid aneuploidy screening by FISH. To obtain a reliable result, 50–100 interphase nuclei have to be evaluated per probe mix; for all five probes, cut-off rates of $\approx$10% were suggested recently *(4)*.

Published studies on rapid aneuploidy screening on, in total, tens of thousands of cases, agree that the results obtained by FISH and banding cytogenetics are in concordance in >99% of the cases *(4–10)*. However, there are single case reports on false-positive or false negative results due to centromeric polymorphisms, the presence of a small supernumerary marker chromosome, dicentric chromosomes, or maternal contamination *(4–14)*. As suggested recently, all these observed pitfalls and misdiagnoses of the prenatal aneuploidy screening (apart from maternal contamination) could easily be omitted by the exclusive use of locus-specific probes *(12)*. The advantage of such probes, like the already applied probes LSI 13 and LSI 21, was proven, e.g., the LSI 21 probe being suited to distinguish between free and translocation trisomy 21 in interphase *(15)*.

As recently stated in **ref.** ***4***, "In summary, the rapid prenatal aneuploidy test is—if applied with the necessary sensibility and by careful explaining its possibilities and limitations—a powerful tool for the clinician in the care for pregnant women. In future it might even be possible to extend this FISH-test to the chorionic villi samples, as a recent paper demonstrates *(16)*."

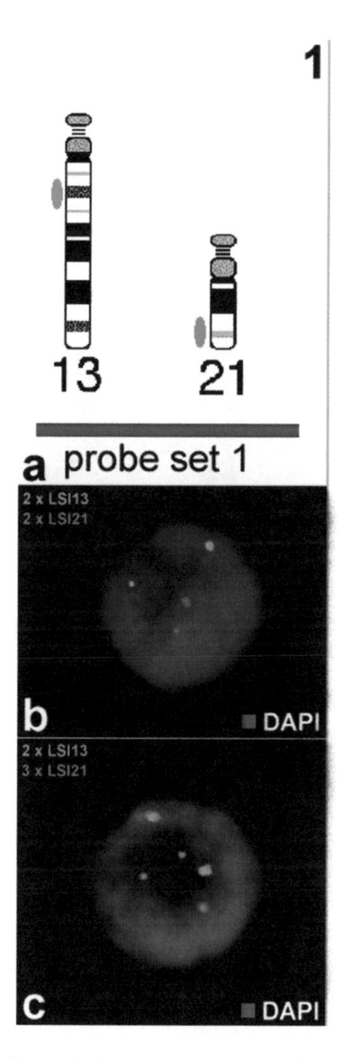

Fig. 1. (**a**) The distribution of the probes LSI 13 (SpectrumOrange) and LSI 21 (SpectrumGreen) is depicted, which comprise probe set 1. (**b**) A nucleus with normal signal distribution of LSI 13 and LSI 21, indicating, if present in at least 45 of 50 evaluated nuclei, the absence of a numerical aberration of chromosomes 13 and 21 with in the foetus. The nucleus is counterstained in blue by DAPI. (**c**) In this nucleus, there are two signals for LSI 13 and three signals for LSI 21 indicative of (free or translocation) trisomy 21.

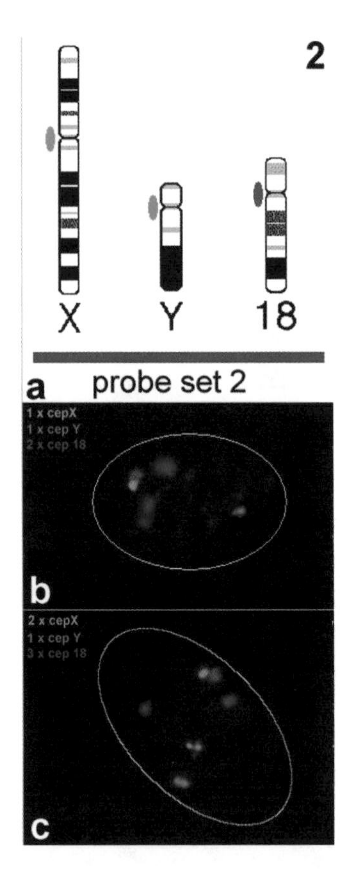

Fig. 2. (**a**) The three centromeric probes cepX, cepY, and cep18 are labeled in SpectrumGreen, SpectrumOrange, and SpectrumBlue, respectively, in probe set 2. (**b**) A nucleus with one signal each for cepX and cepY is seen in this cell of a male fetus. Additionally, two signals only are present for the cep18 probe, and thus with a high probability a free trisomy 18 is excluded. (**c**) Three signals for cep18, and two signals for cepX and one for cepY, were present in each of the analyzed nuclei. Also, three signals each for LSI 13 and LSI 21 were detected (data not shown); thus, a clinically suspected triploidy could be confirmed by FISH, and the pregnancy was terminated.

2. Materials

2.1. Preparation of Amniocytic Cells

1. Trypsin/EDTA (Seromed, Belin, Germany) prewarm 3 ml/case (37°C).
2. Mix 1× phosphate-buffered saline (PBS) with fetal calf serum (Seromed, cat. no. 50113) (4:1) (v/v) and prewarm 5 ml/case (37°C).
3. 0.075 M KCl.
4. Carnoy's fixative, methanol:glacial acetic acid (3:1) (v/v).

2.2. Slide Pretreatment

1. 20× standard saline citrate (SSC) stock solution: 3.0 M NaCl, 0.3 M Na-citrate. Set up with double distilled water. Adjust to pH 7.0 and Autoclave. Store at room temperature (RT).
2. Pepsin stock solution 10% (w/v): Dissolve 100 mg of pepsin (Serva, Heidelberg, Germany) in 1 ml of filtered double-distilled water at 37°C. Aliquot and store at –20°C.
3. Pepsin buffer: Add 1 ml of 1 M HCl to 99 ml of distilled water and incubate at 37°C for ≈20 min. Add 50 μl of the pepsin stock solution 10% (w/v) and leave the coplin jar at 37°C. Make fresh as required.
4. 1× PBS/$MgCl_2$, 5% (v/v): 1 M $MgCl_2$ in 1× PBS.

2.3. FISH and Evaluation

2.3.1. Slide Denaturation

1. Denaturation buffer: 70% (v/v) deionized formamide, 10% (v/v) filtered double-distilled water, 10% (v/v) 20× SSC, 10% (v/v) phosphate buffer. Make fresh as required.
2. Deionized formamide: Add 5 g of ion exchanger Amberlite MB1 (Serva, cat. no. 40701) to 100 ml of formamide (cat. no. 1.09684, Merck, Darmstadt, Germany). Stir for 2 h (RT) and filter twice through Whatman no. 1 filter paper. Aliquot and store at –20°C.
3. Phosphate buffer: Prepare 0.5 M Na_2HPO_4 and 0.5 M NaH_2PO_4. Mix these two solutions (1:1) to get pH 7.0. Aliquot and store at –20°C.

2.3.2. Hybridization

1. Aneu Vision Kit (cat. no. 5J3710, Abbott/Vysis).

2.3.3. Posthybridization Washing

1. 20× SSC (*see* **Subheading 2.2., item 1**).
2. 4,6-Diamidino-2-phenylindole (DAPI) solution: Dissolve 5 μl of DAPI 2 HCl stock solution (cat. no. 18860, Serva) in 100 ml of 4× SSC/0.2% Tween. Make fresh as required.

3. Methods

This section describes how amniocytic cells are prepared from uncultured amniotic fluid cells (*see* **Subheading 3.1.**); how the target of the hybridization, i.e., cytogenetic slides with interphase cells, have to be pretreated (*see* **Subheading 3.2.**); and how FISH itself is performed and evaluated (*see* **Subheading 3.3.**).

3.1. Preparation of Amniocytic Cells

For the rapid aneuploidy screening, it is necessary to prepare the uncultivated amniocytes as recommended in the Aneu Vysion kit protocol.

1. 2–3 ml of amniocytic fluid is put into a 15-ml reaction tube (*see* **Note 1**) and centrifuged (120*g*; 5 min); discard the supernatant.
2. The pellet is suspended in 3 ml of trypsin/EDTA and incubated for 15 min at 37°C.
3. Add 1× PBS/fetal calf serum (4:1), centrifuge (120*g*; 5 min), and discard the supernatant.
4. The pellet is resuspended in 5 ml of 0.075 M KCl and incubated at 37°C for 20 min.
5. Add 2 ml of Carnoy's fixative (methanol:acetic acid [3:1]), centrifuge (120*g*; 5 min), and discard the supernatant.
6. Add 3 ml of Carnoy's fixative, resuspend, and incubate at –20°C for 5 min.
7. Finalize the preparation with a last centrifugation (120*g*; 5 min); the supernatant is discarded, and the cells are diluted in 200 μl of the remaining supernatant.
8. Put whole available suspension on two small regions of one dry and clean slide and air dry for 10 min at RT.
9. Dehydrate in an ethanol series (70, 90, and 100% for 3 min each) and air dry.

3.2. Slide Pretreatment

1. Slides are incubated in 2× SSC for 1 min at RT (in a 100-ml coplin jar on a shaker).
2. Replace 1× PBS with 100 ml of prewarmed pepsin buffer (37°C) and incubate the slides for 5 min at 37°C, without agitation (*see* **Note 2**).
3. Replace fluid with 100 ml of 1× PBS/$MgCl_2$ and incubate at RT for 5 min with gentle agitation ($MgCl_2$ will block the enzymatic activity of pepsin).
4. Postfix nuclei on the slide surface by replacing 1× PBS/$MgCl_2$ with 100 ml of Formalin buffer for 10 min (RT, with gentle agitation).
5. Formalin buffer is replaced by 100 ml of 1× PBS for 2 min (RT, with gentle agitation).
6. Finally, slides are dehydrated by an ethanol series (70, 90, and 100%, 3 min each) and air-dried (*see* **Note 3**).

3.3. FISH and Evaluation

FISH-procedure itself is divided in several steps: denaturation (*see* **Subheading 3.3.1.**), hybridization (*see* **Subheading 3.3.2.**), posthybridization washing (*see* **Subheading 3.3.3.**), and evaluation (*see* **Subheading 3.3.4.**).

3.3.1. Denaturation of Target DNA

1. Add 100 μl of denaturation buffer to the slides and cover them with coverslips (24 × 50 mm).

2. Incubate slides on a heating plate for 3 min at 75°C.
3. Immediately remove the coverslips with forceps and place slides in a coplin jar filled with 70% ethanol (4°C) to conserve target DNA as single strands.
4. Dehydrate slides in ethanol (70, 90, and 100%; 4°C, 3 min each) and air dry.

3.3.2. Hybridization

1. For each slide to be hybridized, add 8 µl of the probe solution 1 (LSI 13 and LSI 21) on one region of the slide and 8 µl of the probe solution 2 (cep 18, cep X, and cep Y) on the dry slide; put 20- × 20-mm coverslips on the drops and seal with rubber cement (Marabu, Tamm, Germany, Fixogum).
2. Incubate slides for one night at 37°C in a humid chamber (*see* **Note 4**).

3.3.3. Posthybridization Washing

1. Take the slides out of 37°C, remove rubber cement with forceps, and let coverslips "swim off" in 4× SSC/0.2%Tween (at RT in a 100-ml coplin jar).
2. Postwash (*see* **Note 5**) the slides one time for 2 min in 0.4× SSC (56–62°C) followed by one time for 1 min in 4× SSC/0.2% Tween (at RT in a 100-ml Coplin jar).
3. Counterstain the slides with DAPI solution (at RT in a 100-ml coplin jar) for 8 min.
4. Wash slides in water for a few seconds and air dry.
5. Add 15 µl of Vectashield antifade (Vector Laboratories via Camon, Wiesbaden, Germany, cat. no. H1000), cover with coverslips, and view under a fluorescence microscope.

3.3.4. Evaluation

1. Fifty interphase nuclei per case and probe combination are evaluated under a fluorescence microscope. This is a semistatistic evaluation counting one, two, three, or four signals for each probe. For a questionable result within the cut-off region, enhance evaluated nuclei to 100 or more (*see* **Note 6**).

4. Notes

1. During the whole preparation, avoid uptake of the cells with a pipette, which would lead to undesired loss of cells; only mix with supernatant by moving the tube.
2. Pepsin pretreatment conditions and denaturation time of the target DNA should be tested in each laboratory on a single slide first. Pepsin concentration can be too stringent, resulting in clean slides without any remaining nuclei.
3. The pretreated slides can be hybridized immediately or stored at RT for up to 3 weeks. If longer storage is necessary slides are stable at –20°C for several months. However, in this special application, this possibility is the exception.

4. Incubation also can be stopped, if necessary, after 3 h; however, especially the signals of the LSI probes may be too weak for evaluation.
5. During the washing steps, it is important to prevent the slide surfaces from drying out, otherwise background problems may arise.
6. As long as not >7–10% of the studied cells present with three specific signals, no trisomy of the corresponding chromosomal region is suspected, and a "normal" report can be issued. Between 7–10% and 20% the evaluation can be regarded as an "unclear result," which is with high probability a normal result; however, in such cases try to evaluate 100 or more nuclei for the probe in question to come to a final decision. A value of >20% is indicative for a trisomy at least as mosaic. We suggested *(4)* differentiation between female and male fetuses and the interpretation of the results. This is because in the amniocytic fluid of male fetuses, maternal cell contamination can be detected, but not in female amniocytic fluid! Thus, the results of female fetuses should be interpreted with more care, and the cut-off rates should be lowered. Similar suggestions are made for the handling of XXX, XYY, or XXY results.

Acknowledgment

This study was supported by Friedrich-Schiller-Universität Jena grant 00469.

References

1. Hulten, M. A., Dhanjal, S., and Pertl, B. (2003) Rapid and simple prenatal diagnosis of common chromosome disorders: advantages and disadvantages of the molecular methods FISH and QF-PCR. *Reproduction* **126,** 279–297.
2. Held, K. R. (2003) QS Zytogenetik Bericht 2002/2003. *Medgen* **15,** 420–421.
3. Claussen, U., Ulmer, R., Beinder, E., and Voigt, H. J. (1993) Rapid karyotyping in prenatal diagnosis: a comparative study of the 'pipette method' and the 'in situ' technique for chromosome harvesting. *Prenat. Diagn.* **13,** 1085–1093.
4. Liehr, T. and Ziegler, M. (2005) Rapid prenatal diagnostics in the interphase nucleus: procedure and cut-off rates. *J. Histochem. Cytochem.* **53,** 289–291.
5. Eiben, B., Trawicki, W., Hammans, W., Goebel, R., Pruggmayer, M., and Epplen, J. T. (1999) Rapid prenatal diagnosis of aneuploidies in uncultured amniocytes by fluorescence in situ hybridization. Evaluation of >3,000 cases. *Fetal Diagn. Ther.* **14,** 193–197.
6. Weremowicz, S., Sandstrom, D. J., Morton, C. C., Niedzwiecki, C. A., Sandstrom, M. M., and Bieber, F. R. (2001) Fluorescence in situ hybridization (FISH) for rapid detection of aneuploidy: experience in 911 prenatal cases. *Prenat. Diagn.* **21,** 262–269.
7. Tepperberg, J., Pettenati, M. J., Rao, P. N., et al. (2001) Prenatal diagnosis using interphase fluorescence in situ hybridization (FISH): 2-year multi-center retrospective study and review of the literature. *Prenat. Diagn.* **21,** 293–301.

8. Witters, I., Devriendt, K., Legius, E., et al. (2002) Rapid prenatal diagnosis of trisomy 21 in 5049 consecutive uncultured amniotic fluid samples by fluorescence in situ hybridisation (FISH). *Prenat. Diagn.* **22,** 29–33.
9. Leung, W. C., Waters, J. J., and Chitty, L. (2004) Prenatal diagnosis by rapid aneuploidy detection and karyotyping: a prospective study of the role of ultrasound in 1589 second-trimester amniocenteses. *Prenat. Diagn.* **24,** 790–795.
10. Caine, A., Maltby, A. E., Parkin, C. A., Waters, J. J., Crolla, J. A., and UK Association of Clinical Cytogeneticists (ACC) (2005) Prenatal detection of Down's syndrome by rapid aneuploidy testing for chromosomes 13, 18, and 21 by FISH or PCR without a full karyotype: a cytogenetic risk assessment. *Lancet* **366,** 123–128.
11. Liehr, T., Beensen, V., Hauschild, R., et al. (2001) Pitfalls of rapid prenatal diagnosis using the interphase nucleus. *Prenat. Diagn.* **21,** 419–421.
12. Liehr, T., Schreyer, I., Neumann, A., et al. (2002) Two more possible pitfalls of rapid prenatal diagnostics using interphase nuclei. *Prenat. Diagn.* **22,** 497–499.
13. Estabrooks, L. L., Hanna, J. S., and Lamb, A. N. (1999) Overwhelming maternal cell contamination in amniotic fluid samples from patients with oligohydramnios can lead to false prenatal interphase FISH results. *Prenat. Diagn.* **19,** 179–181.
14. Skinner, J. L., Govberg, I. J., DePalma, R. T., and Cotter, P. D. (2001) Heteromorphisms of chromosome 18 can obscure detection of fetal aneuploidy by interphase FISH. *Prenat. Diagn.* **21,** 702–703.
15. Liehr, T., Starke, H., Beensen, V., et al. (1999) Translocation trisomy dup(21q) and free trisomy 21 can be distinguished by interphase-FISH. *Int. J. Mol. Med.* **3,** 11–14.
16. Goumy, C., Bonnet-Dupeyron, M. N., Cherasse, Y., et al. (2004) Chorionic villus sampling (CVS) and fluorescence in situ hybridization (FISH) for a rapid first-trimester prenatal diagnosis. *Prenat. Diagn.* **24,** 249–256.

4

Application of Multi-PRINS to Simultaneously Identify Chromosomes 18, X, and Y in Prenatal Diagnosis

Macoura Gadji, Kada Krabchi, Ju Yan, and Régen Drouin

Summary

Since its discovery by Koch in 1989, primed in situ labeling (PRINS) reaction provides an alternative approach for direct detection of human chromosomes. The multiple color (multi)-PRINS technique can simultaneously and specifically display different chromosomes with different colors in the same metaphase or interphase nucleus by using sequential labeling of different chromosome targets. We developed a triple-PRINS reaction on uncultured amniotic cells by omitting the blocking step and taking advantage of mixing two fluorochromes (fluorescein and rhodamine) to create a third color for simultaneous detection in the same amniocytes of three different chromosome targets, e.g., chromosomes 18, X, and Y. Fluorescent signals corresponding to chromosomes 18, X, and Y were shown as yellow, red, and green color spots, respectively. Multi-PRINS is as accurate and reliable as multicolor fluorescent in situ hybridization (multi-FISH) for the detection of aneuploidies involving chromosomes 18, X, and Y. Furthermore, multi-PRINS represents a faster and more cost-effective alternative to FISH for prenatal testing of aneuploidy in uncultured amniocytes.

Key Words: Amniotic fluid; cytogenetic techniques; human chromosomes; multiple color primed in situ labeling (multi-PRINS); molecular cytogenetics; fluorescence in situ hybridization; prenatal diagnosis.

1. Introduction

The routine test for chromosome identification in prenatal diagnosis remains standard karyotyping by using in vitro culture of cells collected during either amniocentesis or chorionic villus sampling. However, there is a real need for

From: *Methods in Molecular Biology, vol. 444: Prenatal Diagnosis*
Edited by: S. Hahn and L. G. Jackson © Humana Press, Totowa, NJ

a rapid method to detect common aneuploidies at prenatal diagnosis. Fluorescence in situ hybridization (FISH) is the method of choice to detect the common autosomal trisomies and aneuploidies of the sex chromosomes ***(1,2)***. This method is particularly useful when a rapid screening for chromosomal abnormality is desired, and the number of analyzable metaphases is limited.

Since its discovery by Koch in 1989 ***(3)***, primed in situ labeling (PRINS) reaction provides an alternative approach for direct detection of human chromosomes. It allows numerous technical variants, and a variety of targets have been detected. PRINS reaction is defined as a nonisotopic in situ labeling method for rapid and efficient detection of interphase and metaphase DNA. This method is based on the annealing of specific oligonucleotide primers, derived from chromosome-specific subsets of DNA sequences, to the denatured DNA, followed by subsequent primer extension by using *Taq* DNA polymerase. Specific labeled nucleotides are incorporated by the polymerase as well as unlabeled nucleotides. Most of the 24 human chromosomes can be identified by means of PRINS reaction by using specific primers ***(4–6)***. However, few chromosomes cannot be uniquely distinguished, because their repetitive centromeric DNA sequences are too similar ***(5,7)***. Initially, PRINS was limited to the detection of repetitive sequences such as those found in centromeric and telomeric DNA. However, some researchers have successfully used PRINS to study single-copy genes ***(8–11)***.

The multiple color primed in situ labeling (multi-PRINS) technique can simultaneously and specifically display different chromosomes with different colors in the same metaphase or interphase nucleus ***(5,6,12–14)*** by using sequential labeling of different chromosome targets. The first technique that was developed to specifically identify different chromosome targets required a blocking treatment between two standard PRINS reactions. The incorporation of dideoxynucleotide triphosphates blocks the free 3′ end of the DNA strands generated during the previous PRINS reaction, thereby avoiding the next PRINS reaction by using it as a primer to perform spurious elongation at nondesired sites ***(12–15)***. In practice, however, the blocking procedure does not always seem effective.

From this finding, we developed a triple-PRINS reactions ***(16)*** by omitting the blocking step and taking advantage of mixing two fluorochromes (fluorescein and rhodamine) to create a third color for detection of three different chromosome targets. By selecting the labeling order, either in bio-dig-bio or in dig-bio-dig order, in the sequential PRINS reaction, then detecting the labeling with a mixture of avidin-fluorescein/anti-dig-rhodamine or a mixture of anti-dig-fluorescein/avidin-rhodamine, the signals at the centromeres of three different chromosomes displayed perfect yellow, red, and green colors, respectively. Because the blocking step is omitted, the entire procedure can be

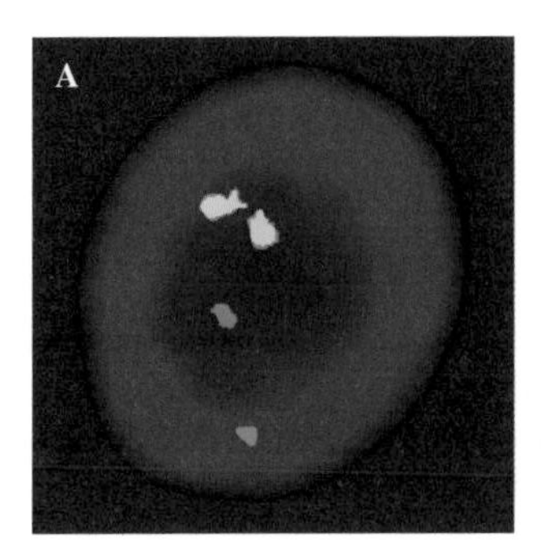

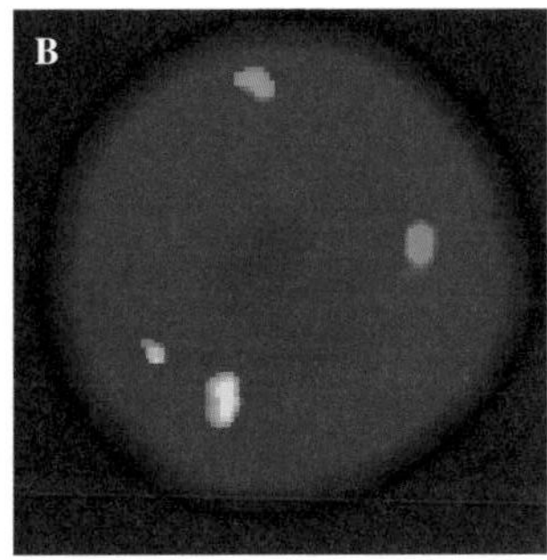

Fig. 1. Fluorescent signals corresponding to chromosomes 18 (yellow), X (red), and Y (green) on a male amniotic cell (A) and on a female amniotic cell (B) by using triple-reaction after direct harvesting. (Images were captured by using a charge-coupled device camera, and they were processed by means of computer analysis; ×1,000 magnification).

completed in <90 min. We showed that this is a practical and efficient way to carry out multi-PRINS so that even more than three chromosome targets, 18, X, and Y, could be detected in the same amniotic cell.

This technique has been successfully used to determine the copy numbers of different chromosome targets from interphase nuclei in peripheral blood and bone marrow ***(17–19)*** and in amniotic fluid cells ***(20,21)*** (*see* **Fig. 1**). Based on the principle of this multi-PRINS technique, four sequential multi-PRINS reactions were successfully performed on spermatozoa, lymphocytes ***(12)***, oocytes, and polar bodies ***(22)***. The fluorochromes fluorescein isothiocyanate, tetramethylrhodamine B isothiocyanate, and Cascade Blue incorporated in sequential PRINS reactions allowed the rapid distinct labeling of four chromosomes displaying distinct color spots (green, red, yellow, and blue).

2. Materials

2.1. Direct Harvesting of Amniotic Cells

1. Amniotic fluid (*see* **Note 1**).
2. Hanks' balanced salt solution (HBSS).
3. Precleaned microscope glass slides and coverslips (20 × 20 mm^2 or 20 × 30 mm^2).
4. Water bath set at 37°C.
5. Freshly prepared trypsin in HBSS and stored at 37°C.
6. KCl (0.56% or 0.075 M) or sodium citrate solution (0.8%). Store both solutions at 4°C and discard after 2 weeks or sooner if the solution seems cloudy or contaminated.
7. Carnoy fixation solution: methanol to glacial acetic acid (3:1) (v/v) kept at –20°C.

2.2. Preparation of Sample

1. Thermotron environmental control unit (CDS-5, Thermotron, Amsterdam, Holland).
2. Formaldehyde (37%).
3. 1 M $MgCl_2$ solution.
4. 10× phosphate-buffered saline (PBS): 1,360 mM NaCl, 20 mM KCl, 106 mM Na_2HPO_4, 15 mM KH_2PO_4, pH 7.2–7.4.
5. 1× PBS, pH 7.2–7.4 (diluted from stock 10× PBS). Store PBS solution at ambient temperature and discard after 6 months or sooner if the solution seems cloudy or contaminated.
6. Formaldehyde, PBS, $MgCl_2$ solution: 92.3 ml of 1× PBS, 5 ml of 1 M $MgCl_2$ for a final concentration of 50 mM $MgCl_2$ and 2.7 ml of 37% formaldehyde.
7. 0.1 M NaH_2PO_4.
8. 0.1 M Na_2HPO_4.
9. 1× PBD (phosphate-buffered detergent) buffer: 0.1 M NaH_2PO_4, 0.1 M Na_2HPO_4, and 0.1% Nonidet P-40 (EM Science, Merck KGaA, Darmstadt, Germany).
10. 70, 80, and 100% ethanol.
11. 20× standard saline citrate (SSC): 3.0 M NaCl, 0.30 M trisodium citrate, pH 7.2–7.4.
12. 2× SSC (diluted from 20× SSC), pH 7.2–7.4: Store the SSC solutions at ambient temperature and discard after 6 months or sooner if the solution seems cloudy or contaminated.
13. Denaturation solution: 70% formamide, 2× SSC. Prepare 50 ml of denaturation solution (70% formamide, 2× SSC) by mixing thoroughly 35 ml of formamide (ultrapure grade; Fluka, Buchs, Switzerland), 5 ml of 20× SSC, and 10 ml of distilled water in a glass coplin jar, pH 7.2–7.4. Prewarm the solution until it reaches 70°C before use. Between uses, store denaturation solution at 2–8°C and discard it after 3 to 5 days.
14. Water bath set at 70°C.
15. Incubator set at 37°C.

2.3. PRINS Reactions

1. TE (Tris-EDTA) buffer: 10 mM Tris-HCl, pH 8.0, and 1 mM EDTA.
2. Primer stock solution: this solution is prepared by dissolving each primer (*see* **Table 1** for the sequence) in TE buffer for a final concentration of 100 μM (*see* **Note 2**).
3. dNTP working solution.

    a. Dilute 1 μl each of 100 mM stock solution of 2′-deoxycytidine 5′-triphosphate (dCTP), 2′-deoxyguanosine 5′-triphosphate (dGTP) and 2′-deoxyadenosine 5′-triphosphate (dATP) (Roche Molecular Biochemicals, Laval, Quebec) with 39 μl of sterile distilled water to a concentration of 2.5 mM.

Table 1
Primers that have been used in multi-PRINS technique for prenatal diagnosis

Name	Location	Sequence	Annealing temp (°C)	Reference
18c	18	ATGTGTGTCCTCAACTAAAG	62.5	***(4)***
Xc	X	GTTCAGCTCTGTGAGTGAAA	65	***(4)***
D599	Y	TGGGCTGGAATGGAAAGGAAT CGAAAC	56	***(6)***
D600	Y paired with D599	TCCATTCGATTCCATTTTTTTC GAGAA	56	***(6)***

b. Dilute 100 mM stock solution of 2′-deoxythymidine 5′-triphosphate (dTTP) (Roche Molecular Biochemicals) to a concentration of 0.25 mM by mixing 1 μl of dTTP with 399 μl of sterile distilled water.

4. 1 mM biotin-16-dUTP or biotin-11-dUTP (Enzo Diagnostics, New York, NY) and digoxigenin-11-dUTP (Roche Molecular Biochemicals).
5. *Taq* DNA polymerase (Roche Molecular Biochemicals).
6. 10× polymerase chain reaction (PCR) buffer (Roche Molecular Biochemicals).
7. Washing buffer.

   a. 4× SSC (diluted from stock 20× SSC), 0.05% Triton X-100, or 0.2% Tween 20.

8. Blocking buffer.

   a. Washing buffer with the addition of 5% skimmed milk powder.

9. Thermal cycling machine equipped with a flat block (PTC-100, MJ Research, Watertown, MA) (*see* **Note 3**).

2.4. Detection

1. 1% Avidin-fluorescein DCS (Vector Laboratories, Burlingame, CA) and 1% anti-dig-rhodamine (Roche Molecular Biochemicals) diluted with blocking buffer.
2. 1% Avidin-rhodamine (Roche Molecular Biochemicals) and 1% anti-digoxigenin-fluorescein (Roche Molecular Biochemicals) diluted with blocking buffer. Either of two detection mixes in **item 1** can be used according to the selected labeling protocol for multiple chromosome target detection (*see* **Note 4**).
3. Counterstaining solution: In 1 ml of 0.1 M Tris-HCl, pH 7.0, dissolve 125 ng of 4′,6-diamidino-2-phenylindole (DAPI) (Sigma, St. Louis, MO), 1 mg *of p* -phenylenediamine. Then mix with 9 ml of glycerol.
4. Fluorescence microscope equipped with appropriate filters and connected to an image system (e.g., ISIS 2, MetaSystems, Belmont, MA).

3. Methods

3.1. Direct Harvesting of Amniotic Cells

1. Centrifuge 10 ml of amniotic fluid for 8 min at 670*g*.
2. Remove the supernatant by hand-suction.
3. Wash the pellet two times by adding 10 ml of HBSS separated by centrifugation for 8 min at 670*g* and hand-suction.
4. Incubate the cell suspension in 10 ml of trypsin solution in HBSS for 35 min at 37°C.
5. To remove the trypsin solution, the cells are then centrifuged for 8 min at 670*g*, and the supernatant is removed by hand-suction.
6. Resuspend the pellet in prewarmed hypotonic KCl solution or sodium citrate solution at 37°C.
7. Incubate for 45 min to 37°C followed by centrifugation for 8 min at 670*g*.
8. Remove the supernatant by hand-suction.
9. Fix the cells by adding slowly drop by drop 10 ml of cold Carnoy (–20°C).
10. Incubate for 10 min at room temperature.
11. Repeat the fixation two times more.
12. Resuspend the pellet in 2 ml of Carnoy and store it at –20°C until use.

3.2. Preparation of Sample

1. Drop 15–25 µl of the cell suspension onto the slide, precleaned and air-dried under the conditions of 28.5°C and 40% humidity in the Thermotron (*see* **Note 5**).
2. Incubate the slide in formaldehyde, PBS, $MgCl_2$ solution for 1 min at room temperature.
3. Immediately after incubation, immerse the slide in 1× PBD buffer for 1 min at room temperature.
4. Dehydrate the slide through a series of ethanol (70, 85, 100%) for 2 min each at room temperature and air dry the slide.
5. Denature the slide in 70% formamide, 2× SSC at 70°C for 2 min and successively pass them through 70, 80, and 100% cold ethanol for 2 min each at –20°C. After air drying, the slide is ready for PRINS procedure (*see* **Note 6**).

3.3. PRINS Reactions

1. Mix 5 µl of primer stock solution with 95 µl of TE buffer or water for a 5 µM primer working solution.
2. For each slide, mix the following in a 1.5-ml microtube:

   a. 5 µl of 10× PCR buffer.
   b. 4 µl of each dNTP working solution, a total of 16 µl.
   c. 2 µl of primer working solution.
   d. 1 µl of 1 mM biotin-dUTP or 1 µl of 1 mM digoxigenin-dUTP.
   e. 2.5 µl of glycerol (*see* **Note 7**).

   f. 23 μl of distilled water.
   g. 0.5 μl of *Taq* DNA polymerase (add just before starting the PRINS reaction).

3. Initiate the PRINS reaction by adding the first reaction solution (50 μl) containing a specific chromosome primer (chromosome 18) and one of the labels onto a slide. Cover the slide with a coverslip and put the slide on the flat block of the thermocycler.
4. Perform annealing and extension steps according to the different primers used for the detection of different chromosome targets (*see* **Note 8**).
5. Wash the slide briefly in 1× PBS solution for 2 min after the first PRINS reaction.
6. Add the second reaction solution containing a primer specific for the second chromosome target (chromosome X) and the other labeled dUTP on the same slide.
7. Perform the second PRINS reaction as in **step 4**.
8. Wash the slide briefly in 1× PBS solution for 2 min and add the third PRINS reaction solution containing a primer specific for the third chromosome target (chromosome Y) on the same slide. Use a labeled dUTP that is the same as the one used in the first PRINS reaction.
9. Perform the third PRINS reaction under an appropriate condition of annealing and strand elongation according to the primer used (*see* **Note 8**).
10. Wash the slide in washing buffer for 5 min with gentle agitation (*see* **Note 9**).
11. Drain the washing buffer from the slide and quickly go to detection steps.

3.4. Detection

1. Mount the slide with 100 μl of blocking buffer and a coverslip. Leave the slide at room temperature for 5 min.
2. Remove the coverslip, drain the blocking buffer of the slide, and apply 100 μl of a detection solution mix (*see* **Note 10**) to the slide.
3. Place a coverslip and incubate the slide in a moist chamber at 37°C for 30 min.
4. Remove the coverslip and wash the slide three times for 5 min each in washing buffer at room temperature or two times at 37°C with gentle agitation.
5. As soon as the slide has been air-dried in the dark, apply 10–20 μl of DAPI counterstaining solution to the slide and cover the slide with a coverslip.
6. Examine the slide under the fluorescence microscope and image system.

4. Notes

1. Do not use amniotic fluid contaminated with maternal blood because this leads to the risk for false-negative results.
2. The lyophilized oligonucleotide is stable at –20°C for 1 year or longer. It is generally accepted that oligonucleotides dissolved in TE are stable at least 6 months at –20 or 4°C. Oligonucleotides dissolved in water are stable at least 6 months at –20°C. Do not store oligonucleotides in water at 4°C. TE is

recommended over deionized water, because the pH of the water is often slightly acidic and can cause hydrolysis of the oligonucleotides.

3. In our experience, the MJ Research Thermocycler (PTC-100) gives the best results. The Hybrite from Vysis can be used as well, but the temperature program variation should be validated to perform multi-PRINS adequately.
4. In the detection solution, a relatively weak fluorochrome, e.g., avidin-fluorescein or anti-digoxigenin-fluorescein (green), should be used for the last PRINS target labeling to minimize the color mixing in the previous PRINS target. Therefore, whereas the first labeling uses biotin-dUTP and the second labeling uses digoxigenin-dUTP (labeling order bio-dig) for two different chromosome targets, respectively, an appropriate fluorochrome mix should be avidin-rhodamine/anti-digoxigenin-fluorescein. In contrast, if the labeling order is dig-bio, the fluorochrome mix should be anti-digoxigenin-rhodamine/avidin-fluorescein. The principle of selecting a relatively weak fluorochrome for the last PRINS target detection is critical for double-PRINS. For triple-PRINS reaction, the weak fluorochrome should be used to detect the first and last PRINS targets.
5. The efficiency of the PRINS relies highly on the conditions of the temperature and the humidity when making the slides. The optimal conditions for dropping cells onto slides can be reached by using a Thermotron (CDS-5) ***(23)*** or in a temperature/humidity-adjustable chamber. The best conditions may vary from laboratory to laboratory and should be determined by pretesting. Ideally, the nuclei on the slide should show a gray color and no reflection type of the nuclei themselves or bright ring around the nuclei should be seen.
6. The PRINS reaction works best with freshly prepared slides. Just let the slides dry completely, and then they are ready for PRINS. Alternatively, slides stored in –20°C up to 1 month old can be used for PRINS reaction.
7. Glycerol (5%) can effectively prevent the drying of the PRINS reaction solution during the annealing and extension steps of PRINS on the heating block of the thermocycler. Higher concentration of glycerol may cause the DNA polymerase inactivation.
8. A single step of primer annealing and strand extension for most primers was performed at 62.5°C for 15 min (example chromosome 18). For X chromosome detection, the PRINS program is as follows: annealing at 65°C for 10 min and extension at 72°C for 10 min. For Y chromosome, annealing is at 56°C for 10 min and extension is at 72°C for 10 min.
9. Usually, the last wash at room temperature for 5 min should be sufficient. In a high background, the slide can be washed once at 45°C for 5 min and twice at room temperature for 5 min each.
10. When the labeling order is bio-dig-bio, a 100-µl mix of 1% avidin-fluorescein DCS and 1% anti-dig-rhodamine is used, whereas when the labeling order is dig-bio-dig, a 100-µL mix of 1% anti-dig-fluorescein and 1% avidin-rhodamine is used. In triple-PRINS, a new color, yellow, is created for the first PRINS target, because it mixes twice the green labeling and once the red labeling (*see* **Fig. 1**).

Acknowledgments

We are grateful to Marc Bronsard for technical help. This study was supported by a grant from the Canadian Institute for Health Research to Régen Drouin. M.G. was the recipient of a studentship from the "The Foundation for Research into Children's Diseases". J.Y. is a research scholar (junior I level) of the "Fonds de la Recherche en Santé du Québec" program. R.D. holds the Canada Research Chair in Genetics, Mutagenesis and Cancer.

References

1. Cremer, T., Lichter, P., Borden, J., Ward, D. C., and Manuelidis, L. (1988) Detection of chromosome aberrations in metaphase and interphase tumor cells by in situ hybridization using chromosome-specific library probes. *Hum. Genet.* **80,** 235–246.
2. Eiben, B., Trawicki, W., Hammans, W., Goebel, R., Pruggmayer, M., and Epplen, J. T. (1999) Rapid prenatal diagnosis of aneuploidies in uncultured amniocytes by fluorescence in situ hybridization. Evaluation of >3,000 cases. *Fetal Diagn. Ther.* **14,** 193–197.
3. Koch, J. E., Kolvraa, S., Petersen, K. B., Gregersen, N., and Bolund, L. (1989) Oligonucleotide-priming methods for the chromosome-specific labelling of alpha satellite DNA in situ. *Chromosoma* **98,** 259–265.
4. Pellestor, F., Girardet, A., Lefort, G., Andreo, B., and Charlieu, J. P. (1995) Selection of chromosome-specific primers and their use in simple and double PRINS techniques for rapid in situ identification of human chromosomes. *Cytogenet. Cell Genet.* **70,** 138–142.
5. Koch, J., Hindkjaer, J., Kolvraa, S., and Bolund, L. (1995) Construction of a panel of chromosome-specific oligonucleotide probes (PRINS-primers) useful for the identification of individual human chromosomes in situ. *Cytogenet. Cell Genet.* **71,** 142–147.
6. Speel, E. J., Lawson, D., Hopman, A. H., and Gosden, J. (1995) Multi-PRINS: multiple sequential oligonucleotide primed in situ DNA synthesis reactions label specific chromosomes and produce bands. *Hum. Genet.* **95,** 29–33.
7. Hindkjaer, J., Koch, J., Brandt, C., Kolvraa, S., and Bolund, L. (1996) Primed in situ labeling (PRINS). A fast method for in situ labeling of nucleic acids. *Mol. Biotechnol.* **6,** 201–211.
8. Cinti, C., Stuppia, L., and Maraldi, N. M. (2002) Combined use of PRINS and FISH in the study of the dystrophin gene. *Am. J . Med . Genet .* **107,** 115–118.
9. Kadandale, J. S., Wachtel, S. S., Tunca, Y., Wilroy, R. S., Jr., Martens, P. R., and Tharapel, A. T. (2000) Localization of SRY by primed in situ labeling in XX and XY sex reversal. *Am. J. Med. Genet.* **95,** 71–74.
10. Kadandale, J. S., Wachtel, S. S., Tunca, Y., Martens, P. R., Wilroy, R. S., and Tharapel, A. T. (2002) Deletion of RBM and DAZ in azoospermia: evaluation by PRINS. *Am. J. Med. Genet.* **107,** 105–108.

11. Tharapel, A. T., Kadandale, J. S., Martens, P. R., Wachtel, S. S., and Wilroy, R. S. Jr. (2002) Prader Willi/Angelman and DiGeorge/velocardiofacial syndrome deletions: diagnosis by primed in situ labeling (PRINS). *Am. J . Med. Genet.* **107,** 119–122.
12. Pellestor, F., Girardet, A., Coignet, L., Andreo, B., and Charlieu, J. P. (1996) Assessment of aneuploidy for chromosomes 8, 9, 13, 16, and 21 in human sperm by using primed in situ labeling technique. *Am. J. Hum. Genet.* **58,** 797–802.
13. Gosden, J. R. and Lawson, D. (1997) Multiple sequential oligonucleotide primed in situ DNA syntheses (MULTI-PRINS). In: Gosden, J. R. (ed.). PRINS and in situ PCR protocols. Humana Press, Totowa, NJ. pp 39–44.
14. Hindkjaer, J., Koch, J., Terkelsen, C., Brandt, C. A., Kolvraa, S., and Bolund, L. (1994) Fast, sensitive multicolor detection of nucleic acids in situ by PRimed IN Situ labeling (PRINS). *Cytogenet. Cell Genet.* **66,** 152–154.
15. Mennicke, K., Yang, J., Hinrichs, F., Muller, A., Diercks, P., and Schwinger, E. (2003) Validation of primed in situ labeling for interphase analysis of chromosomes 18, X, and Y in uncultured amniocytes. *Fetal Diagn. Ther.* **18,** 114–121.
16. Yan, J., Bronsard, M., and Drouin, R. (2001) Creating a new color by omission of 3' end blocking step for simultaneous detection of different chromosomes in multi-PRINS technique. *Chromosoma* **109,** 565–570.
17. Yan, J., Marceau, D., and Drouin, R. (2001) Tetrasomy 8 is associated with a major cellular proliferative advantage and a poor prognosis. Two cases of myeloid hematologic disorders and review of the literature. *Cancer Genet. Cytogenet.* **125,** 14–20.
18. Krabchi, K., Gros-Louis, F., Yan, J., et al. (2001) Quantification of all fetal nucleated cells in maternal blood between the 18th and 22nd weeks of pregnancy using molecular cytogenetic techniques. *Clin. Genet.* **60,** 145–150.
19. Krabchi, K., Gadji, M., Yan, J., and Drouin, R. (2006) Dual-color primed *in situ* labeling (PRINS) for *in situ* detection of fetal cells in maternal blood. *Methods Mol. Biol.* **334,** 141–149.
20. Gadji, M., Krabchi, K., and Drouin, R. (2005) Simultaneous identification of chromosomes 18, X and Y in uncultured amniocytes by using multi-primed in situ labelling technique. *Clin. Genet.* **68,** 15–22.
21. Yan, J., Gadji, M., Krabchi, K., and Drouin, R. (2006) New rapid multicolor PRINS protocol. *Methods Mol. Biol.* **334,** 3–13.
22. Pellestor, F., Anahory, T., Andreo, B., et al. (2004) Fast multicolor primed in situ protocol for chromosome identification in isolated cells may be used for human oocytes and polar bodies. *Fertil. Steril.* **81,** 408–415.
23. Spurbeck, J. L., Zinsmeister, A. R., Meyer, K. J., and Jalal, S. M. (1996) Dynamics of chromosome spreading. *Am. J. Med. Genet.* **61,** 387–393.

5

Prenatal Diagnosis Using Array CGH

Catherine D. Kashork, Aaron Theisen, and Lisa G. Shaffer

Summary

Microarray-based comparative genomic hybridization (array CGH) is a fast and high-resolution approach to the diagnosis of a large number of genetic syndromes associated with loss or gain of the human genome. This technology has proven to be useful in several clinical settings, including postnatal diagnosis of mental retardation, developmental delay, and congenital malformation syndromes. We describe the use of array CGH for prenatal diagnosis of a range of chromosomal syndromes associated with congenital malformations visible by ultrasound. The procedure is reproducible in a clinical setting and provides reliable results in a short period (~5 days). Thus, depending on the array used, array CGH may develop into an excellent tool for prenatal diagnosis.

Key Words: Array CGH; microarray; congenital malformations; prenatal diagnosis; comparative genomic hybridization.

1. Introduction

Pre- and postnatal cytogenetic testing is used for the detection of structural and numerical chromosomal abnormalities that often cause an imbalance of dosage-sensitive genes in a particular chromosome or chromosomal segment; these imbalances can result in complex and specific phenotypes. Although traditional cytogenetic techniques may detect abnormalities as small as ~5 Mb, the identification of clinically significant submicroscopic chromosome rearrangements by using molecular techniques (e.g., fluorescence in situ hybridization [FISH]) emphasizes the need for higher resolution screening techniques for the characterization of chromosome abnormalities.

Microarray-based comparative genomic hybridization (array CGH) provides a higher resolution, more comprehensive means of scanning the genome for

From: *Methods in Molecular Biology, vol. 444: Prenatal Diagnosis*
Edited by: S. Hahn and L. G. Jackson © Humana Press, Totowa, NJ

copy-number imbalances. Array CGH uses two genomes, a control (reference) and a patient (test) genome, that are fluorescently labeled and competitively hybridized to a solid support containing hundreds or thousands of DNA targets. The advantages to a microarray with numerous overlapping targets compared with traditional molecular cytogenetic techniques are twofold. First, because a microarray can quickly scan the all the targets for imbalances, there is no need for suspicion of a specific phenotype before testing. Second, because the fluorescence ratios can be compared between overlapping targets, array CGH can narrow the region of copy-number change to within a fraction of a target length; thus, the resolution is limited only by the size and distance between targets. Coupled with a potential turnaround time less than that of chromosome analysis, array CGH is well suited for prenatal diagnosis of chromosomal abnormalities.

Although we restrict our comments to array CGH by using DNA from cultured amniotic fluid or chorionic villi, methods have been developed to prepare directly from prenatal specimens, further decreasing the processing time.

Cultured prenatal samples may be received for array CGH in a diagnostic laboratory after routine cytogenetics has excluded visible numerical and structural abnormalities and balanced rearrangements, which are not detectable by array CGH. On the day the sample is received, DNA can be extracted and sonicated in preparation for labeling and hybridization. Extraction is performed by a lysis procedure using a commercial extraction kit. Sonication is performed to fragment the genomic DNA into segments approx 500 bp to 4 kb.

After extraction and sonication, the genomic DNA of both the patient and the control specimen is labeled by random priming by using a commercial kit. Each patient sample and control may be labeled twice, with different fluorochromes for a dye reversal hybridization strategy. After labeling, the genomic DNA is purified, precipitated, and denatured before being placed on the array. The arrays are hybridized overnight with the two differentially labeled genomes. The arrays are then washed to remove any unbound genomic DNA. Slides are scanned using a two-color scanner and analyzed.

Herein, we report common array CGH procedures used to take a sample from receipt to results in a clinical diagnostic setting.

2. Materials

2.1. Preparation of Samples

1. AmnioMax Medium (Invitrogen, Carlsbad, CA).
2. 30-mm petri dishes with coverslips, sterile (MatTek, Ashland, MA).
3. T-25 vented flasks (Falcon; BD Biosciences Discovery Labware, Bedford, MA).

2.2. DNA Extraction

1. Puregene Genomic DNA Purification kit for whole blood (Gentra Systems, Inc., Minneapolis, MN).
2. Glycogen solution, 20 mg/ml.
3. Proteinase K solution, 20 mg/ml.

2.3. Quantification of Genomic DNA

1. Hoechst Dye 33258 (Sigma-Aldrich, St. Louis, MO): Prepare by dissolving 10 mg of Hoechst Dye 33258 in 10 ml of Millipore water (Millipore Corporation, Bedford, MA). Store in the dark.
2. 10× TNE buffer: Dissolve 12.11 g of Tris-HCl, 3.72 g of EDTA-$Na_2 \cdot 2H_2O$, and 116.89 g of NaCl in 800 ml of Millipore water. Adjust pH to 7.4 with concentrated HCl. Bring total volume to 1,000 ml. Filter before use (0.45 μm).
3. Calf thymus DNA standard (100 ng/μl): Adding 250 μl of sterile H_2O to the vial of concentrated calf thymus (Sigma-Aldrich). Yields 1 ng/μl concentration.

2.4. Blocking

1. Salmon sperm DNA.
2. 20× standard saline citrate (SSC): 3.0 M NaCl and 0.3 M Na-citrate. Prepare with double-distilled water; adjust to pH 7.0. Autoclave and store at room temperature.
3. Fraction V bovine serum albumin (BSA) (Sigma-Aldrich): Combine 10 g of fraction V BSA, 150 ml of 20× SSC, and 840 ml of double distilled water.
4. 10% sodium dodecyl sulfate (SDS): Prepare by dissolving 100 g of SDS in 900 ml of Millipore water at 68°C until dissolved. Adjust to pH 7.2 with HCl or NaOH, then bring volume to 1,000 ml.
5. Formamide.
6. Dextran sulfate.
7. Hybridization solution: 50% formamide, 10% dextran sulfate, and 2× SSC.

2.5. Labeling

1. BioPrime Labeling lit (Invitrogen).
2. Cyanine 3-dCTP (PerkinElmer Life and Analytical Sciences).
3. Cyanine 5-dCTP (PerkinElmer Life and Analytical Sciences).
4. DNA triphosphate set (Roche Diagnostics, Indianapolis, IN).

2.6. Purification of Labeled Genomic DNA

1. Micron PCR filter (Millipore Corporation).

2.7. Hybridization of Genomic DNA

1. Human Cot I DNA.
2. 5 M NaCl: Dissolve 292.2 g of NaCl in 800 ml of sterile water. Adjust the volume to 1 liter after the NaCl is dissolved.
3. ULTRAhyb hybridization buffer (Ambion, Austin, TX).
4. Deionized formamide (VWR, West Chester, PA).
5. 20× SSC.

2.8. Washing the Microarray

1. 20× SSC.
2. Deionized formamide (VWR).
3. 10% SDS.
4. Phosphate-buffered saline (PBS).
5. Washing solution: 50% formamide, 2× SSC, and 0.1% SDS. Warm to 45°C before use.

3. Methods

3.1. Preparation of Samples

1. Amniotic fluid specimens are usually received in 50-ml tubes. Divide the fluid between two 15-ml conical tubes. Centrifuge at 168*g* for 10 min.
2. After centrifugation, evaluate the pellet. If the sample looks clear (pellet should look white), then set up one to two coverslips for chromosomes or FISH. Set up the other pellet in a T-25 flask.
3. Working under a hood, label the dishes and flasks with the patient information as well as the set-up date and coverslip number (e.g., 1A, 2B, and so on). Using a 10-ml pipette, aseptically remove most of the supernatant from the tube, being careful not to disturb the pellet.
4. Using a 2-ml sterile pipette, add enough Amniomax medium to the cell pellet to bring the volume to 0.5 ml per petri dish. Triturate the suspension gently but thoroughly, without creating bubbles. Dispense 0.5 ml carefully to each coverslip to form a bead.
5. For the flask, add ~2–3 ml of Amniomax to the pellet. Triturate the suspension gently and carefully add to the bottom of the T-25 flask. Carefully rock the flask to cover the entire surface area.
6. Allow the cultures to attach overnight and feed the coverslips with 1.5 ml of Amniomax to a total volume of 2 ml. Continue to monitor the cultures for attachment and growth. Feed the cultures as necessary.
7. When the coverslips show sufficient colony growth (5–10 independent colonies), harvest the coverslips using standard cytogenetic harvesting techniques *(1)*.
8. When the flask has sufficient growth for subculturing, trypsinize the culture to lift the cells and spread to another flask, feeding both of the flasks. Continue growth until three T-25s or one T-75 at 70% confluence is produced. When there is sufficient growth for extraction, trypsinize the cells and proceed to DNA extraction.

3.2. DNA Extraction

DNA extraction is performed using the Puregene DNA Purification kit (Gentra Systems, Inc., www.gentra.com) according to manufacturer's instructions.

3.2.1. Cell Lysis

1. After lifting the cells from the flask surface, centrifuge for 10 min at 671*g* in a 15-ml conical tube. Remove supernatant with a pipette, leaving behind the visible white pellet and ≈10–20 μl of liquid.
2. Vortex tube vigorously to resuspend cells.
3. Add 900 μl of cell lysis solution, resuspend, and lyse cells either by vortexing or by pipetting, and divide among three 1.5-ml tubes. Incubate samples for 5 min at 37°C to ensure sample resuspension.

3.2.2. RNase A Treatment

1. Add 1.5 μl of RNase A solution to the cell lysate in all tubes within the rack.
2. Mix sample by inverting tube 25 times and incubate at 37°C for 5 min.

3.2.3. Protein Precipitation

1. Cool sample to room temperature by placing on ice for 8 min.
2. Add 100 μl of protein precipitation solution to the RNase A-treated cell lysate.
3. Vortex two tubes vigorously at high speed for 25 s to mix protein precipitation solution uniformly with cell lysate (required to completely remove contaminants).
4. Centrifuge at 13,000–16,000*g* for 3 min. A tight, dark brown pellet should form from the precipitated protein (*see* **Note 1**).

3.2.4. DNA Precipitation

1. Pour supernatant from the protein precipitation into a clean 1.5-ml microfuge tube containing 300 μl of 100% isopropanol at room temperature.
2. Mix sample by gently inverting 50 times.
3. Centrifuge at 13,000–16,000*g* for 1 min. DNA should be visible as a small white pellet.
4. Carefully aspirate any liquid from the pellet. Add 300 μl of 70% ethanol at room temperature and invert tube several times to wash DNA pellet.
5. Centrifuge at 13,000–16,000*g* for 1 min. Carefully aspirate any liquid from the pellet.
6. Recap samples, pulse centrifuge, and remove any additional supernatant. Let the pellet dry for 3–5 min.

3.2.5. Standard DNA Hydration

1. Add 50 µl of DNA hydration solution to each tube (*see* **Note 2**).
2. Rehydrate DNA by incubating at 37°C for 60 min. Vortex the sample and spin down in the centrifuge for 3 s. Combine the suspension of all three tubes into one tube.
3. Store DNA at 4°C. For long-term storage, store at –20 or –80°C.

3.3. Quantification of Genomic DNA

1. Turn on the fluorometer and let the UV lamp warm up for at least 15 min before use.
2. Make fresh low-range assay solution by adding 90 ml of sterilized Millipore water, 10 ml of 10× TNE buffer, and 10 µl of Hoechst dye stock solution at least 30 min in advance of measurements to let it warm up to room temperature.
3. Aliquot 2 ml of assay solution in each disposable cuvette.
4. Place the cuvette with assay solution in the fluorometer, close door, and press the "zero" button.
5. Add 2 µl of the standard DNA (100 µg/ml) to the assay solution; mix with a disposable pipette. Place cuvette in the machine, close door, and press "calibrate."
6. For each sample to be measured, use the cuvette with assay solution only to "zero," and add 2 µl of DNA to the same cuvette, mix, and measure by closing the door. The measurement that appears on fluorometer is the final concentration (nanograms per microliter).
7. Calibrate the fluorometer after measuring every 10 samples by repeating **steps 4** and **5**.

3.4. Blocking of the Microarray

3.4.1. Reagent Preparation

For each slide blocked, prepare salmon sperm DNA mix by combining 10.7 µl of salmon sperm with 0.3 µl of 10% SDS and 7.7 µl of hybridization solution (50% formamide, 10% dextran sulfate, and 2× SSC).

3.4.2. Procedure

1. Denature the salmon sperm DNA mix in the thermoblock at 72°C for 10 min.
2. Add 10 µl of sterile Millipore water to the holes in the chamber. Place the hybridization chambers along with the slides to be blocked on the hot-plate at 37°C.
3. Pipette 6.8 µl of the BSA onto the upper right edge of the printed area (rather than the middle) of the slide. Then, add 8.7 µl of the denatured salmon sperm DNA onto the lower right edge of the printed area.
4. Cover with a 22- × 22-mm^2 coverslip.

5. Place the slide in a Corning hybridization chamber. Assemble the incubation chamber with the top and clips.
6. Place the incubation chamber in a Ziploc bag with moistened paper towels to keep the environment humid.
7. Place the Ziploc bag in a shaking incubator at 45°C for 2 h at 220 rpm.
8. Disassemble the hybridization chamber. To remove the coverslip, place the slide in a petri dish of sterile Millipore water until the coverslip lifts off.
9. Rinse the slide in sterile Millipore water. Move the slides to a staining jar rack.
10. Heat sterile Millipore water in an oven at 80°C. Pour the hot water in staining jars and submerge the slides for 2 min.
11. Quickly transfer the slides to cold 95% ethanol for 1 min.
12. Dry the slides in the centrifuge with the plate spinner. Centrifuge for 4 min at 329*g*.
13. Slides are ready for hybridization or can be stored in the desiccator for later use.

3.5. Labeling Genomic DNA

1. For each sample (patient and control), prepare two labeled tubes: one tube labeled cyanine 3 (Cy3) and the other tube labeled cyanine 5 (Cy5). Use the safe lock tubes to prevent the lids from opening when boiling. Put 500 ng of purified, digested DNA into each tube and add H_2O to bring the total volume to 12 μl. Then, add 10 μl of 2.5× random primer/reaction buffer mix.
2. Briefly vortex the tube and do a quick spin to collect all the liquid at the bottom of the tube. Place the tubes into a foam or plastic float and set them into a beaker with boiling water for 5 min.
3. Remove the beaker from the boiling water. Immediately place the tubes on ice for 5 min in a darkened room.
4. While keeping the samples on ice, add 1.25 μl of the nucleotide mix to each tube.
5. Add 0.75 μl of cyanine 3-dCTP to the normal control Cy3 and the patient Cy3 tubes. Add 0.75 μl of cyanine 5-dCTP to normal control Cy5 and the patient Cy5 tubes.
6. Add 1 μl of Klenow fragment to each tube, mix the sample by brief vortexing, and recollect the samples by brief centrifugation.
7. Incubate the tubes in the oven at 37°C for 2 h.
8. Stop the reaction by adding 2.5 μl of stop buffer.
9. Incubate the tubes in a heating block well filled with water at 72°C for 5 min.

3.6. Purification of Genomic DNA

1. Use the Millipore Micron-PCR filter units that are made up of two parts: the purple filter column and the clear 1.5-ml collection tube. Assemble one unit for every sample to be purified by placing one purple filter column in a labeled clear collection tube (purple side up and white side down).
2. Add 472 μl of autoclaved water to each of the sample tubes.

3. Pipet the 500 µl of diluted sample onto the filter. Close the lid gently but securely and spin the tubes for 15 min at 4,890*g*.
4. After spinning the tubes, transfer the purple filter column (purple side up) to a clean, labeled collection tube and discard the original collection tube and its remaining contents.
5. Place 30 µl of autoclaved water onto the center of the filter. Gently take the filter out of the tube, invert it, and place it back into the collection tube with the white side facing up (purple side down). Once the tubes have been inverted, the lids cannot be closed.
6. Place the tubes in the microfuge and spin the tubes for 2 min at 4,890*g*.
7. After spinning, the filter column can be removed. The two 30-µl samples can be combined before hybridization.

3.7. Hybridization of Genomic DNA

1. Combine the 30 µl of Cy3-labeled normal control DNA sample with the 30 µl of Cy5-labeled patient sample and combine the 30 µl of Cy5-labeled normal control DNA sample with the 30 µl of Cy3-labeled patient sample. Add 50 µg of human Cot I to each of the two tubes.
2. Calculate the total volume (X µl of Cot + 30 µl of Cy3 + 30 µl of Cy5). Add 10% of the total volume of 5 M NaCl and vortex the tubes to mix. Then, add 100% of the total volume of room temperature isopropanol to the tube. Mix the samples well by gently inverting the tubes 30× and incubate in the dark at room temperature for 15 min.
3. Centrifuge the samples at 29,200*g* for 20 min.
4. Using a micropipettor, remove the supernatant, avoiding the pellet.
5. Rinse the pellets once with 500 µl of 70% room temperature ethanol and spin for 7 min at 29,200*g*. Using a micropipettor, remove the supernatant and allow the pellets to air dry for 2–5 min.
6. Add 30 µl of ULTRAhyb hybridization buffer to each tube. Vortex and centrifuge before adding to the pellet. Incubate in a 37°C water bath for 60 min.
7. Denature the samples by incubating in a heat block well filled with water at 72°C for 5 min.
8. After the denaturation of the samples, incubate the samples in the oven at 37°C for 60 min. During this time, prepare the hybridization container. Dampen two paper towels with distilled water and place them in a heavy-duty sealable freezer bag. Place the freezer bag into the oven along with the samples to warm the bag. Get out the plastic hybridization chambers. Using a cotton-tipped applicator, dry the wells in the bottom of the chamber. Near the end of the 60-min incubation time, add 10 µl of 2× SSC + 50% formamide to each well.
9. Remove the tubes from the oven and use a marker slide to locate the array.
10. Pipette the contents of the tube (30 µl) onto the middle of the array, starting at the upper end and moving down to the lower end, forming a line of hybridization solution.

11. CAREFULLY lay a 22- × 40-mm^2 coverslip onto the hybridization solution. Align with a marker slide to ensure that the hybridization solution covers the array.
12. Place the slide in a hybridization chamber with the array side up. Gently lay the top of the chamber over the bottom and attach the black clips.
13. Wrap with aluminum foil, being very careful not to invert the chamber. Put the chambers in a heavy-duty plastic bag with wet paper towels and close the bag. Fold down the top and secure.
14. Incubate the bag in a 37°C shaking incubator for 14–16 h.

3.8. Washing the Microarray

3.8.1. Preparation

1. Prewarm 50 ml of the 50% formamide, 2× SSC, and 0.1% SDS solution. Warm at 45°C in the FISH water bath for at least 30 min.
2. Prepare 1× PBS.
3. Prepare ~200 ml of 0.2× SSC.

3.8.2. Procedure

1. Fill petri dishes with 1× PBS. In the dark, remove the slides from the hybridization chamber. Place slide with coverslip in the 1× PBS solution and carefully slide off the coverslip.
2. Fill a new dish with the 50% formamide wash and place slides in the dish. Cover and place in the 45°C Shake-n-Bake oven for 20 min.
3. Pour off the formamide wash. Add 1× PBS to cover the slides. Agitate dishes vigorously. Pour off into waste. Add 1× PBS to cover the slides for a second wash and repeat as before. Move to a fresh dish containing approximately 25 ml of 1× PBS. Place on a rocking shaker at room temperature for 20 min.
4. Prepare three staining dishes of 1× PBS, one jar of 0.2× SSC, and one jar of Millipore water.
5. When the 20× PBS wash is complete, move the slides to a staining rack of 1× PBS and dunk up and down for 1 min. Repeat this step through the other two jars of PBS. Then, move the slides to 0.2× SSC and dunk 10 times up and down. Move the slides to another staining jar filled with Millipore water. Allow the slides to stay in the water as each slide is dried with a whoosh duster or centrifuge in a plate spinner at 1,400 rpm for 4 min.

3.9. Scanning and Analysis of Microarray

Although the procedures for scanning and analysis of microarray data will vary depending upon the platform used, image analysis entails four primary steps: scanning, image analysis, data normalization, and visualization. Images are scanned using a dual-laser scanner. Two simultaneous scans of each array at two different wavelengths are required; if Cy3 and Cy5 are used for labeling, the arrays will be scanned at 532- and 635-nm wavelengths, respectively. After scanning, an image analysis program defines the positions of each DNA spot in the image, based on the file generated from the printer. The fluorescence intensity of each spot is then calculated, and the ratio of these spots for each subject are averaged and analyzed. The background noise is subtracted and the ratio of fluorescence intensities derived from hybridized test and reference DNA are calculated and normalized (*see* **Note 3**). The results are then converted to $\log_2$ scale (*see* **Note 4**) and plotted using an image plotting program.

4. Notes

1. If a tight dark pellet does not form, repeat **step 3**, followed by incubation on ice for 5 min, and then continue to **step 4**.
2. Samples are stable in DNA hydration solution for at least 9 years at 4°C.
3. Because there will be variations between the relative Cy3 and Cy5 signals from one slide to another, the hybridization ratios must be normalized. Normalization can be either linear or nonlinear. The most common method of linear normalization uses the total signal, which is the sum of all signals from the array. The most common method of nonlinear normalization calculates the normalization for subsets of clones in the same way that normalization for the whole array is performed. Nonlinear normalization is usually preferable to linear normalization, because it accounts for variations in the normalization factor over the whole signal range, whereas linear normalization does not.
4. When dye-reversal experiments are performed, $\log_2$ ratios are preferable to raw ratios for plotting hybridization results, because, when Cy5/Cy3 experiments are plotted concurrently with Cy3/Cy5 experiments by using raw ratios, a single-copy loss would give a value of 0.5 (1:2 ratio) on one experiment but a value of 2 (2:1 ratio) on the dye-reversed experiment, producing a nonsymmetrical plot. Plotting the log_2 ratio corrects for this bias, resulting in a value of 1 for a 2:1 ratio and a value of –1 for a 1:2 ratio. Furthermore, a normal 2:2 ratio will fluctuate around a value of 1 using raw ratios, but the $\log_2$ of these spots will fluctuate around 0 *(2,3)*.

Acknowledgments

We thank the technical staff at Signature Genomic Laboratories who developed and refined these protocols for use in a clinical laboratory.

References

1. Barch, M. J., Knutsen, T., and Spurbeck, J. L. (eds.) (1997) The AGT cytogenetics laboratory manual. Lippincott Williams & Wilkins, Philadelphia, PA.

2. Hui, A. B., Lo, K. W., Yin, X. L., Poon, W. S., and Ng, H. K. (2001) Detection of multiple gene amplifications in glioblastoma multiforme using array-based comparative genomic hybridization. *Lab. Invest.* **81,** 717–723.
3. Vissers, L. E., de Vries, B. B., Osoegawa, K., et al. (2003) Array-based comparative genomic hybridization for the genomewide detection of submicroscopic chromosomal abnormalities. *Am. J. Hum. Genet.* **73,** 1261–1270.

6

Prenatal Detection of Chromosome Aneuploidy by Quantitative Fluorescence PCR

Kathy Mann, Erwin Petek, and Barbara Pertl

Summary

Autosomal chromosome aneuploid pregnancies that survive to term, namely, trisomies 13, 18, and 21, account for 89% of chromosome abnormalities with a severe phenotype. They are normally detected by full karyotype analysis of cultured cells. The average reporting time for a prenatal karyotype analysis is approximately 14 days, and in recent years, there has been increasing demand for more rapid prenatal results with respect to the common chromosome aneuploidies, to relieve maternal anxiety and facilitate options in pregnancy. The rapid tests that have been developed negate the requirement for cultured cells, instead directly testing cells from the amniotic fluid or chorionic villus sample, with the aim of generating results within 48 h of sample receipt. Interphase fluorescence in situ hybridization is the method of choice in some genetic laboratories, usually because the expertise and equipment are readily available. However, a quantitative fluorescence (QF)-PCR–based approach is more suited to a high-throughput diagnostic service. This approach has been investigated in a small number of pilot studies and reported as a clinical diagnostic service in many studies. It may be used as a stand-alone test or as an adjunct test to full karyotype analysis, which subsequently confirms the rapid result and scans for other chromosome abnormalities not detected by the QF-PCR assay.

Key Words: quantitative fluorescence-polymerase chain reaction (QF-PCR); chromosome aneuploidies; rapid prenatal test.

1. Introduction

Autosomal chromosome aneuploid pregnancies that survive to term, namely, trisomies 13, 18, and 21, account for 89% of chromosome abnormalities with a severe phenotype *(1)*. They are normally detected by full karyotype analysis

From: *Methods in Molecular Biology, vol. 444: Prenatal Diagnosis*
Edited by: S. Hahn and L. G. Jackson © Humana Press, Totowa, NJ

of cultured cells. The average UK reporting time for a prenatal karyotype analysis is approximately 14 days ***(2)***, and in recent years, there has been increasing demand for more rapid prenatal results with respect to the common chromosome aneuploidies, to relieve maternal anxiety and facilitate options in pregnancy. The rapid tests that have been developed negate the requirement for cultured cells, instead directly testing cells from the amniotic fluid (AF) or chorionic villus sample (CVS), with the aim of generating results within 48 h of sample receipt. Interphase-fluorescence in situ hybridization (FISH) ***(3,4)*** is the method of choice in many genetic laboratories, usually because the expertise and equipment are readily available. However, a quantitative fluorescence-polymerase chain reaction (QF-PCR)-based approach is more suited to a high-throughput diagnostic service. This approach has been investigated in a small number of pilot studies ***(5–9)*** and reported as a clinical diagnostic service ***(10–14)***. It may be used as a stand-alone test or as an adjunct test to full karyotype analysis, which subsequently confirms the rapid result and scans for other chromosome abnormalities not detected by the QF-PCR assay.

1.1. Principle

QF-PCR refers to the amplification of chromosome-specific polymorphic microsatellite markers by using fluorescence-labeled primers, followed by quantitative analysis of the products on a genetic analyzer to determine copy number of specific chromosomal material. Tetranucleotide repeat markers are used to minimize PCR-generated "stutter bands" (amplified sequences that are one to three repeat units smaller than the true allele size, and both mitotic and meiotic allele instability). Where a microsatellite marker is heterozygous, the ratio of its allele peak areas represents a disomic (1:1) or trisomic (2:1, 1:2 or 1:1:1) chromosome complement (*see* **Fig. 1**). A marker is uninformative if only a single peak is observed.

Due to allele size heterogeneity and differences in sample type and quality, the amplification of a single marker relative to other markers in the assay may vary greatly. Thus, a comparison of allele peak areas between markers, as an indicator of chromosome copy number, is not recommended. Furthermore, because only peak areas within a single locus are compared, allele dosage ratios are more resilient to the effects of the plateau phase of the PCR than other dosage assays ***(15)***.

The procedure described here uses a "one-tube test," where 16 markers are coamplified in one multiplex reaction (*see* **Table 1** and **Fig. 1**). Five markers are used for both chromosomes 13 and 21, and six markers are used for chromosome 18. A separate sex chromosome multiplex can be used for sexing purposes and to screen for sex chromosome aneuploidy (*see* **Table 2** and **Fig. 2**) ***(16)***.

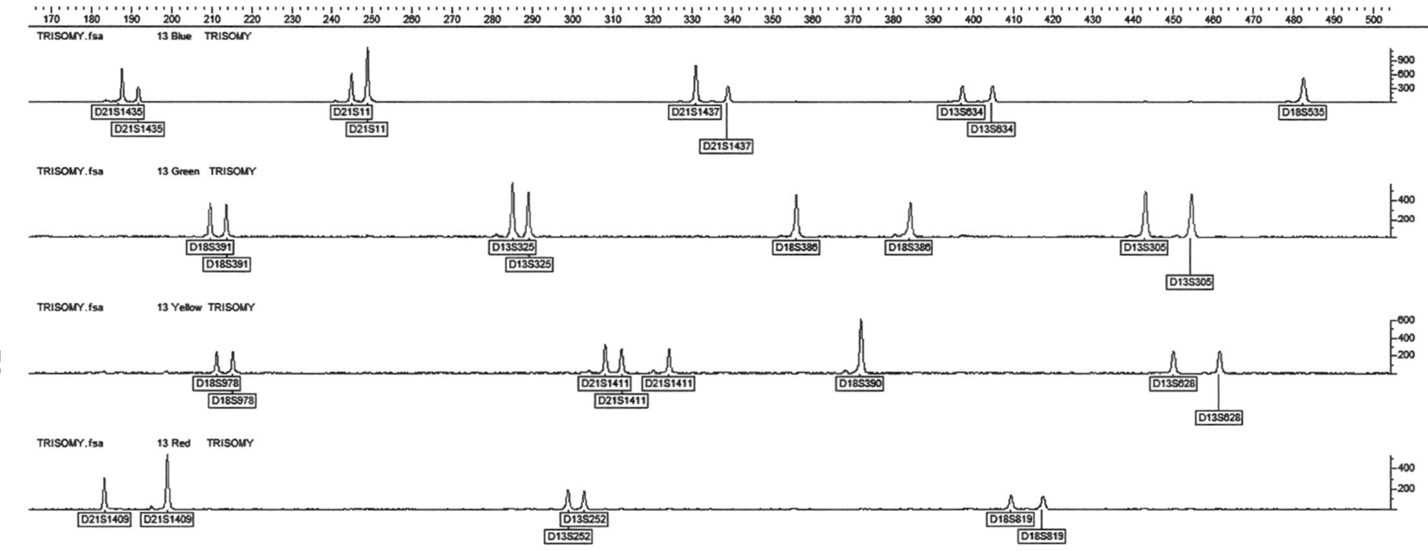

Fig. 1. A Genotyper profile of a trisomy 21 sample amplified in the QF-PCR multiplex and analyzed on a 3100 genetic analyzer. Size in base pairs is shown on the horizontal axis, fluorescent units on the vertical axis. Peaks are labeled with marker name. Heterozygous markers for chromosomes 13 and 18 exhibit 2 allele peaks with peak areas in a 1:1 ratio. Markers represent trisomy 21 as either 3 allele peaks in a 1:1:1 ratio (D21S1411), or 2:1 (D21S1435 and D21S1437) or 1:2 (D21S1409 and D21S11) allele ratios.

Table 1
Details of primers used in the trisomy multiplex. Size ranges given are those used in Genotyper, version 3.7 (Applied Biosystems). Heterozygosity values are based on our cohort and may vary in other populations. All markers are tetranucleotide repeats

Marker name	Location	Heterozygosity	Size range (bp)	Primer sequences (5′-3′)	Final conc. (μM) of each primer
D13S252-F D13S252-R	13q12.1	0.85	260–330	PET-GCAGATGTACTGTTTTCCTACCAA AGATGGTATATTGTGGGACCTTGT	0.675
D13S305-F D13S305-R	13q13.3	0.75	418–482	VIC-TGGTTATAGAGCAGTTAAGGCAC GCCTGTTTGAGGACCTGTCGTTA	1
D13S628-F D13S628-R	13q31.1	0.688	425–474	NED-TGGATGAATACGCCACTTTTC TGGTTAAAAGATTGCCAAGGAG	1.7
D13S634-F D13S634-R	13q21.33	0.812	355–440	6-F-GGCAGATTCAATAGGATAAATAGA GTAACCCCTCAGGTTCTCAAGTCT	0.38
D13S325-F D13S325-R	13q12.12	0.75	235–315	VIC-CTGTGCTATCTCCTCCAACG GTTTGAAAGATAGGCCATGCAG	0.175
D18S386-F D18S386-R	18q22.1	0.875	320–417	VIC-TGAGTCAGGAGAATCACTTGGAAC CTCTTCCATGAAGTAGCTAAGCAG	0.7
D18S390-F D18S390-R	18q22.3	0.75	340–415	CTCCAACCTCACTTGAGAGTA NED-GGTCAATAGTGAATATTTGGATAC	0.2
D18S391-F D18S391-R	18p11.31	0.75	190–235	CTGGCTAATTGAGTTAGATTACAA VIC-GGACTTACCACAGGCAATGTGACT	0.08

D18S535-F D18S535-R	18q12.3	0.92	450–500	6-F-CAGCAAACTTCATGTGACAAAAGC CAATGGTAACCTACTATTTACGTC	0.5
D18S819-F D18S819-R	18q11.2	0.78	370–450	PET-CTTCTCACCTGAATTACTATGGT TTTGTAATCGATCTACCACAGTT	1
D18S978-F D18S978-R	18q12.3	0.667	180–230	NED-GTAGATCTTGGGACTTGTCAGA GTCTCCCATGGTCACAATGCT	0.18
D21S11-F D21S11-R	21q21.1	0.9	220–283	6-F-TTTCTCAGTCTCCATAAATATGTG GATGTTGTATTAGTCAATGTTCTC	0.45
D21S1437-F D21S1437-R	21q21.1	0.84	283–351	6-F-CTACCACTGATGGACATTTAG GTGGAGGGTGTACCTCCAGAA	0.88
D21S1409-F D21S1409-R	21q21.2	0.81	160–220	PET-GGAGGGGAATACATTTGTGTA TTGCCTCTGAATATCCCTATC	0.5
D21S1411-F D21S1411-R	21q22.3	0.933	256–345	ATAGGTAGATACATAAATATGATGA NED-TATTAATGTGTGTCCTTCCAGGC	0.75
D21S1435-F D21S1435-R	21q21.3	0.75	160–200	6-F-CCCTCTCCAATTGTTTGTCTACC ACAAAAGGAAAGCAAGAGATTTCA	0.3

Table 2
Details of primers used in the sex chromosome multiplex. Size ranges given are those used in Genotyper, version 3.7 (Applied Biosystems). Heterozygosity (Het) values are based on our cohort and may vary in other populations. DYS448 is a hexanucleotide repeat. DX6807, DXS981, DXS1187, XHPRT, DXS7423, and DXYS267 are tetranucleotide repeats, whereas DXS1283E is a dinucleotide repeat. AMEL and SRY are not polymorphic

Name	Location	Het	Allele size range (bp)	Sequence 5′-3′	Final conc. (μM) of each primer
DXS6807-F DXS6807-R	Xp22.3	0.7	300–380	**6-FAM**-TCTCCCTTATTTGTGGTTTTGC AGCAGTTCTCCCTTATCCAC	1.6
DXS1283-F DXS1283-R	Xp22.3	0.89	295–340	**NED**-AGTTTAGGAGATTATCAAGCTG CCCATACACAAGTCCTCAAAGTGA	0.8
DXS981-F DXS981-R	Xq13.1	0.86	225–260	**6-FAM**-CTCCTTGTGGCCTTCCTTAAATG TTCTCTCCACTTTTCAGAGTCA	0.22
DXS1187-F DXS1187-R	Xq26.2	0.72	125–170	**VIC**-CAGCTACTCAATGAAAAGCC TGATGGAGAAAGTCACTGAAC	0.22
XHPRT-F XHPRT-R	Xq26.2	0.78	265–300	**VIC**-ATGCCACAGATAATACACATCCCC CTCTCCAGAATAGTTAGATGTAGG	0.6
DXS7423-F DXS7423-R	Xq28	0.74	350–420	**VIC**-TACTGGAGGTGAGGGTTGTG TGGGCTGCCCAGATACAACT	1.6
DXYS267-F DXYS267-R	Xq21.31Yp11.31	0.87	240–280	**PET**-ATGTGGTCTTCTACTTGTGTCA GTGTGTGGAAGTGAAGGATAG	0.8
AMEL-F AMEL-R	Xp22.2/Yp11.2		106/112	**NED**-CCCTGGGCTCTGTAAAGAATAGTG ATCAGAGCTTAAACTGGGAAGCTG	0.2
SRY-F SRY-R	Yp11.31		248	**NED-**AGTAAAGGCAACGTCCAGGAT TTCCGACGAGGTCGATACTTA	0.4
DYS448-F DYS448-R	Yq11.223		323–370	**PET-**CAAGGATCCAAATAAAGAACAGAGA GGTTATTTCTTGATTCCCTGT G	0.4

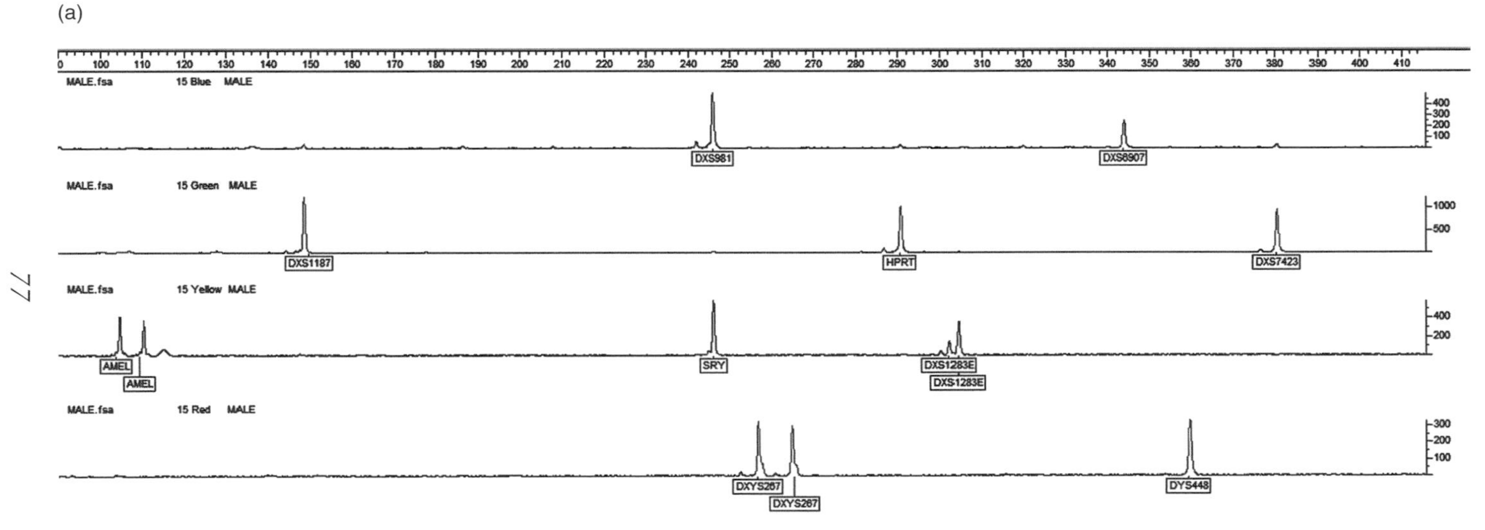

Fig. 2. (Continued)

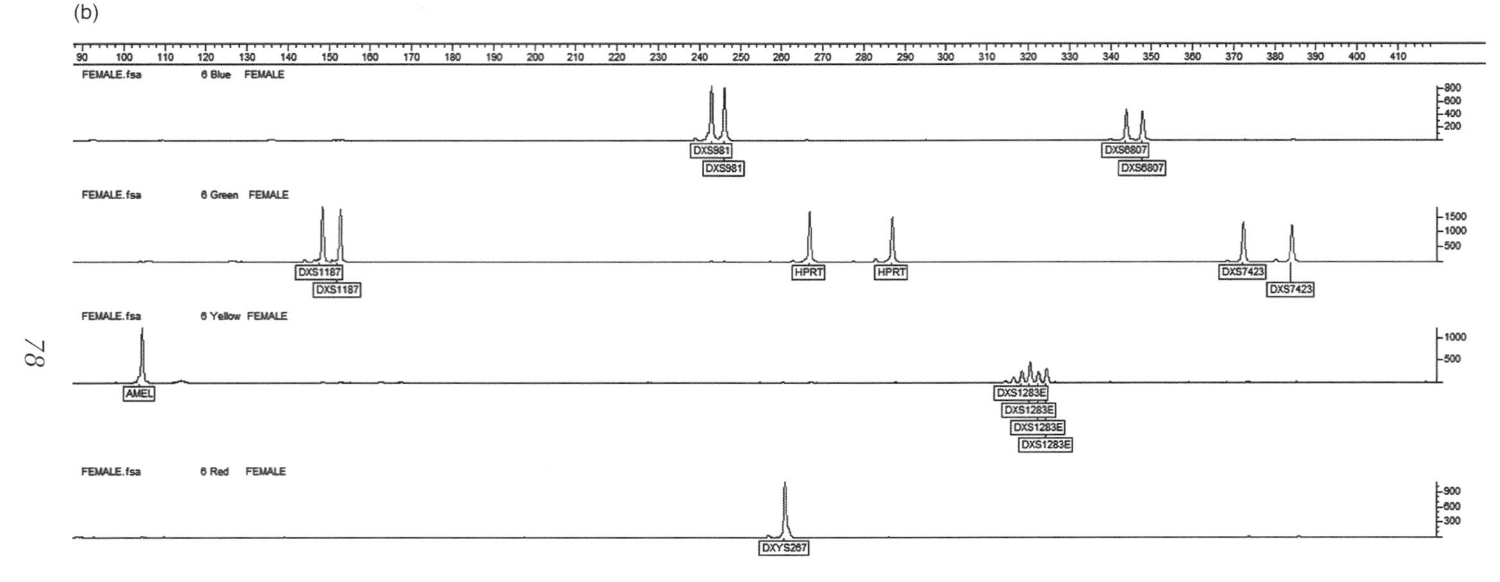

Fig. 2. **(a)** Genotyper profile of a normal male sample. Y-specific sequences are present (AMEL-112 bp, SRY, DYS448), X and Y sequences are present in an equal ratio (AMEL and DXYS267), and all X chromosome markers show a single allele consistent with, although not diagnostic of, a single X chromosome. **(b)** A Genotyper profile of a normal female sample. Y-specific sequences are absent, and heterozygous markers for the X chromosome exhibit 2 allele peaks with peak areas in a 1:1 ratio.

Markers are located along the length of each chromosome to increase the chance of detecting unbalanced chromosome rearrangements.

1.2. Potential Problems

1.2.1. Maternal Cell Contamination (MCC)

Evidence of a second genotype, as shown by inconsistent dosage ratios for each chromosome, extra allele peaks, or both, usually indicates contamination of the sample by maternal cells (*see* **Fig. 3**), although it may represent a twin or chimera. Maternal cell contamination is usually associated with blood-stained AF samples, although the degree of blood staining should not in itself be used as an indicator of maternal cell contamination; blood cells may be fetal or maternal in origin. Usually, samples are accompanied by some degree of blood staining, although this ranges from a pale pink cell pellet and clear liquor to a deep red color of the whole fluid.

When the majority genotype shows consistent normal or abnormal results with no inconclusive allele ratios, then the result may be reported, although it may be advisable to confirm the origin of the majority genotype by analysis of a maternal blood sample. We have found the level of blood staining in the AF cell pellet to broadly correlate with the level of maternal genotype; the majority genotype from pellets with fewer red blood cells is consistently fetal in origin. However, when the presence of two genotypes causes allele ratios to skew outside of the normal or abnormal range (*see* **Subheading 3.5.1.**), it is recommended that the QF-PCR results are not interpreted due to the increased risk of a misdiagnosis. In these cases, if a rapid result is required in these cases and one of the genotypes is determined as fetal in origin, either by sexing or by genotype analysis of a maternal sample, interphase-FISH may then be used if the analysis takes into account the fetal to maternal ratio. The number of analyzed cells can be increased to account for those that are maternal, or a sex chromosome probe can be cohybridized with an autosome probe and only the male cells analyzed.

In our London sample set, approximately 10% of AF samples are found to have two genotypes, although allele ratios vary considerably, proportional to the relative contribution of each cell line. The majority of these samples exhibit a very low level second genotype and can be reported as normal, whereas approximately 2% of AF samples in both the London and Graz cohorts exhibit second genotypes that prevent confident interpretation of allele ratios; therefore, they are reported as unsuitable due to MCC.

The detection of maternal cells in an amniotic fluid sample should not discredit the karyotype analysis of cultured cells. Subsequent genotype analysis of cultured cells from samples showing MCC, normally demonstrates a single

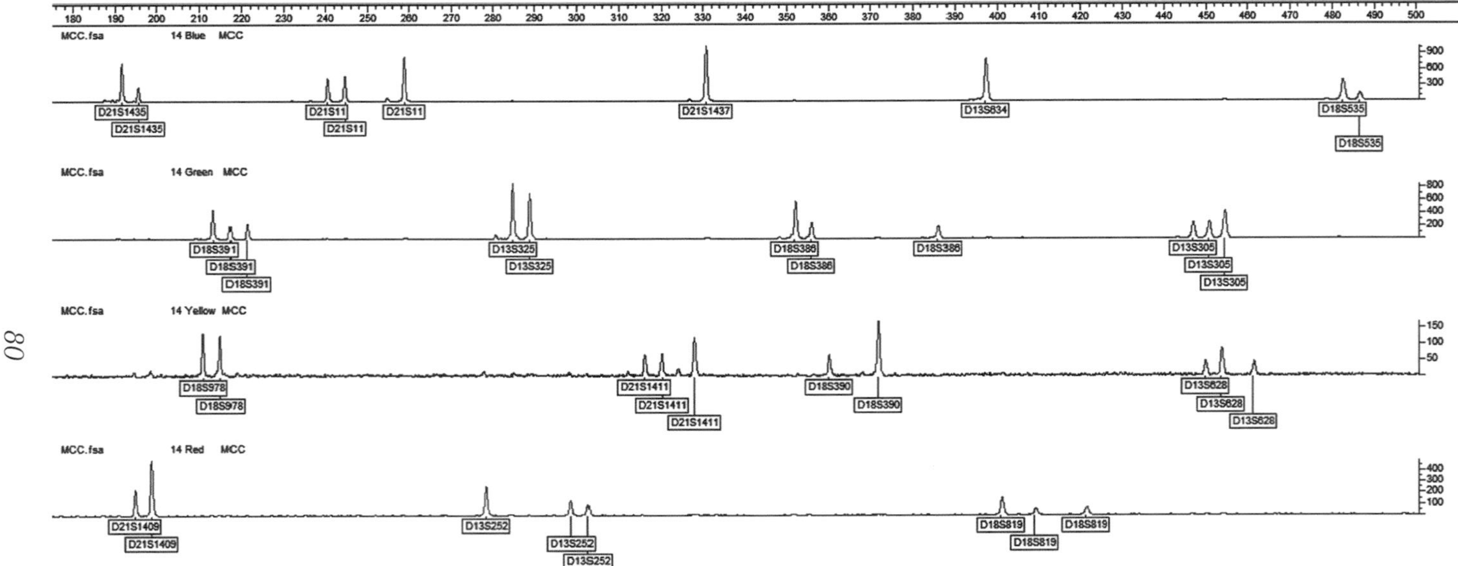

Fig. 3. A Genotyper profile of a sample exhibiting a high level of maternal cell contamination. The characteristic triallelic pattern, where the fetal-specific and maternal-specific allele peak areas combine to equal the shared fetal/maternal allele, is observed in markers D21S11, D18S391, D18S386, D13S305, D21S1411, D13S628, D13S252, and D18S819.

genotype, consistent with the selection and growth of fetal cells, and loss of maternal cells during the culture process *(17)*. In those samples with a mixed female/male cell population evident on the direct analysis, a single male genotype is usually detected on the follow-up test. However, the presence of two genotypes in samples where no blood staining is evident may indicate a maternal tissue plug. This may grow in culture; therefore, genotype analysis of cultured cells may be useful.

1.2.2. Mosaicism

The problem of mosaic genotypes and karyotypes in prenatal samples is well documented, particularly in CVS (*see* **Fig. 4**). With respect to QF-PCR, two issues are relevant; the levels of mosaicism detectable by the QF-PCR technique, and the degree of concordance between a direct test result and the fetal genotype.

The first of these can only be determined by the analysis of samples (both postnatal and prenatal) exhibiting mosaicism for one of the tested regions. The generation of "artificial mosaics" by the mixing of two genotypes in known measures represents a chimera rather than a mosaic genotype. The presence of a triallelic result is consistent with a meiotic nondisjunction event generating the trisomy cell line, whereas the absence of a triallelic result is evidence, although not diagnostic, of a normal conception followed by a mitotic nondisjunction event. Analysis of mosaic cases in our sample set found that a minimum level of 15% trisomy mosaicism could be detected, if a triallelic allele pattern was observed and 20% trisomy mosaicism if only dialleic ratios were present *(18)*. Thus, a mitotic error occurring in a disomic fetus may be harder to detect, due to the absence of a third allele. Indeed, QF-PCR identified only one of three trisomy 18 or 21 mosaics described by Pertl et al. *(8)*; this mosaic case was also triallelic, in this case for chromosome 21.

Discrepancies between the QF-PCR and karyotype result have recently been described *(19,20)*. These have been shown to be due to mosaicism and confinement of cell lines to different regions of the tested sample. To minimize such discrepant results it is recommended that DNA be prepared from dissociated cells prepared from 5–15 mg of cleaned villi. Although only a small aliquot of this cell suspension is required, all major cell lines in both the mesoderm and cytotrophoblast should be represented and detected by QF-PCR. Karyotype analysis of cell populations that are subsequently cultured from this cell suspension should minimise discrepancies between the two techniques.

Chorionic villi consist of an outer cytotrophoblast layer and internal mesoderm. The mesoderm layer is derived from a later fetal cell lineage, whereas the cytotrophoblast is derived from a much earlier lineage; thus, it is

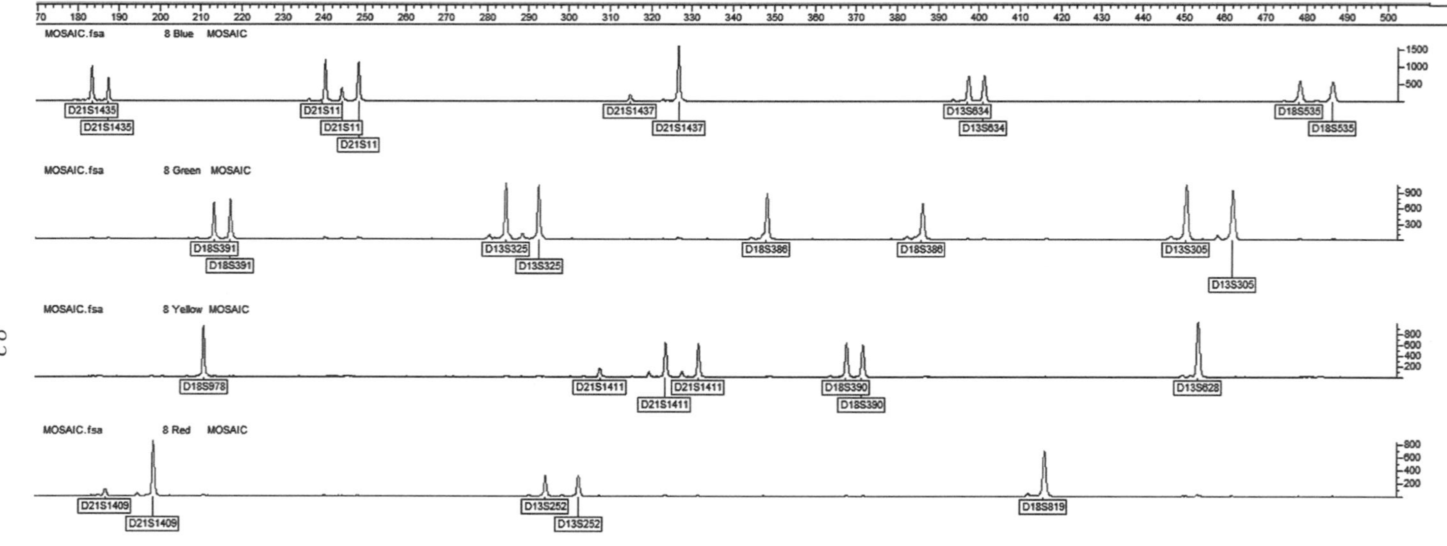

Fig. 4. A Genotyper profile of a sample exhibiting trisomy 21 mosaicism. All informative chromosome 21 markers show either an unequal triallelic pattern (D21S11 and D21S1411) or unequal diallelic ratios (D21S1435, D21S1437, and D21S1409). Chromosome 13 and 18 markers are normal. The presence of triallelic chromosome 21 markers is consistent with a meiotic nondisjunction event followed by trisomy rescue to generate the normal cell line.

less representative of fetal tissue ***(21)***. DNA prepared from dissociated cells represents cells from both the cytotrophoblast layer and mesenchymal core ***(22)***. Interphase-FISH results also are thought to represent both cell layers. In contrast, karyotype analysis of direct CVS preparations concerns only the cytotrophoblastic line, whereas culture conditions primarily lead to expansion of the mesoderm cell line, resulting in a final karyotype that is more representative of the fetus. In summary, care should be taken in the interpretation of trisomic prenatal results derived from CVS material, in the absence of a triallelic result demonstrating a meiotic origin to the trisomy cell line.

1.2.3. Submicroscopic Duplications

Partial chromosome duplication may be identified by QF-PCR analysis by the presence of both normal and abnormal marker results on one chromosome. This pattern may indicate a cytogenetically visible abnormality ***(11)*** or one that is submicroscopic. If the most distal or proximal markers are duplicated, then this may indicate the unbalanced product of a reciprocal translocation. However, in our experience, the presence of a single abnormal marker result, all other informative results are normal, is most likely to represent a submicroscopic duplication (SMD) ***(12)***. In the majority of cases, analysis of parental samples shows these SMDs to be inherited. It is necessary to establish the inheritance of SMDs in each case, because the extent of the duplicated region may vary between families. Inherited submicroscopic duplications are unlikely to be clinically significant, and they can be categorized as copy number variants ***(23)***.

1.2.4. Primer Site Polymorphisms (PSPs)

Primer site polymorphisms are a known phenomena of PCR assays ***(24)***. Sequence differences between the genomic DNA and the primers can result in complete or partial allele dropout (ADO) due to reduced or absent hybridization of the primers to genomic DNA. Partial ADO in a normal sample can either give an abnormal diallelic ratio consistent with trisomy for that region, or an inconclusive ratio (*see* **Subheading 3.5.**). Complete ADO in an abnormal sample can result in a normal diallelic ratio at that locus. In all cases of suspected ADO caused by PSPs, it is recommended to repeat the PCR at a lower annealing temperature (for example, 4°C lower than the standard temperature). This provides a less stringent environment for primer hybridization, resulting in reduced ADO, as represented by a change in the allele ratio. If the follow-up tests are consistent with the presence of a PSP, the marker result should be failed and not used as part of the QF-PCR analysis, even if it shows a normal ratio. A PSP may cause an abnormal diallelic result to seem normal at a lower annealing temperature.

1.2.5. Somatic Microsatellite Mutations (SMMs)

Somatic changes in the length of a microsatellite sequence, due to DNA replication and proof-reading errors, may be visible as an unequal triallelic result, where the areas of the two lowest alleles combine to equal the highest allele (**Fig. 5**), or skewed diallelic ratios. The characteristic triallelic pattern represents two cell lines that have one common allele and a second allele of different lengths.

QF-PCR analysis of AF samples found <0.1% to have an SMM at a single locus, whereas SMMs were observed at a much higher frequency (4%) in individual whole villi *(25)*. This high frequency is probably due to the small clonal cell population being analyzed. QF-PCR analysis of digested/chopped villi containing a more diverse cell population (*see* **Subheading 3.1.2.**) is therefore likely to result in the identification of fewer SMMs. Interestingly, the majority of SMMs are single repeat unit expansion/contraction events.

These somatic changes in repeat length are specific to the microsatellite and are not clinically significant. Thus, a single marker result, showing a characteristic triallelic pattern consistent with a SMM, where all other markers on that chromosome are normal, does not require further clarification. However, it is more difficult to class skewed diallelic results as SMMs; PSPs (*see* **Subheading 1.2.3.**) and SMDs (*see* **Subheading1.2.4.**) may be alternative explanations. It is important to distinguish between these events if a 1:2 or 2:1 allele ratio is observed, because these ratios may represent a clinically significant imbalance. Once PSPs have been excluded, analysis of cultured cell populations may help to distinguish between an SMD and SMM. In an SMM, the proportion of the two cell lines may change between uncultured and cultured cells thus altering the allele ratio.

2. Materials

2.1. DNA Preparation

1. InstaGene Matrix (Bio-Rad, Hercules, CA). Store at 4°C.

2.2. PCR

1. T.1E buffer: 10 mM Tris-HCl, pH 7.6, and 0.1 mM EDTA. Filter, sterilizes, and store at room temperature.
2. 10× Multiplex PCR kit (QIAGEN, Hilden, Germany). Store at –20°C. Once thawed store at 4°C. Kit contains *Taq* polymerase and dNTPs.
3. 5′-labeled fluorescent oligonucleotide primers: We use Applied Biosystems (Foster City, CA) primers, 5′ labeled with 6-FAM, VIC, NED, or PET. Fluorescence-labeled primers should not be exposed to light for prolonged periods or repeatedly freeze-thawed (*see* data sheet). These are supplied as precipitates, and they are resuspended in T.1E buffer to give a working concentration of 100 μM. The primers are stored at –20°C. Aliquots in use are stored at 4°C.

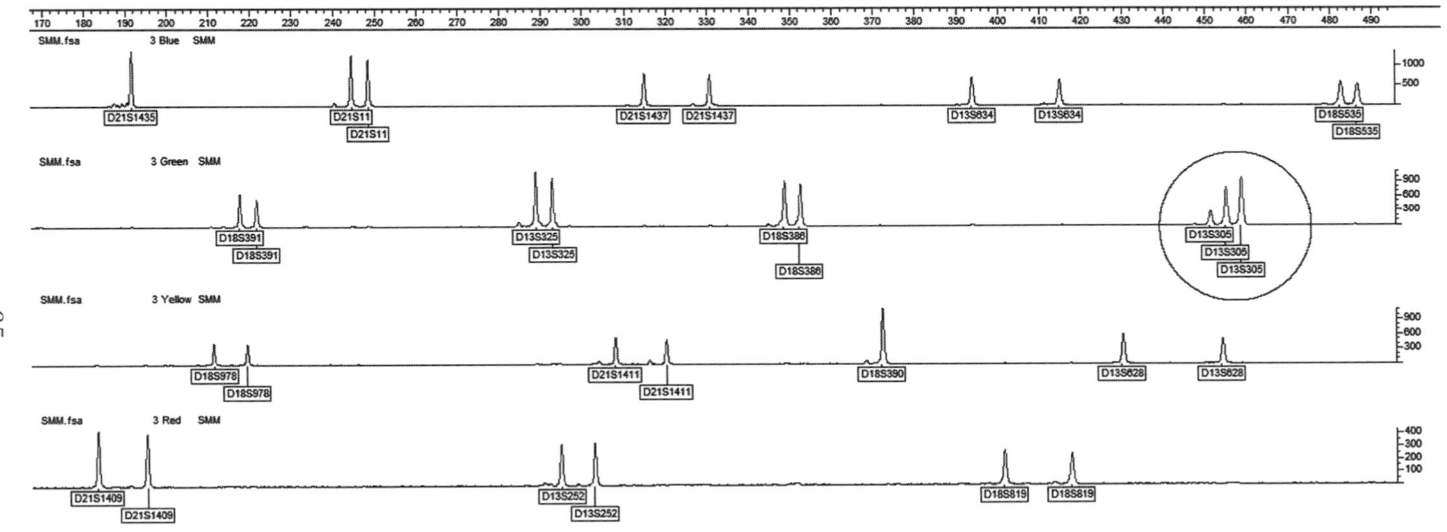

Fig. 5. A Genotyper profile of a sample exhibiting an SMM at marker D13S305. The characteristic triallelic result is evident where the two lowest allele peaks representing the two mosaic cell lines combine to equal the higher allele present in both cell lines. All other informative markers are normal.

4. Unlabeled oligonucleotide primers: We use Invitrogen (Carlsbad, CA) primers. These primers are supplied as precipitates, and they are resuspended in T.1E buffer to a working concentration of 100 μM and stored at –20°C. Although unlabeled primers are more stable than fluorescent-labeled primers, freeze-thawing should be avoided. Aliquots in use are stored at 4°C.

2.3. Analysis

1. We use a capillary-based genetic analyser (3100, Applied Biosystems, Foster City, CA). All consumables (10× EDTA buffer, POP6 polymer, 36cm capillaries, plate septa) are purchased from Applied Biosystems.
2. Deionized formamide (Toxic: refer to the material safety data sheet): HI-DI formamide (Applied Biosystems) is highly deionized, to minimize the breakdown of the fluorescent label. It is stored at –20°C in appropriate aliquots.
3. Genescan-500 LIZ size standard: we use Applied Biosystems size standard, designed for 100–500-bp fragment analysis.
4. 96-well plates: We use those recommended for the 3100 genetic analyzer (Thermo-Fast 96 Detection Plates, ABgene, Epsom, Surrey, UK).
5. Software: For fragment analysis, GeneScan Analysis, version 3.7 is used. For allele size calling and labeling, Genotyper, version 2.5 (Applied Biosystems) is used. Microsoft Excel (Microsoft, Redmond, WA) is used to calculate and tabulate allele dosage ratios.

3. Methods

The processing of a number of prenatal samples at one time, and the risk of sample mixup, necessitates stringent quality control procedures (*see* **Note 1**). In addition, care must be taken, as with all PCR-based tests, to avoid contamination of tested material with amplified products of previous reactions and external DNA (*see* **Note 2**).

3.1. Sample Preparation

Sample and DNA preparation procedures are carried out in a class II biological containment cabinet, up to **step 5, Subheading 3.2.** (*see* **Note 3**).

3.1.1. AF

Between 10 and 20 ml of AF is normally received in a 20-ml sterile universal container. The sample is centrifuged at 200*g* for 10 min. Fluid is carefully removed using a 20-ml syringe to leave approximately 1 ml of fluid (equal to the top of the conical section of the Universal). The cell pellet is resuspended in the fluid by using a plastic Pasteur pipette, and approximately 100 μL (2 to

3 drops) is transferred to a 0.5-ml Eppendorf tube. This aliquot (approx 1/10 of the original AF sample) can be stored at 4°C until processed. The remainder can be cultured for subsequent karyotype analysis, as required.

*3.1.2. Chorionic Villus (see **Note 4**) and Tissue Samples (see **Note 5**)*

Maternal deciduas are removed from 5 to 15 mg of chorionic villi. Cleaned villi are digested with collagenase (final concentration 2.5 mg/ml) at 37°C for 35 min, followed by trypsin digestion (final concentration 2.5 mg/ml) at 37°C for 25 min. The addition of 2 ml of medium prevents further digestion. A single drop (approx 100 μl) of the dissociated cell slurry is removed for QF-PCR analysis. This may be stored at 4°C. The rest of the sample is cultured for karyotype analysis. For tissues, a small section of whole tissue sample is isolated and can be stored at 4°C.

3.2. DNA Preparation (see Note 6)

This is a quick and simple procedure and it can be successfully applied to a number of sample types and different sample qualities (*see* **Note 7**). Because it is based on a boiling lysis protocol, aided by vigorous vortexing, the safety issues associated with phenol/chloroform extractions also are avoided. After cell lysis, a commercial resin (InstaGene Matrix, Bio-Rad) removes trace metal contaminants that may inhibit the PCR.

Before the DNA extraction, place the InstaGene Matrix on a magnetic stirrer and set at a medium speed for at least 5 min.

1. Pellet the cells/villus/tissue at 12,000*g* for 1 min in a microcentrifuge and carefully remove the liquid, leaving enough to resuspend the pellet (approx 10–20 μl). For amniotic fluid samples, any blood staining in the cell pellet is noted as a percentage of whole pellet. If >40% of the pellet is red, follow **steps 2** and **3**. For all other samples, proceed to **step 4**.
2. Vortex to resuspend the cell pellet and add 200 μl of H_2O to wash the sample (*see* **Note 8**).
3. Vortex the sample, pellet the cells and remove the wash solution as described above. Resuspend the cells in the remaining wash solution by vortexing.
4. Add between 100 and 400 μl (*see* **Note 9**) of InstaGene Matrix to the cells/villus by using a wide-bore pipette tip, e.g., a Gilson p1000 tip and vortex.
5. Incubate at 100°C for 8 min.
6. Vortex again at high speed for 10 s, and pellet the InstaGene Matrix at 12,000*g* for 3 min in a microcentrifuge.
7. Place the samples on ice to cool.
8. The DNA preparation should be stored at –20°C (*see* **Note 10**).

3.3. PCR Setup

Batches of PCR assays can be prepared in advance, tested and stored at –20°C. These are 20-μl aliquots of a master mix that contains all components except DNA, which is added immediately before temperature cycling, to give a total volume of 25 μl. The final concentrations of the reaction components are 1× Multiplex PCR kit and 2.5–42.5 pmol of each primer (*see* **Tables 1 and 2**), in a total volume of 25 μl (Note: DNA is added according to volume rather than concentration, as concentration is not measured).

1. To a thin-walled PCR tube containing 20 μl of the master mix, add 5 μl of DNA solution (*see* **Note 11**), taking care not to disturb the InstaGene Matrix pellet. Mix by pipetting.
2. Add 1 drop of mineral oil, if a heated lid is not being used, and place in the PCR machine.
3. PCR cycling conditions: *Taq* polymerase activation and initial denaturation at 95°C for 15 min followed by 25 cycles of 94°C for 30 s, 58°C for 90 s, 71°C for 90 s (*see* **Note 12**). Final synthesis: 72°C for 20 min followed by storage at 10°C (*see* **Note 13**).

3.4. Analysis

3.4.1. PCR Product Preparation

Post-PCR cleanup to remove excess primers and free dye molecules is not carried out, due to time constraints (*see* **Note 14**). We use 3100 Applied Biosystems genetic analyzers, and conditions specific to this instrument are described. Standard use of this analyzer is not detailed here. Other genetic analyzers capable of fragment resolution, fluorescence detection, and quantification also can be used, and they include Applied Biosystems capillary-based analyzer model 310, 3730, and 3700; and Amersham-Pharmacia ALF and MegaBase systems, although fluorescent labels may have to be substituted depending on the filter sets of each analyzer.

1. Prepare PCR products for analysis by the addition of 3 μl of product to 15 μl of HI-DI formamide in 96-well plates (*see* **Note 15**).
2. Denature at 95°C for 2 min and snap-chill on ice.

3.4.2. 3100 Analysis

Separate PCR products through a 36-cm capillary array filled with POP6 (*see* **Note 16**). A 10-s injection time is suitable for most samples (*see* **Note 17**). The running conditions are 60°C for 3,000 s.

3.4.3. Genotyper Analysis

Macros are used to label allele peaks with marker name, size, and peak area (**Fig. 1**) (*see* **Note 18**). The Genotyper table is transferred to an Excel spreadsheet for allele ratio analysis.

3.5. Result Interpretation

The criteria listed below are based on >19,000 QF-PCR prenatal tests, all of which were followed by karyotype analysis of cultured cells (*see* **Note 19**). For additional information, see the UK CMGS Best Practice Guidelines for the Diagnosis of Aneuploidy at http://cmgs.org/BPG/Guidelines/2004/QFPCR.htm.

1. Normal allele dosage ratios range between 0.8 and 1.4 (*see* **Note 20**). For alleles separated by >24 bp, ratios up to 1.5 are acceptable. Trisomy is indicated by an allele ratio of between 1.8 and 2.4 or between 0.65 and 0.45 or by the presence of three alleles of equal areas (*see* **Note 21**). All of these results are described as informative.
2. Two informative markers per chromosome are required for confident interpretation. This minimizes the risk of misdiagnosis due to primer-site polymorphisms and/or somatic repeat instability (*see* **Note 22**). Single marker abnormal results should not be reported.
3. If both normal and abnormal marker results are obtained for a single chromosome, follow-up studies should be carried out. Such results may represent polymorphisms or a clinically significant partial chromosome imbalance (*see* **Subheadings 1.2.3.**, **1.2.4.**, and **1.2.5**). Single marker assays, additional markers, lowering the PCR annealing temperature and analysis of cultured cell populations and parental samples can clarify these results.

4. Notes

1. To prevent sample mix-up a minimal number of tube-to-tube transfers should be used (three transfers are required for this protocol). Each sample transfer and analysis should be checked by another laboratory member, and the use of two identifiers per tube, such as sample number and name, aids sample tracking.
2. Contamination of a PCR by external DNA is evident by the appearance of allele peaks in the negative (no DNA) PCR control; a critical part of any PCR procedure and a reaction that should be setup last in a series of samples. Separation of the PCR setup and post-PCR analysis areas should help to prevent contamination.
3. To ensure that DNA is prepared from the correct sample, it is advisable to prepare the initial sample aliquot (*see* **Subheadings 3.1.1.** and **3.1.2.**) one sample at a time in a class II biological containment cabinet, with only one sample in the cabinet during the procedure. Because the subsequent DNA preparation (*see* **Subheading 3.2.**) is carried out without further tube transfers, DNA from

a number of samples can be prepared simultaneously. This DNA can then be used for subsequent PCR tests.

4. It is recommended that DNA is prepared from dissociated cells prepared from 5 to 15 mg of cleaned villi. Although only a small aliquot of this cell suspension is required, all major cell lines in both the mesoderm and cytotrophoblast should be represented and detected by QF-PCR. The analysis of whole villi has rarely been associated with discrepant results between QF-PCR and karyotype analysis due to mosaic cell lines confined to one region of the sample ***(19,20)***. Analysis of cell populations that are subsequently cultured for karyotype analysis should minimize discrepancies between the two techniques.
5. The QF-PCR procedure can be applied effectively to solid tissue samples (e.g., skin and cartilage). This is particularly useful for confirming a prenatal diagnosis or for a poor-quality sample where chromosome analysis of cultured cells may not be possible.
6. DNA prepared using the protocols described here also may be used for other molecular prenatal tests. The protocol has benefits over the traditional phenol/chloroform-based approach in terms of labor and time savings and reduced safety risks. However, as the extracted DNA may contain residual contaminants, its suitability for use in each test should be determined.
7. Although the procedure is generally successful in extracting DNA of sufficient quality for use in the multiplex, in our experience, DNA extracted from blood-stained or discolored AF fluid may contain PCR inhibitors. These inhibitors can be removed by a subsequent extraction (*see* **Note 9**).
8. Deionized water lyses red blood cells, and it also may aid lysis of cells in villus and tissue samples. A deionized water wash is used for blood-stained/discolored AF.
9. It is beneficial to adjust the volume of InstaGene Matrix to balance removal of all cell lysis products with excessive dilution of the DNA. A 300-μl volume of InstaGene Matrix is generally used for CVS and tissue samples and the larger AF cell pellets (those that cover the base of the 0.5-ml microcentrifuge tube). Only 100 μl of InstaGene Matrix is required for average and small AF pellets. It is important that the DNA extraction is not overloaded with too much starting material. This leads to a failure by the InstaGene Matrix to chelate all metal ions and can result in inhibition of the PCR. In particular, the larger sized markers in the multiplex may fail to amplify. If inhibition is observed, a further extraction can be used to remove the contaminants; 100 μl of the DNA extract is added to 100 μl of InstaGene Matrix and treated as per the extraction protocol (*see* **Subheading 3.2., steps 5–8**).
10. Because the DNA prepared here is relatively crude, and cell lysis products that damage DNA may remain, the DNA should be stored at –20°C. However, the DNA does seem to be stable for at least 3 d at room temperature, which allows some flexibility, such as transfer of the sample to another laboratory.
11. As well as the necessary inclusion of the negative (no DNA) PCR control for the reasons given in **Note 2**, the use of a DNA control trisomic for one

of the chromosomes and exhibiting 2:1 or 1:2 dosage for at least one marker is recommended. Normal allele dosage is exhibited by the other two nontrisomic chromosomes. The control not only demonstrates that the amplified DNA represents allele copy number for that procedure but also can be used as a standard against which any spurious background bands or free-dye peaks can be compared.

12. Samples that do not generate sufficient amplified sequences for analysis, due to low initial DNA concentration (usually evident by a very small original cell pellet) can be amplified with a greater number of cycles. Any change in PCR cycling conditions should always be accompanied by a trisomy control, to ensure the reaction is still quantitative.
13. The 20-min incubation at 72°C is necessary, because *Taq* polymerases lacking exonuclease activity add a template-independent dATP to the 3′ end of amplified sequences ***(26)***. Without the 72°C incubation, a single base-pair size difference in the amplified sequences can resolve as a "split peak" on the profile and hinder analysis. This is a particular problem here due to the size of the fragments generated (100–500 bp). In this size range, the analysis systems efficiently resolve single base-pair differences especially in smaller alleles.
14. Free-dye peaks are caused by the detachment of the fluorescent molecule from the labelled primer. These molecules are resolved as broad peaks, usually up to 180 bp (*see* **Note 16**); as such, they can be distinguished from allele peaks. The breakdown of fluorescent primers can be minimized by the use of deionized formamide stored at –20°C and reduced exposure of labeled primers to temperatures above –20°C. Free-dye molecules can be removed, along with unincorporated primers, by standard post-PCR cleanup protocols if required.
15. Accurate transfer of samples to the wells of a 96-well plate can be difficult. The risk of error can be minimized not only by the use of a multichannel pipette but also by the addition of loading buffer containing dextran blue, which is visible but does not interfere with the fluorescent analysis. Transparent piercable sheets also are available (ABGene) that can be sealed onto the plate, or rubber septa can be placed over wells that are not in use.
16. The POP6 polymer can be used on the 3100 genetic analyzer if greater resolution is required, although this requires longer run times. Resolution of free-dye molecules (*see* **Note 14**) is not linear in respect to fragment size, but it is influenced by both temperature and the separating matrix. A different polymer may be used to resolve free-dye molecules that coincide with allele peaks.
17. Allele peak heights >6,000 fluorescent units on the 3100 genetic analyzer are not analyzed. The charge-coupled device camera becomes saturated in this range, and peak fluorescence may be underrepresented. If a sample is overloaded, injection times can be reduced to accommodate differences in DNA concentration and the corresponding amplification. This is one of the advantages of the capillary-based genetic analyzers, where a repeat injection does not require repeat sample preparation.

18. Tetranucleotide alleles demonstrate few visible stutter bands (*see* **Subheading 1.2.**), and only the main allele peak is labeled. However, some microsatellite markers contain a mix of both tetranucleotide and dinucleotide repeats and generate significant stutter bands. For dinucleotide alleles, the larger alleles generally exhibit more significant stutter effects than smaller alleles, due to the longer repeat. It is therefore necessary to recognize and label at least the first stutter peak and include it in the allele peak area measurement.
19. Although there are now several published studies describing the use of QF-PCR as a diagnostic test ***(5–14)***, it is important to validate the QF-PCR strategy in the laboratory in which it is to be used. Control samples are required, and a pilot study is recommended before the implementation of a QF-PCR based aneuploidy diagnostic service, especially if primer sets are used that are not described in the published literature.
20. The large normal range is necessary due to the use of tetranucleotide repeats. These can result in widely spaced alleles (up to 50 bp apart), and marked preferential amplification of the smaller allele, which in turn results in skewed allele dosage ratios. However, closely spaced alleles should exhibit less allele specific preferential amplification and would be expected to have dosage ratios closer to 1.0.
21. As more than one sample is usually processed, the sample identity of abnormal results should be confirmed. This can be done by a repeat QF-PCR test or by another technique such as interphase-FISH or karyotype analysis of uncultured or cultured cells. Alternatively, genotype analysis of a maternal blood sample by using the same markers can be used to confirm sample identification.
22. Polymorphisms in the primer-binding site can result in partial or complete amplification failure of an allele (*see* **Subheading 1.2.4.**). This can result in a misdiagnosis if the result is used in isolation. In addition, SMMs (*see* **Subheading 1.2.5.**) ***(25)*** and submicroscopic imbalance (*see* **Subheading 1.2.3.**) ***(12)*** also could result in misdiagnosis if a single marker was used independently.

References

1. Lewin, P., Kleinfinger, P., Bazin, A., Mossafa, H., and Szpiro-Tapia, S. (2000) Defining the efficiency of fluorescence in situ hybridization on uncultured amniocytes on a retrospective cohort of 27 407 prenatal diagnoses. *Prenat. Diagn.* **20,** 1–6.
2. Waters, J. J. and Waters, K. S. (1999) Trends in cytogenetic prenatal diagnosis in the UK: results from UKNEQAS external audit, 1987–1998. *Prenat. Diagn.* **19,** 1023–1026.
3. Spathas, D. H., Divane, A., Maniatis, G. M., Ferguson-Smith, M. E., and Ferguson-Smith M. A. (1984) Prenatal detection of trisomy 21 in uncultured amniocytes by fluorescence in situ hybridisation: a prospective study. *Prenat. Diagn.* **14,** 1049–1054.

4. Klinger, K., Landes, G., Shook, D., et al. (1992) Rapid detection of chromosomal aneuploidies in uncultured amniocytes by using fluorescence *in-situ* hybridization (FISH). *Am. J. Hum. Genet.* **51,** 55–65.
5. Pertl, B., Yau, S. C., Sherlock, J., Davies, A. F., Mathew, C. G., and Adinolfi, M. (1994) Rapid molecular method for prenatal detection of Down's syndrome. *Lancet* **343,** 1197–1198.
6. Pertl, B., Kopp, S., Kroisel, P. M., Häusler, M., et al. (1997) Quantitative fluorescence polymerase chain reaction for the rapid prenatal detection of common aneuploidies and fetal sex. *Am. J. Obstet. Gynecol.* **177,** 899–906.
7. Verma, L., Macdonald, F., Leedham, P., McConachie, M., Dhanjal, S., and Hulten, M. (1998) Rapid and simple prenatal DNA diagnosis of Down's syndrome. *Lancet* **352,** 9–12.
8. Pertl, B., Kopp, S., Kroisel, P. M., Tului, L., Brambati, B., and Adinolfi, M. (1999) Rapid detection of chromosome aneuploidies by quantitative fluorescence PCR: first application on 247 chorionic villus samples. *J. Med. Genet.* **36,** 300–303.
9. Schmidt, W., Jenderny, J., Hecher, K., et al. (2000) Detection of aneuploidy in chromosome X, Y, 13, 18 and 21 by QF-PCR in 662 selected pregnancies at risk. *Mol. Hum. Reprod.* **6,** 855–860.
10. Levett, L. J., Liddle, S., and Meredith, R. (2001) A large-scale evaluation of amnio-PCR for the rapid prenatal diagnosis of fetal trisomy. *Ultrasound. Obstet. Gynecol.* **17,** 115–118.
11. Mann, K., Fox, S. P., Abbs, S. J., et al. (2001) Development and implementation of a new rapid aneuploidy diagnostic service within the UK National Health Service and implications for the future of prenatal diagnosis. *Lancet.* **358,** 1057–1061.
12. Mann, K., Donaghue, C., Fox, S. P., Docherty, Z., and Ogilvie, C. M. (2004) Strategies for the rapid prenatal diagnosis of chromosome aneuploidy. *Eur. J. Hum. Genet.* **12,** 907–915.
13. Cirigliano, V., Voglino, G., Canadas, M. P., et al. (2004) Rapid prenatal diagnosis of common chromosome aneuploidies by QF-PCR. Assessment on 18,000 consecutive clinical samples. *Mol. Hum. Reprod.* **10,** 839–846.
14. Ogilvie, C. M., Donaghue, C., Fox, S. P., Docherty, Z., and Mann, K. (2005) Rapid prenatal diagnosis of aneuploidy using quantitative fluorescence-PCR (QF-PCR) *J. Histochem. Cytochem* . **53,** 285–288.
15. Lubin, M. B., Elashoff, J. D., Wang, S.-J., Rotter, J. I., and Toyoda, H. (1991) Precise gene dosage determination by polymerase chain reaction: theory, methodology, and statistical approach. *Mol. Cell. Probes* **5,** 307–317.
16. Donaghue, C., Roberts, A., Mann, K., and Ogilvie, C. M. (2003) Development and targeted application of a rapid QF-PCR test for sex chromosome imbalance. *Prenat. Diagn.* **3,** 201–210.
17. Stojilkovic-Mikic, T., Mann, K., Docherty, Z., and Ogilvie, M. C. (2005) Maternal cell contamination of prenatal samples assessed by QF-PCR genotyping. *Prenat. Diagn.* **25**, 79–83.

18. Donaghue, C., Mann, K., Docherty, Z. and Ogilvie, C. M. (2005) Detection of mosaicism for primary trisomies in prenatal samples by QF-PCR and karyotype analysis. *Prenat. Diagn.* **25**, 65–72.
19. Waters, J. J., Walsh, S., Levett, L. J., Liddle, S. and Akinfenwa, Y. (2006) Complete discrepancy between abnormal fetal karyotypes predicted by QF-PCR rapid testing and karyotyped cultured cells in a first-trimester CVS. *Prenat. Diagn.* **26,** 892–897.
20. Waters, J. J., Mann, K., Grimsley, L., et al. Complete discrepancy between QF-PCR analysis of uncultured villi and karyotyping of cultured cells in the prenatal diagnosis of trisomy 21 in three CVS. *Prenat Diagn.* 2007 Apr; **27**(4): 332–339.
21. Gardner, R. J. M. and Sutherland, G. R. (1996) Chromosome Abnormalities and Genetic Counseling, Oxford University Press, Oxford, UK, pp. 336–355.
22. Mann, K., Kabba, M., Donaghue, C., Hills, A., and Ogilvie, C. M. Analysis of a chromosomally mosaic placenta to assess the cell populations in dissociated chorionic villi: implications for QF-PCR aneuploidy testing. *Prenat Diagn.* 2007 Mar; **27**(3): 287–289. No abstract available.
23. Sharp, A. J., Locke, D. P., McGrath, S. D., et al. (2005) Segmental duplications and copy-number variation in the human genome. *Am. J. Hum. Genet.* **77,** 78–88.
24. Andrew, S. E., Whiteside, D., Buzin, C., Greenberg, C., and Spriggs, E. (2002) An intronic polymorphism of the hMLH1 gene contributes toward incomplete genetic testing for HNPCC. *Genet. Test* . **6,** 319–322.
25. Mann, K., Donaghue, C., and Ogilvie, C. M. (2003) In vivo somatic microsatellite mutations identified in non-malignant human tissue. *Hum. Genet.* **114,** 110–114.
26. Clark, J. M. (1988) Novel non-templated nucleotide addition reactions catalyzed by procaryotic and eucaryotic DNA polymerases. *Nucleic Acids Res.* **16,** 9677–9686.

7

Real-Time Quantitative PCR for the Detection of Fetal Aneuploidies

Bernhard G. Zimmermann and Lech Dudarewicz

Summary

In prenatal analysis, one of the major concerns is the detection of fetal aneuploidies. Several molecular methods have been described recently for the rapid analysis of amniotic fluid and chorionic villi. Fluorescence in situ hybridization (FISH) and polymerase chain reaction (PCR) of short tandem repeats are already implemented in prenatal laboratories and permit the evaluation of the major chromosomal aberrations within 24 h. However, both methods have their disadvantages.

The advent of real-time PCR has revolutionized the measurement of nucleic acid copy numbers in recent years. Quantitative PCR, a term that was only 10 years ago considered to be an oxymoron, is now widely accepted. We demonstrated previously the feasibility of detection of trisomy 21 by real-time PCR, and here we describe the modified test that permits simultaneous analysis of trisomies 18 and 21. This approach has been demonstrated in a recent large-scale analysis, and it is presented in detail in this chapter.

Key Words: Multiplex real-time polymerase chain reaction (PCR); relative quantification; prenatal diagnosis; amniotic fluid; trisomy 21; Down syndrome; trisomy 18; aneuploidy; DNA analysis; gene copy number; molecular probes; PCR primers; PCR kinetics.

1. Introduction

1.1. Background: Prenatal Screening and Cytogenetic Analysis

The detection of gross chromosomal abnormalities is a major focus of prenatal diagnostics, and it constitutes the most frequent indication for a prenatal invasive diagnostic procedure. The most common cytogenetic anomaly in live

From: *Methods in Molecular Biology, vol. 444: Prenatal Diagnosis*
Edited by: S. Hahn and L. G. Jackson © Humana Press, Totowa, NJ

births is trisomy 21, corresponding to the clinical diagnosis of Down syndrome. Other fetal aneuploidies frequently detected involve chromosomes 13, 16, and 18 and both sex chromosomes (X and Y). The cytogenetic analysis requires intact, living cells that currently may only be obtained by an invasive procedure that carries a low, but not negligible risk to the fetus and mother *(1)*. This risk, together with the high cost of the cytogenetic study is the reason why only the fetuses at the highest risk of the cytogenetic anomaly may be tested.

Conventional cytogenetics using Giemsa banding of metaphase preparations in a single run detects a wide range of aberrations with high reliability, deserving the position of a gold standard in prenatal diagnosis. Nevertheless, classical cytogenetics is subject to technical problems such as culture failure, external contamination, and sometimes even selective growth of maternal cells. Quantitative interpretation of banded karyotypes also can be limited by cell-to-cell variability in chromosome condensation and staining characteristics. Moreover, the nature of the karyotype analysis by banding methods is to some degree subjective. Probably, the principal drawback of the most widespread method of invasive prenatal diagnosis—genetic amniocentesis—is the long time to result: it returns the final answer after approximately 2 weeks!

There have been two main avenues to overcome the drawbacks of classical cytogenetics: screening was devised to avoid as many unnecessary invasive procedures as possible and rapid molecular tests were invented to shorten the turnaround time.

At present, the risk of the most common and significant fetal aneuploidies (trisomies 21, 18, and 13, triploidy, both maternal and paternal and monosomy X) may be estimated with noteworthy accuracy. Indeed, >95% of fetuses with the above-mentioned aneuploidies may be detected at <5% of false-positive results in the general population *(2)*. Pregnancies are considered to be at high risk after performing a screening test for fetal aneuploidy. Historically, the first screening test for Down syndrome that was introduced in the late 1960s, almost instantly when prenatal karyotyping became feasible, was a questionnaire based on maternal age. In this test, after prenatal counseling, the woman was asked her age, and if it was advanced enough to justify an (much more risky than nowadays) amniocentesis, it was considered positive. In the 1980s, the first markers of fetal aneuploidy were discovered in maternal serum, and the concept of risk screening was coined *(3)*. The screening emerged with the risk as "the common currency" into which different markers may be converted. Because of the number and the different nature of used markers for fetal aneuploidy—quantitative, qualitative, sonographic, and biochemical—there seems to exist no other simple and elegant way to interpret multiple markers in a given patient, in conjunction with her age and demographic data than the risk. Currently, the most efficient way to screen for fetal aneuploidies is risk calculation on the basis of maternal age in conjunction with the following first trimester ultrasound and biochemical markers, but other markers, also in the second trimester of pregnancy, are still widely used in different countries:

1. Sonographically measured nuchal translucency, which is a physiologic fluid-filled space at the back of the neck of the fetus, that is, on average, increased in the most common aneuploidies in the first trimester of pregnancy ***(4)***.
2. Absence of nasal bones, which are not visible on ultrasound in approximately 65% of the fetuses with trisomy 21 and approximately 55% of the fetuses with trisomy 18 ***(5)***.
3. Presence of tricuspid regurgitation, which is detected in approximately 65–74% of the fetuses with trisomy 21 ***(6)***.
4. Maternal serum concentration of pregnancy-associated plasma protein A, which is, on average, decreased by 50–60% in trisomy 21 ***(7)***.
5. Maternal serum concentration of free β-subunit of human chorionic gonadotrophin, which is, on average, elevated by 100% in trisomy 21 ***(8)***.

In clinical settings, the individual risk figure calculated on the basis of any combination of parameters is used not as an absolute threshold, but as a guideline describing the magnitude of risk to the patient. In all cases, detailed and skilled, nondirective genetic counseling is necessary, to enable the patient arrive at her final decision of whether to an undergo invasive genetic procedure or not.

The detection rate of the most common aneuploidies in the order of 90–95% may be perceived as impressive, but the false-positive rate of 1 to 5% is not much so, because it means that even in the ideal situation, approximately 1 to 5% of all pregnant women are assigned "high risk" and proposed invasive procedure with a lengthy turnaround time. From this group, only a minority will finally turn out to have a baby affected by aneuploidy. Particularly in this group of highly anxious patients there is the need for the fast diagnostic result. The rapid molecular tests are a blessing for those women and their families.

1.2. Rapid Molecular Methods

To address these needs, more rapid methods for the prenatal diagnosis of fetal chromosomal aneuploidies have recently been developed and implemented. The first of these methods to be widely applied is multicolor fluorescence in situ hybridization (FISH). Although FISH is very reliable, it is a time- and labor-intensive procedure ***(9,10)***. The next method that has seen widespread clinical application, particularly in the UK, is quantitative fluorescent polymerase chain reaction (PCR) analysis of short tandem repeats (STRs) ***(11,12)***. This method is rapid and reliable, but it needs to be optimized for the target population due to the use of polymorphic sequences. An important point to bear in mind regarding both of these rapid diagnostic tests is that they currently only permit a result regarding the most common fetal aneuploidies (chromosomes X, Y, 13, 18, and 21). Hence, it is still necessary to resort to the normal 2-week cell culture-based analysis to obtain a full karyotype. Nevertheless, it is being discussed whether the karyotype analysis can be replaced by molecular methods for pregnancies at moderate risk ***(13)***. In the past years, several alternative molecular methods have been shown to be feasible for the analysis of aneuploidies ***(14)***.

The recent development of real-time PCR has rapidly emerged as a powerful tool for the accurate and precise determination of template copy numbers, and it has found widespread applicability in the analysis of gene expression, cell free DNA in body fluids, and the measurement of gene duplications or deletions in cancer research ***(15,16)***. The precision of current technology, however, needs to be optimized to determine less than twofold differences of target template concentrations. After successful completion of a pilot study ***(17)***, we extended this method for the simultaneous detection of trisomy 18 and trisomy 21. In this analysis, the two chromosomes being interrogated are quantified relative to each other. In the past years, we performed two blinded feasibility studies of this real-time PCR approach. First we tested a cohort of almost 100 archived DNA samples that had been extracted from amniotic fluid with chelex resin. We achieved a sensitivity of 100% for the detection of trisomies 18 and 21. However, three normal samples also were scored as trisomic. The reason for the false-positive results was found to be adsorption of the DNA to the resin during the prolonged storage period.

Learning from this experience, we initiated a study with >400 samples, where we extracted previously frozen amniotic fluid in our laboratory and analyzed the fresh DNA. We were able to conclude the study with 471 correctly analyzed samples and only 14 inconclusive results, in which the analysis was either hindered by inadequate concentrations of template DNA or by the presence of inhibitors in the original DNA preparation. Importantly, no sample was misdiagnosed.

These results clearly demonstrate that the real-time PCR analysis of aneuploidies is feasible with very high reliability. Problems for the widespread application are the differences between instruments and between batches of reagents. Further caveats are DNA concentration, purity and sample storage. However, these can easily be avoided, and the method offers several control possibilities without additional work.

It is worth noting that the use of real-time quantitative (q)PCR does have a few advantages in comparison to the established and clinically implemented quantitative fluorescent PCR analysis of STRs: the setup of the test is very straightforward, and it requires no postamplification handling. Thus, it facilitates high throughput and it is less susceptible to carryover contamination. Also, because the real-time qPCR targets nonpolymorphic sequences, it produces valid results regardless of the patient population. In clinical application, the analysis of several targets per chromosome would be warranted and further reduces the potential error rate considerably.

Thus, in the long run, making use of further technological developments, the real-time approach might prove to be the method of choice in the future.

1.3. The Principle of Relative Quantification by Real-Time PCR for Determination of Chromosome Ploidy

Real-time PCR is based on the detection and quantitation of a fluorescent signal directly proportional to the PCR product generated. In our assay, the fluorescence is generated by the 5′ nuclease method: A so-called TaqMan probe hybridizes specifically to one strand of the DNA sequence between the two primers. The probe is labeled with a fluorescent detector dye at its 5′ end and with a quencher dye at its 3′ end. The melting temperature of the probe is 10°C higher than the annealing temperature (T_A) of the PCR reaction, which is the melting temperature of the primers. This ensures that a probe binds to every target sequence before primers anneal. When a primer binds to the target, the Taq polymerase extends the primer and by its nuclease activity cleaves the 5′ end of the probe, thereby separating the detector dye from the quencher dye. Only now, the detector dye emits a characteristic fluorescence signal when excited by an appropriate light source. The cleavage of each probe molecule is the result of one template amplification; the amount of fluorescent signal is therefore directly proportional to the amount of PCR product synthesized.

One parameter has been devised for the quantitative analysis of such real-time PCR assays: The C_T, or threshold cycle value, is the cycle number at which point the amplification curve crosses the (user-defined) fluorescence threshold of the amplification plot (*see* **Fig. 1**). This C_T is a function of the initial template amount and the PCR efficiency: the higher the initial amount of target sequence, the lower the C_T . In an optimally efficient reaction, 1 cycle difference means a twofold difference in starting material.

The main advantage of the TaqMan probe based detection is that the real-time PCR can be multiplexed. The real-time PCR test for fetal aneuploidy entails the simultaneous amplification of two chromosomal loci (chromosome 21 versus chromosome 18) in one reaction vessel, and their product formation is detected by two different fluorescent dyes. Because each of these reporter dyes has a discretely different emission spectrum, and because the amount of each dye measured is proportional to the respective product, the simultaneous quantification of both chromosomes relative to each other is possible.

It is important that the coamplification of two sequences occurs in the same reaction vessel, because this guarantees that no well to well variation occurs

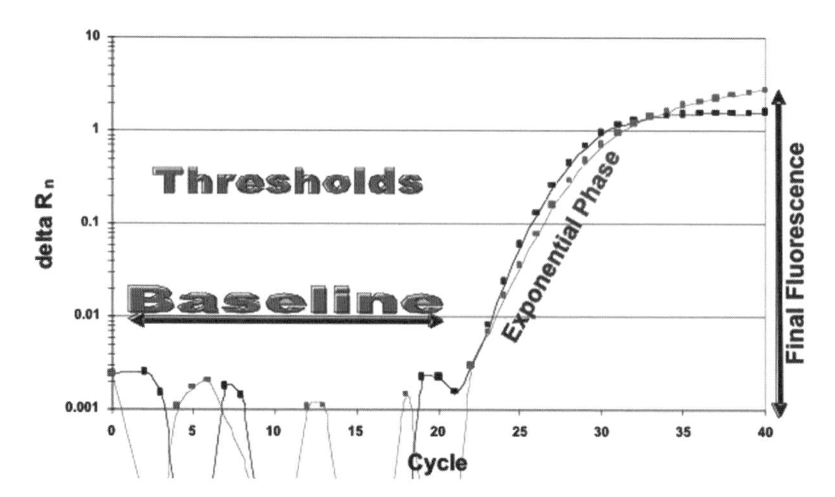

Fig. 1. Amplification plot of the real-time PCR data. Labeled in the plot are the baseline (phase of background fluorescence); the three threshold lines at 0.2, 0.3, and 0.45; the exponential phase; and the final fluorescence. Displayed is an amplification of a sample with trisomy 18. The gray curves represent data from chromosome 18 (FAM dye), and the black curves represent data from chromosome 21 (VIC dye).

between the two amplified targets *(18)*. In this regard, the analysis carried out in a singleplex manner would be more prone to error. Many influences, such as pipetting differences, alterations in polymerase activity, nonuniform heating block temperature, illumination intensities of different wells, and unequal reagent depletion result in well-to-well deviations, can add up to an inaccurate assessment of the target chromosome ploidy.

In a similar manner, it is less reliable to determine the ploidy of a sample by referring to a standard curve. Variability of the measurements of standards can lead to a skewed standard curve *(19)*. Even if this effect is small, the resulting deviations can introduce too much error. The relative amount of the target sequences is calculated by using the ΔC_T , the difference of the C values of the first (chromosome 18) and the second amplified sequence (chromosome 21) in one reaction well (*see* **Fig. 2**):

$$\Delta C_T = C_T(\text{target A}) - C_T(\text{target B}) = C_T(\text{chromosome 18}) - C_T(\text{chromosome 21})$$

The ΔC_T can be converted into a ratio *(20)*:

$$\text{target A}/\text{target B} = 2^{-(\Delta CT)} = \text{chromosome 18}/\text{chromosome 21}$$

In theory, the fluorescent signals from both amplifications are detected simultaneously if the sample is karyotypically normal, i.e., both chromosomes being

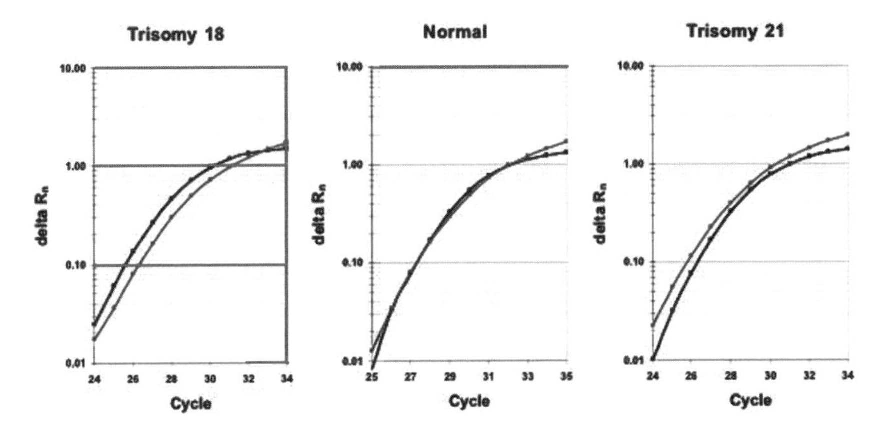

Fig. 2. Real-time PCR amplification plots of three amniotic fluid samples with trisomy 18, normal, and trisomy 21 ploidy. Plotted are the cycle number against the normalized fluorescence (ΔR_n). The amplification of chromosome 18 is in black, and amplification of chromosome 21 in gray.

interrogated have the same copy number. Thus, the difference in threshold cycle number (ΔC_T) for a normal (and for a triploid) karyotype would be 0 cycles. In a trisomy, the chromosome present at three copies per cell would be detected at a lower (earlier) C_T value than the other ($\Delta C_T = \pm 0.58$ cycles, as $2^{-(0.58)} = 1.5$).

However, to date it is not possible to produce perfectly parallel signals from two different assays. To correct for slight differences of signal intensities of the two probes and differences of amplification and cleavage efficiencies, the $\Delta\Delta C_T$ method with a calibrator should be used ***(20)***. In general, the calibrator is a sample of normal karyotype. The ΔC_T of the sample is corrected by the ΔC_T of the calibrator, and the ratio of the target sequences in the sample can be calculated as follows:

$$\Delta\Delta C_{T\ \text{calibrated}} = \Delta C_T(\text{sample}) - \Delta C_T(\text{calibrator})$$
$$= 2^{-(\Delta\Delta CT\ \text{calibrated})}$$

However, even though this method is better than solely relying on the ΔC_T value of the analyzed sample, the use of a single calibrator sample can introduce error. As with every sample, measurement precision is not ideal for a calibrator sample either, and deviations from the "true" calibrator value can cause misinterpretation of the sample analyzed. In this regard, it is worth noting that a seemingly small shift in the reference ΔC_T value of only +0.2 cycles from its true value results in the false assessment of a sample with a similarly small and common deviation of –0.2 cycles. Thus, we recommend an adaptation of the $\Delta\Delta C_T$ method, whereby the samples being analyzed are quantified relative to a reference ΔC_T value determined from the average ΔC_T values of many karyotypically normal samples tested previously. At the beginning of a study, we advise to measure at least 20 samples of normal karyotype. The ΔC_T values of these samples are averaged to generate the calibrator value for the consecutive analysis.

Finally, to balance small fluctuations of the fluorescence readings and to ensure that the amplification of both target sequences has proceeded with similar efficiency, we analyzed, instead of a solitary C_T value, three to four points with equal distances (i.e., equal -fold increase of R_n) along the exponential phase of the amplification (*see* **Fig. 1**). Those samples in which a deviation is found to occur are reanalyzed.

For aneuploidy detection it is not necessary to determine the ratio, because the $\Delta\Delta C_T$ values themselves are indicative of the karyotype, i.e., close to 0 for normal and –0.58 for trisomy 18 and +0.58 for trisomy 21. In this manner, it is possible to determine the samples' karyotype with a minimum of postrun work by setting a cut-off around the average $\Delta\Delta C_T$ of a given karyotype. In our analyses, we set a cut-off range of 0.32 cycles around the average measured $\Delta\Delta C_T$ of normal karyotype (as $2^{0.32} = 1.25$). In this regard, the $\Delta\Delta C_{Tcalibrated}$ of karyotypically normal samples has a range of –0.32 to +0.32 cycles.

2. Materials

2.1. DNA Extraction

1. DNA can be extracted with many methods and kits. In our large study, we used the extraction of genomic DNA with the MagnaPure extraction instrument as described in Chap 21.

2.2. Performing real-time PCR

1. Real-time qPCR analysis is performed on an ABI7000 Sequence Detection System (SDS7000) (*see* **Note 1**).
2. PCR reactions are carried out in MicroAmp Optical 96-well reaction plates (Applied Biosystems, Basel, Switzerland, cat. no. 4306737) with the ABI Prism Optical adhesive covers (cat. no. 4311971).
3. The reactions are carried out using the TaqMan Universal PCR master mix (cat. no. 4304437) containing AmpliTaq Gold DNA Polymerase and AmpErase uracil-*N*-glycosylase (UNG), dNTPs with dUTP and passive reference 1 (ROX) fluorescent dye (*see* **Note 2**).
4. Primer (high-performance liquid chromatography [HPLC] purified, Microsynth, Balgach, Switzerland) (*see* **Note 3**) and minor groove binder (MGB) TaqMan probe (Applied Biosystems, cat. no. 43160324) (*see* **Note 4**) sequences are listed

Table 1
Primer and probe sequences for the real-time amplification of chromosomes 18 and 21. Probes were with MGB

Chromosome location	Gene[a]	Name	Sequence
18p11.32	TYMS	Chr_18F	TGACAACCAAACGTGTGTTCTG
		Chr_18R	AGCAGCGACTTCTTTACCTTGATAA
		Chr_18P_MGB	VIC-GGTGTTTTGGAGGAGTT
21q21.3	APP	Chr_21F	CCCAGGAAGGAAGTCTGTACCC
		Chr_21R	CCCTTGCTCATTGCGCTG
		Chr_21P_MGB	FAM-CTGGCTGAGCCATC

[a]TYMS, thymidylate synthetase; APP, amyloid precursor protein

in **Table 1**. The Chromosome 18 probe is 5′ labeled with the fluorescent dye VIC, and the chromosome 21 probe is 5′ labeled with 5-carboxyfluorescein (FAM). It is advisable to store aliquots with a stock concentration of 5 μM at –20°C, where they are stable for over 1 year. If used rapidly (within 1 week), the probes can be stored at 4°C. In this case, aliquots of both probes should be used in parallel. Avoid light exposure. Primers stored at 4°C are stable for several weeks.

5. The reference sample for control purposes and for quantification of the DNA content of the samples can be any genomic DNA (gDNA) of known concentration. We use DNA extracted from buffy coat with the same method as the samples to be analyzed. Alternatively gDNA from a commercial source can be used. The concentration should be such that between 100 and 1,000 genome equivalents (GE; 1 GE = 6.6 pg of human gDNA) is added to one reaction. Reference DNA should be stored frozen in aliquots in nonstick Microfuge tubes (Ambion, Austin, TX, cat. no. 12450). Once thawed, it can be used for 2 weeks when kept at 4°C.

3. Methods

3.1. DNA Extraction of Amniotic Fluid Samples

Automated DNA extraction using the MagnaPure system is described in Chap 21 (*see* **Note 5**).

3.2. Performing Real-Time PCR

1. The suggested total reaction volume is 25 μl (*see* **Note 6**), containing 2 μl of the sample DNA solution (*see* **Note 7**), 300 nM of each primer (*see* **Note 8**) and 200 nM of each probe (*see* **Note 9**) at 1× concentration of the Universal PCR reaction mix.
2. Each sample is analyzed in triplicate.

3. The preparation of the real-time PCR reactions should be carried out on a PCR cooler block (Eppendorf, Hamburg, Germany, catalog no. 022510525) or on ice (*see* **Note 10**).
4. Include at least one control sample of normal karyotype and known concentration in the analytic run (*see* **Note 11**).
5. First pipette the PCR reaction mixture into the reaction wells. Then, add 2 μl of the sample DNA solution. Try to avoid generating air bubbles. Seal the reaction plate with the optical adhesive cover. Centrifuge at 2000*g* at 4ºC for 1 min to spin down any droplets and remove air bubbles.
6. Immediately transfer the plate to the real-time PCR thermocycler and start the following cycling protocol with the emulation mode off (*see* **Note 12**).
7. Incubation at 50°C for 2 min to permit AmpErase activity, 10 min at 95ºC for activation of AmpliTaq Gold and denaturation of the genomic DNA, and 40 cycles of 1 min at 60°C and 15 s at 95°C.

3.3. Analysis of Real-Time PCR Data

1. The experimental analysis is performed with the automatic baseline setting (*see* **Note 13**). Replicate curves are checked for uniformity in the amplification plot view (*see* **Note 14**).
2. Analyze three different fluorescence thresholds (R values) at 0.2/0.3/0.45, as indicated in **Fig. 1** (*see* **Note 15**). Export the C_T result files and the ΔR_n file and analyze the data as shown in the sample spreadsheet (*see* **Table 2**).
3. As quality controls for the entire plate, compare the C_T values of the reference sample to previous runs. The deviation should not be greater than 0.2 cycles.
4. As a first quality control for the individual curves, determine the input copy number for both chromosomes. The minimal input copy number per reaction is 100 GE (*see* **Note 7**).
5. Second, examine the final fluorescence (ΔR_n). The minimal acceptable ΔR_n is 1.5 for the signals of either probe (*see* **Note 16**).
6. The third control for the individual curves is the determination of the fluorescence increase; This is the slope of the linear regression when plotting the C_T values against the logarithm of the fluorescence. The "efficiency of probe cleavage" within the examined threshold interval can be calculated in Excel by the following formula:

$$E_{\text{fluo inc}} = 10^{(\text{SLOPE})-1}$$

   SLOPE = [log(0.45) [minus] log(0.2)]/[C_T (0.45) [minus] C_T (0.2)].
   The minimal acceptable fluorescence increase is 50%.
7. For every replicate calculate the ΔC_T for each of the three thresholds (*see* **Table 2**).
8. Calculate the $\Delta\Delta C_T$ for each of the three thresholds:

$$\Delta\Delta C_{T\text{ calibrated}} = \Delta C_T(\text{sample})[\text{minus}]\Delta C_T(\text{calibrator})$$

Table 2
Experimental data and analysis of one trisomy 18 sample (single replicate). The $\Delta\Delta C_T$ values and the determined karyotype are in bold. For illustrative purposes, the measurements of the reference are used to calculate the ΔC_T (calibrator). The reference copy number was 500 GE

	Thr = 0.2	Thr = 0.3	Thr = 0.45	ΔR_n	Fluorescence increase
C_T (chr 18) reference	28.23	28.89	29.62		
C_T (chr 21) reference	28.31	29.04	29.81		
ΔC_T (calibrator)	–0.08	–0.15	–0.19	minimum: 1.5	minimum: 50%
C_T (chr 18) sample	26.50	27.18	27.96		
C_T (chr 21) sample	27.27	28.00	28.78		
Chr 18	1659 GE	1636 GE	1580 GE	1.62	74%
Chr 21	1028 GE	1028 GE	1021 GE	2.85	71%
Control passed	Y	Y	Y	Y	Y
ΔC_T	–0.77	–0.82	–0.82		
$\Delta\Delta C_T$	–0.69	–0.67	–0.63		
Karyotype	Tris 18	Tris 18	Tris 18		

The $\Delta\Delta C_T$ is indicative for the chromosomal status. It is between –0.32 and +0.32 for normal karyotype and outside of this range for trisomy 18 (< –0.32) and 21 (> +0.32).

9. If the results of all three replicates are indicative of the same chromosomal balance, the ploidy of the sample is that of the matched reference ΔC_T calibrators; otherwise the result is not conclusive.

4. Notes

1. One feature of the SDS7000 is that the fluorescence data are normalized to the passive reference dye (ROX). Any instrument that offers such feature should be suited for the test.
2. The use of AmpErase UNG is important to avoid amplification of any missprimed nonspecific products formed before specific amplification. If this step is omitted, the balance of the two multiplexed reactions can be affected

and one of the sequences can be amplified with a greater efficiency, resulting in false results. For the same reason, use of a Taq polymerase activated by a hot start is required.

3. It is recommended to use HPLC-purified primers, because these reduce the probability of artefacts that form unspecific products. These have an adverse effect on the reaction efficiency. Additionally, nonpurified primers are subject to greater batch-to-batch variation than HPLC-purified primers.
4. MGB probes emit lower background fluorescence than dual-labeled probes. This is mainly due to more efficient quenching, because the probes are shorter. As a result, MGB probes have a longer observable exponential phase, and this permits the use of a wider range for the thresholds. It is also important to be aware, that the use of different batches of the same probes can result in differences in measurement due to minor differences in quality.
5. Any genomic DNA extraction method from amniotic fluid can be used. We also tested manual extraction with the high-pure PCR template kit from Roche (Basel, Switzerland) and with the QIAamp blood kits from QIAGEN (Hilden, Germany). However, for resin-based extraction methods, it is necessary to separate the supernatant from the resin directly after the extraction, because the resin can absorb the DNA during storage.
6. Reaction volumes tested were 15 to 50 μl. Using 15 μl works, but the deviations are increased. A volume of 50 μl works fine, but it is not necessary to use such a big volume. The same volume should always be used, because small differences in collected data occur if the reaction volume is different from the volume for which the assay is characterized.
7. The amount of DNA needed for highly reproducible results is at least 50 GE (330 pg). One milliliter of amniotic fluid extracted into 100 μl usually yields DNA in the required concentration range. It is not necessary to determine the DNA concentration (C) of the sample before analysis, because template copy number will be determined by coamplifying a reference sample of known amount of DNA:

$$C_{sample} = 2^{-(C_{T,sample} - C_{T,reference})} \times C_{reference}$$

   If the concentration of the DNA is too low, up to 8 μl of DNA can be used in a 25-μl reaction. If the concentration is still too low, DNA has to be concentrated. Also higher concentrations of genomic DNA disturb the reaction.
8. Higher primer concentrations result in a decrease of the reaction efficiency in the multiplex assay. This can be attributed to an increase of unspecific reactions.
9. Probe concentrations of 100 μM also work fine, but precision may be reduced. Probe concentrations of 400 μM work slightly better. However, it will be necessary to use a new batch more quickly, which is a cost factor and in addition requires to test the assay again.
10. Sample setup on ice is important to reduce nonspecific product formation to a minimum. The reaction plate should be kept on ice at all times before starting the real-time PCR.

11. Use of a reference sample of known quantity allows quantification as mentioned in **Note 7**. If a dilution series of the reference is amplified (template input between 100 and 10,000 GE), the sequence detection software can determine the concentration of all samples tested with standard curves automatically. In theory, these curves would permit karyotyping of the samples; however, the results are not good enough.
12. Run the program always at the same heating and cooling rates. Differences in the temperature program result in slight differences of the final C_T and ΔC_T values. We use the emulation off setting, because it is faster.
13. Optimal baseline setting for individual wells is now performed automatically. In previous versions of the software, the baseline had to be adjusted manually: it is used to separate unspecific fluorescent signal from the signal generated by the reaction. If the baseline is too far away from the amplification curve, results can be affected by noise. Setting the baseline too close skew the amplification curves and change the ΔC_T to smaller values.
14. In the amplification plot view, check whether the amplification curves of both dyes look normal and parallel for all three replicates. If the three replicates do not look very similar, something might be wrong with at least one of the reactions. This can be due to contamination with DNA, pipetting errors, dirt or fluorescence source other than the dyes, leakage of the well, inhibitors, incorrect automatic baseline setting, or other reasons. If the curves do not resemble each other closely, the test should be rerun.
15. The use of several thresholds increases the accuracy of the method. By using the four proposed thresholds, fluorescence data collected during three or four cycles is used compared with data from only 2 cycles when using only one threshold. A threshold of 0.1 sometimes contains a good amount of unspecific signal. To exclude that possibility, the minimal threshold setting used is 0.2. The maximum threshold of 0.45 still gives good data, although the reaction is reaching the end of the exponential phase and efficiencies decreasing. The optimal thresholds can be determined empirically and should not change between experiments if the same stock of reagents is used and the instrument performance is stable. Recalibration of the instrument requires reestablishment of calibrator values and optimal thresholds for analysis.
16. In our large analysis, the ΔR_n after 50 cycles was usually between 2.0 and 3.0. Runs with a final ΔR_n of <1.5 were disregarded and repeated.

References

1. Thorp, J. A., Helfgott, A. W., King, E. A., King, A. A., and Minyard, A. N. (2005) Maternal death after second-trimester genetic amniocentesis. *Obstet. Gynecol.* **105,** 1213–1215.
2. Cicero, S., Spencer, K., Avgidou, K., Faiola, S., and Nicolaides, K. H. (2005) Maternal serum biochemistry at 11–13(+6) weeks in relation to the presence or absence of the fetal nasal bone on ultrasonography in chromosomally abnormal

fetuses: an updated analysis of integrated ultrasound and biochemical screening. *Prenat. Diagn.* **25,** 977–983.

3. Wald, N., Cuckle, H., Wu, T. S., and George, L. (1991) Maternal serum unconjugated oestriol and human chorionic gonadotrophin levels in twin pregnancies: implications for screening for Down's syndrome. *Br. J. Obstet. Gynaecol.* **98,** 905–908.
4. Nicolaides, K. H., Azar, G., Byrne, D., Mansur, C., and Marks, K. (1992) Fetal nuchal translucency: ultrasound screening for chromosomal defects in first trimester of pregnancy. *BMJ* **304,** 867–869.
5. Cicero, S., Longo, D., Rembouskos, G., Sacchini, C., and Nicolaides, K. H. (2003) Absent nasal bone at 11–14 weeks of gestation and chromosomal defects. *Ultrasound Obstet. Gynecol.* **22,** 31–35.
6. Falcon, O., Auer, M., Gerovassili, A., Spencer, K., and Nicolaides, K. H. (2006) Screening for trisomy 21 by fetal tricuspid regurgitation, nuchal translucency and maternal serum free beta-hCG and PAPP-A at 11 + 0 to 13 + 6 weeks. *Ultrasound Obstet. Gynecol.* **27,** 151–155.
7. Brambati, B., Macintosh, M. C., Teisner, B., et al. (1993) Low maternal serum levels of pregnancy associated plasma protein A (PAPP-A) in the first trimester in association with abnormal fetal karyotype. *Br. J. Obstet. Gynaecol.* **100,** 324–326.
8. Macri, J. N., Spencer, K., Aitken, D., et al. (1993) First-trimester free beta (hCG) screening for Down syndrome. *Prenat. Diagn.* **13,** 557–562.
9. Witters, I., Devriendt, K., Legius, E., et al. (2002) Rapid prenatal diagnosis of trisomy 21 in 5049 consecutive uncultured amniotic fluid samples by fluorescence in situ hybridisation (FISH). *Prenat. Diagn.* **22,** 29–33.
10. Klinger, K., Landes, G., Shook, D., et al. (1992) Rapid detection of chromosome aneuploidies in uncultured amniocytes by using fluorescence in situ hybridization (FISH). *Am. J. Hum. Genet.* **51,** 55–65.
11. von Eggeling, F., Freytag, M., Fahsold, R., Horsthemke, B., and Claussen, U. (1993) Rapid detection of trisomy 21 by quantitative PCR. *Hum. Genet.* **91,** 567–570.
12. Levett, L. J., Liddle, S., and Meredith, R. (201) A large-scale evaluation of amnio-PCR for the rapid prenatal diagnosis of fetal trisomy. *Ultrasound Obstet. Gynecol.* **17,** 115–118.
13. Hulten, M. A., Dhanjal, S., and Pertl, B. (2003) Rapid and simple prenatal diagnosis of common chromosome disorders: advantages and disadvantages of the molecular methods FISH and QF-PCR. *Reproduction* **126,** 279–297.
14. Dudarewicz, L., Holzgreve, W., Jeziorowska, A., Jakubowski, L., and Zimmermann, B. (2005) Molecular methods for rapid detection of aneuploidy. *J. Appl. Genet.* **46,** 207–215.
15. Gingeras, T. R., Higuchi, R., Kricka, L. J., Lo, Y. M., and Wittwer, C. T. (2005) Fifty years of molecular (DNA/RNA) diagnostics. *Clin. Chem.* **51,** 661–671.
16. Bustin, S. A. and Mueller, R. (2005) Real-time reverse transcription PCR (qRT-PCR) and its potential use in clinical diagnosis. *Clin. Sci.* **109,** 365–379.

17. Zimmermann, B., Holzgreve, W., Wenzel, F., and Hahn, S. (2002) Novel real-time quantitative PCR test for trisomy 21. *Clin. Chem.* **48,** 362–363.
18. Hahn, S., Zhong, X. Y., Burk, M. R., Troeger, C., and Holzgreve, W. (2000) Multiplex and real-time quantitative PCR on fetal DNA in maternal plasma. A comparison with fetal cells isolated from maternal blood. *Ann. N Y Acad. Sci.* **906,** 148–152.
19. Zimmermann, B., El-Sheikhah, A., Nicolaides, K., Holzgreve, W., and Hahn, S. (2005) Optimized real-time quantitative PCR measurement of male fetal DNA in maternal plasma. *Clin. Chem.* **51,** 1598–1604.
20. Livak, K. J. and Schmittgen, T. D. (2001) Analysis of relative gene expression data using real-time quantitative PCR and the 2(-Delta Delta C(T)) method. *Methods* **5,** 402–408.

8

MLPA for Prenatal Diagnosis of Commonly Occurring Aneuploidies

Jan Schouten and Robert-Jan Galjaard

Summary

Multiplex ligation-dependent probe amplification (MLPA) is a new method to determine the copy number of up to 45 genomic DNA sequences in a single multiplex polymerase chain reaction (PCR)-based reaction. In contrast to standard multiplex PCR, only one pair of PCR primers is used. MLPA reactions with currently commercial available kits result in very reproducible gel patterns with fragments of 130 to 480 bp that can be analyzed by sequence type electrophoresis. Comparison of this gel pattern to that obtained from a control sample indicates which sequences show an aberrant copy number.

Key Words: Aneuploidy; multiplex ligation-dependent probe amplification (MLPA); trisomy; multiplex polymerase chain reaction (PCR); amniotic fluid; gene dosage.

1. Introduction

Multiplex ligation-dependent probe amplification (MLPA) is a new multiplex method to detect abnormal copy numbers of genomic DNA sequences *(1)*. MLPA reactions are easy to perform, and they require little hands-on time. Up to 96 samples can be handled simultaneously, and results can be obtained within 24 h. The thermocycler and sequencing-type electrophoresis equipment that are required are present in most DNA diagnostic laboratories. One of the currently available MLPA kits (P095, MRC-Holland, Amsterdam, The Netherlands, www.MLPA.com) contains eight independent probes for each of the chromosomes 13, 18, 21, and X, and four Y-specific probes, and it is used by several medical centers on a large scale as a rapid test for aneuploidies

From: *Methods in Molecular Biology, vol. 444: Prenatal Diagnosis*
Edited by: S. Hahn and L. G. Jackson © Humana Press, Totowa, NJ

of these chromosomes ***(3)***. Although MLPA reactions require a minimum of only 20 ng of human DNA, the use of 1 ml of amniotic fluid at 16 weeks of pregnancy results in test failures ascribed to low DNA input in 2–8% ***(5)***. The use of 2 ml of amniotic fluid and DNA isolation reduces this to 0.2% (***Unpublished data***).

MLPA allows discrimination of sequences that differ only in a single nucleotide; therefore, MLPA also can be used for the detection of known mutations. A variation on the MLPA technique can be used to determine the methylation status of DNA sequences ***(4)***. Although performing an MLPA reaction is easy, the development of new MLPA assays is complex and time-consuming. MLPA kits for >100 different applications are commercially available from the MRC-Holland (SALSA MLPA kits, www.MLPA.com).

These applications include kits for detection of the following:

1. Aneuploidy of chromosomes 13, 18, 21, X, and Y.
2. Deletions or duplications of complete genes or chromosomal areas, e.g., the 22q11 region involved in DiGeorge syndrome and various microdeletion syndromes. Besides, MLPA kits are available to detect copy number changes of all subtelomeric regions, or all centromeric regions, in a single reaction.
3. Deletions or duplications involving only one or more exons of a gene. Small chromosomal rearrangements can be detected because the sequences analyzed by MLPA probes are only 60 nucleotides in length.

MLPA is basically a method to make a nucleic acid sample suitable for a multiplex polymerase chain reaction (PCR) with the use of only one pair of PCR primers. In the currently available kits, the amplification products generated are separated by sequence-type electrophoresis (*see* **Fig. 1**).

Although the MLPA amplification reaction is very reproducible, some sequences are amplified with an efficiency that is 1–2% higher or lower during each PCR cycle, resulting in different peak areas for the different amplification products. Therefore, a single MLPA amplification profile will not provide the information required. MLPA profiles must be compared with a similar profile obtained from a control DNA sample. Compared with a control reaction, the relative peak area of each amplification product reflects the relative copy number of the target sequence of that probe in the analyzed sample.

An aberrant copy number of one or more of the sequences detected by the MLPA probes can therefore be detected by a decrease or increase in relative peak area of the amplification products of the probes detecting those sequences.

MLPA tests are designed in such a way that the length of the amplification product of each probe is different. In the currently available probe mixes (SALSA MLPA kits, MRC-Holland), length difference between different amplification products is usually six to nine nucleotides, with MLPA amplification products that range in size between 130 and 480 nucleotides. This

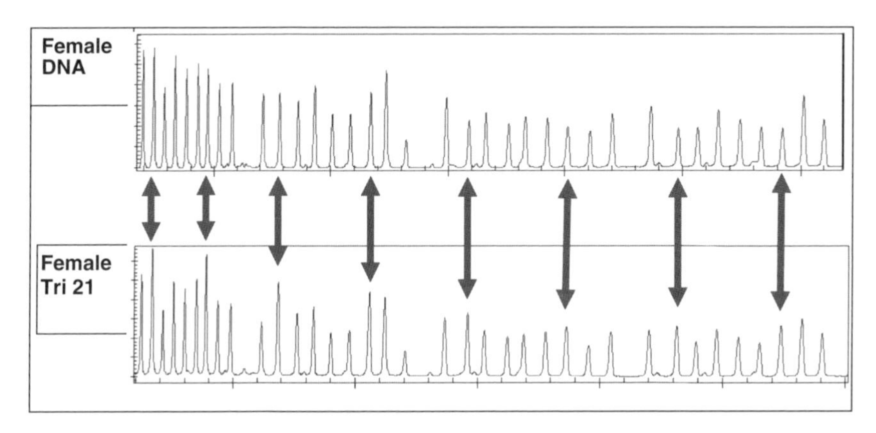

Fig. 1. Detection of trisomy 21 with MLPA. Arrows indicate alterations. An increase of the signal from the chromosome 21-specific probes is seen.

is the size range that provides an optimal separation and low background on sequencing type electrophoresis gels.

Standard multiplex PCR requires one pair of PCR primer oligonucleotides for every fragment to be amplified. These PCR primers are present in high concentrations during the PCR, and the presence of high numbers of different primers often causes problems. Because the efficiency of different primer pairs is not equal, small differences in reaction conditions can result in large differences in the results obtained. MLPA reactions are more robust, because all fragments are amplified with only one pair of PCR primers.

In the PCR of an MLPA reaction, no sample nucleic acids are amplified. In MLPA, a copy is made of each target sequence (*see* **Fig. 2**). These copy sequences are amplified in a multiplex PCR reaction with the use of a single primer pair.

To make one copy of each target sequence, a specific MLPA probe is added to the nucleic acid sample for each of the sequences to be quantified. Each MLPA probe consists of two oligonucleotides that both hybridize to the sequence to be detected at sites immediately adjacent to each other. Each of these two oligonucleotides contains one of the two sequences recognized by the PCR primer pair. The two parts of each probe can be ligated to each other by a specific ligase enzyme, but only when both probe oligonucleotides are perfectly hybridized to adjacent sites of the target sequence.

The resulting ligated products have both PCR primer sequences, and they can be amplified exponentially by PCR. Thus, each probe consists of two oligonucleotides, and up to 45 different probes can be present in an MLPA probe mix. The concentration of probes and the duration of the hybridization reaction are sufficient to allow near complete hybridization of target sequences to their corresponding probe oligonucleotides. Prolonging the hybridization reaction

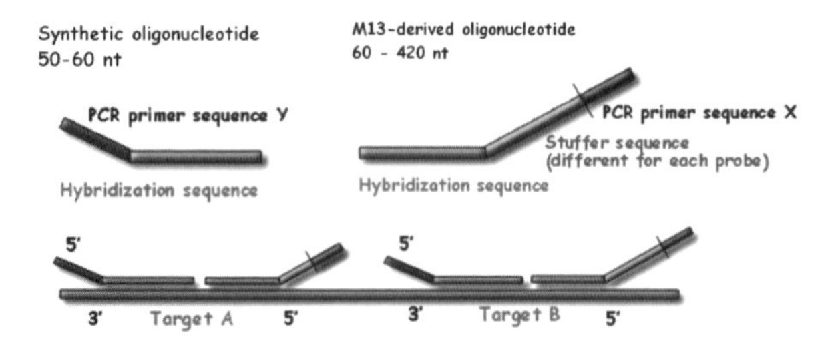

1. Denaturation and hybridization to the target sequence

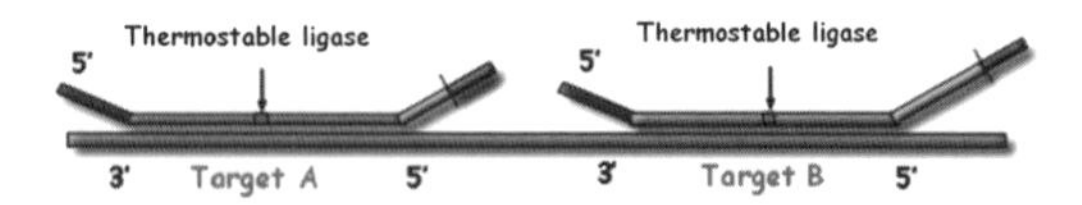

2. Ligation of the two probe parts

3. Amplification of probes by PCR

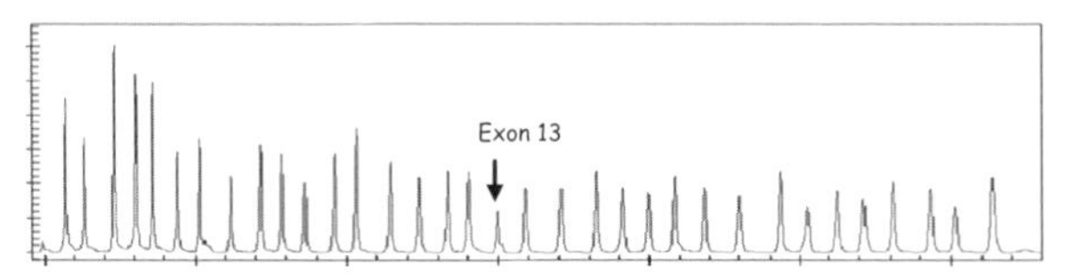

4. Separation of amplification products by capillary sequencer

Fig. 2. Outline of the MLPA technology.

or addition of larger amounts of the probes will not influence the results. In a typical MLPA reaction, approx 500,000,000 copies of each probe oligonucleotide are present, whereas only 20,000 copies of most target sequences are present in a 60-ng human DNA sample.

Only probes hybridized to a target sequence are ligated, and they are exponentially amplified by PCR. The relative signal strength of each amplification product is determined primarily by the copy number of the target sequence in the sample. MLPA probes that do not find a target sequence cannot be amplified by PCR, and they do not have to be removed; therefore, the protocol for an MLPA reaction is very simple.

1. Denature 20–500 ng of DNA by heating to 98°C in a thermocycler.
2. Add the MLPA probes and leave overnight at 60°C for hybridization.
3. Add the ligase and ligase buffer and ligate at 54°C for 15 min.
4. Inactivate the ligase by heating to 98°C
5. Add PCR primers, dNTPs, and polymerase and start the PCR.
6. Analyze the products by electrophoresis.

In MLPA it is important that all fragments are amplified with the use of only one pair of PCR primers. Similar to amplified fragment-length polymorphism and differential display techniques, the amplification reaction of MLPA slows down after approx 30 cycles due to primer depletion. The effect of extra PCR cycles on the relative amounts of the amplification products is very small. Thus, MLPA is very reproducible, and the results do not vary with the amount of input DNA, which should be between 20 and 500 ng.

For aneuploidy testing on capillary electrophoresis systems, the SALSA P095 probe mix is available from the MRC-Holland (www.MLPA.com). This probe mix contains 36 probes: eight probes each that are specific for chromosome 13, 18, 21, and X sequences, and four probes for human chromosome Y target sequences.

MLPA is not able to detect all chromosomal abnormalities seen with karyotyping. It is designed for detection of specific copy number changes as mentioned. MLPA analysis is expected to detect high level of chromosomal mosaicism, because it will give the average copy number per cell. The P095 probe mix, however, has been shown not to detect low-level mosaicism (*3* and ***Unpublished data***). The P095 probe mix is specifically designed to detect common occurring aneuploidies, except for detection of triploidies *(2)*. It will not detect cases of 69,XXX *(3,5)*. Cases of 69,XXY and 69,XYY are difficult to discriminate from maternal DNA contamination, as was shown for a case of 69,XXY, which was assigned as a sample contaminated by maternal DNA *(3)*. However, we have correctly diagnosed two cases (***Unpublished data***). Keep in mind that many triploidies result in fetal ultrasound abnormalities. The MLPA aneuploidy kit is not designed for detection of balanced chromosomal rearrangements, such as translocations and inversions.

MLPA, however, provides several opportunities that are not possible with other techniques such as karyotyping. For example, when a quick diagnostic result is warranted, e.g., in cases of fetal abnormalities visualized by advanced ultrasound examination, amniotic fluid samples can first be rapidly tested for the copy number of the most commonly occurring aneuploidies. If normal, further selective testing based on clinical preselection could be done, e.g., for all subtelomeric regions and the most common microdeletion syndromes. MLPA products for these applications are commercially available. Apart from (prenatal) diagnostics of chromosome abnormalities already mentioned, single-gene defects also can be diagnosed using currently available MLPA kits. Probe mixes can include probes specific for certain frequent point mutations. In northwestern Europe 40% of the cystic fibrosis (CF) patients can be detected by inclusion of a single probe for the cystic fibrosis transmembrane conductance regulator ΔF508 mutation sequence. Most spinal muscular atrophy (SMA) patients could be detected by inclusion of a single MLPA probe. The inclusion of eight probes for selected regions of the DMD gene in a MLPA probe set is expected to detect 40–50 % of all Duchenne muscular dystrophy (DMD) patients. Whether the aim for the prenatal detection of chromosomal abnormalities should include sex chromosome abnormalities, balanced translocations, or, e.g., single-gene defects such as in DMD, SMA, and CF, is still debated in different countries.

2. Materials

2.1. Contents of SALSA Kit for 100 MLPA Reactions

1. SALSA probe mix (mixture of up to 45 pairs of probe oligonucleotides).
2. SALSA MLPA buffer. Contains 1.5 M salt + additives. Does not always freeze at –20°C.
3. Ligase-65 enzyme solution.
4. Ligase-65 buffer A. Contains cofactor NAD required for the Ligase-65 enzyme.
5. Ligase-65 buffer B.

   a. Contains the salts required by the Ligase-65 enzyme.

6. SALSA PCR buffer.
7. SALSA PCR primers. Contains one fluorescently labeled and one unlabeled PCR primers + dNTPs.
8. SALSA polymerase enzyme solution (similar to Taq polymerase).
9. SALSA enzyme dilution buffer. Stabilizes the polymerase/primers dilution.

SALSA MLPA kits are stable for at least 1 year when stored in the dark at –20°C. All enzymes, nucleic acids, and buffer constituents are nonhazardous. SALSA MLPA kits are available from MRC-Holland (www.MLPA.com).

Mix the contents of the tubes after thawing and centrifuge the tubes for a few seconds before opening.

3. Methods

3.1. Amniotic Fluid Sample Preparation Protocol 1: Cell Lysate Preparation

1. Mix the amniotic fluid gently, just before a sample is removed. Most cells might be at the bottom of the tube. Blood-contaminated amniotic fluid samples require DNA purification such as used in protocol 2 (*see* **Subheading 3.2.**).

2. Centrifuge 1.5 ml of amniotic fluid for 5 min at 10,000*g*. Remove the supernatant carefully.
3. Centrifuge again for 15 s at 10,000*g*. Remove the last traces of supernatant with a micropipette.
4. Resuspend the pellet in 10–15 μl of proteinase K (ProtK*) solution and incubate 30 min at 55°C.
5. Transfer 5 μl to a 0.2-ml vial for the MLPA reaction. Store the remainder at –20°C.

* ProtK solution: 1 mg/ml proteinase K in 20 mM Tris-HCI, pH 7.6, and 50 mM NaCl. We use a 20-fold dilution of the recombinant ProtK solution sold by Roche (Mannheim, Germany; cat. no. 745723).

3.2. Amniotic Fluid Sample Preparation Protocol 2: DNA Isolation with QIAamp DNA Mini kit (QIAGEN, Hilden, Germany)

1. Mix the amniotic fluid gently, just before a sample is removed. Most cells might be at the bottom of the tube.
2. Centrifuge a 2-ml sample of amniotic fluid for 5 min at 10,000*g*. Remove the supernatant carefully. Wash the pellet with 1 ml of phosphate-buffered saline (PBS) and centrifuge for 5 min at 10,000*g*. Remove supernatant carefully and resuspend the pellet in 200 μl of PBS. DNA isolation is done according to the Blood and Body Fluid Spin Protocol (QIAGEN). DNA is eluted with 50 μl of buffer AE instead of 200 μl to increase the DNA concentration. Transfer 5 μl to a 0.2-ml vial tube for the MLPA reaction. Store the remainder at –20°C.

3.3. DNA Denaturation and Hybridization of SALSA Probes

1. Heat 5 μl of DNA sample (20–500 ng of DNA) (*see* **Notes 1** and **2**) for 5 min at 98°C in a 0.2-ml vial in a thermocycler with heated (105°C) lid (*see* **Note 3**).
2. Cool to 25°C before opening the thermocycler.
3. Prepare a mixture of equal volumes SALSA probe mix and MLPA buffer at room temperature. Mix well.
4. Add 3 μl of this mixture to each sample. Mix with care by repeated pipetting.
5. Incubate for 1 min at 95°C, followed by a 16-h incubation (*see* **Note 4**) at 60°C (*see* **Note 5**).

3.4. Ligation Reaction

1. Prepare a ligase mixture containing 3 μl of Ligase-65 buffer A, 3 μl of Ligase-65 buffer B, 25 μl of water, and 1 μl of Ligase-65 enzyme for each reaction (*see* **Note 6**). Mix well by repeated pipetting.
2. Reduce the temperature of the thermal cycler to 54°C.
3. Add 32 μl of ligase mixture to the MLPA reaction (*see* **Note 7**) and mix by repeated pipetting.

4. Incubate 15 min at 54°C and then heat 5 min at 98°C. Ligated reactions are stable and can be stored at –20 °C (*see* **Note 8**).

3.5. PCR

1. Prepare a PCR mixture (*see* **Note 9**) containing 4 μl of SALSA PCR buffer (*see* **Notes 10** and **11**), 26 μl of water, and 10 μl of the MLPA-ligated reaction in a 0.2-ml vial for each reaction.
2. Prepare a polymerase mixture for each reaction containing 2 μl of SALSA PCR primers, 2 μl of SALSA enzyme dilution buffer, and 5.5 μl of water.
3. Mix and add 0.5 μl of SALSA polymerase. Mix well but do not vortex. Store on ice until used.
4. While the vials are in the thermal cycler at 60°C, add 10 μl of polymerase mix to each vial. Mix by pipetting and thenimmediately start the PCR reaction (*see* **Note 12**).

3.5.1. PCR Conditions

1. 30 s at 95°C (*see* **Note 13**).
2. 30 s at 60°C.
3. 60 s at 72°C; 35 cycles (*see* **Note 14**).
4. End with 20-min incubation at 72°C (*see* **Note 15**).

3.6. Separation and Quantification of MLPA Amplification Products by Capillary Electrophoresis

The amount of the MLPA PCR reaction required for analysis by capillary electrophoresis depends on the apparatus and fluorescent label used. For example, conditions for the Beckman CEQ apparatus (Beckman Coulter, Fullerton, CA) are provided below:

Following the PCR reaction, mix 0.7 μl of the PCR reaction, 0.4 μl of the Beckman D1-labeled 60–600 molecular weight marker (Beckman Coulter, no. 608095), 40 μl of deionized formamide (Applied Biosystems, Foster City, CA, no. 4311320, or Beckman sample loading solution).

Settings: capillary temperature, 50°C; denaturation, 90°C for 90 s; injection time, 2.0 kV for 60 s; and runtime, 60 min at 4.8 kV. Analysis settings: include peaks >3%; Size standard-600. Slope threshold 1 (*see* **Note 16**).

3.7. Thermal Cycler Program for Complete MLPA Reaction

1. Heated is lid is at 105°C at all steps.
2. 5 min at 98°C; 25°C pause.
3. 1 min at 95°C; 60°C pause.
4. 15 min at 54°C.

5. 5 min at 98°C; 4°C pause.
6. 60°C pause.
7. 30 s at 95°C; 30 s at 60°C; at 60 s 72°C, 35 cycles.
8. 20 min at 72°C; 4°C pause.

4. Notes

1. If necessary, dilute DNA with TE (10 mM Tris-HCl, pH 8.0, and 1 mM EDTA).
2. The volume of the reaction is important for the hybridization speed, which is probe and salt concentration dependent. Do not use >5-µl sample DNA.
3. The use of a thermocycler with heated lid (105°C) is essential. Use 0.2-ml vials.
4. Minimum recommended hybridization period is 14 h; maximum is 20 h.
5. Excessive evaporation can be a source of problems. After the overnight incubation at 60°C, the volume at the bottom of the tubes should be at least 5.5 µl. Excessive evaporation can be due to poor-quality plastics or insufficient temperature or pressure from the heated lid, resulting in a much lower peak area of the longer fragments.
6. Diluted Ligase-65 can be stored at 4°C for at least 1 h.
7. The ligase mixture can be added to the samples as soon as the temperature of the samples has dropped to 54°C. Ligation is almost complete after 2 min at 54°C. We recommend incubation at 54°C for 10–15 min after the last samples have been mixed with the ligase mixture.
8. After the 98°C ligase inactivation treatment, samples can be stored at 4°C for 1 week or at –20°C for longer periods. Usually, the PCR reaction will be started instantly.
9. Never use micropipettes for performing MLPA reactions that have been used for handling MLPA PCR products. After PCR, the tubes should not be opened in the vicinity of the thermal cycler.
10. All volumes used in the PCR reaction can be reduced twofold to save reagents.
11. In each SALSA probe mix, the DNA quantity (DQ) control fragments are included. These DQ fragments generate amplification products that are much smaller than the probe amplification products. The purpose of these DQ fragments is to give a warning when the amount of sample DNA used is lower than the 20 ng of human DNA that is required for reliable MLPA results. These MLPA DQ fragments will generate amplification products of 64, 70, 76, and 82 bp, even when ligation failed or when no sample DNA was present. In addition, a 94-bp amplification product is made that is sample DNA dependent and ligation dependent. The relative peak height of the four 64–82-bp amplification products depends on the amount of DNA present in the sample. The peak size of the 94-bp amplification product reflects the copy number of a 2q14 DNA sequence. This amplification product is generated by an MLPA probe that consists of two synthetic probe oligonucleotides. The peak area of the 94 fragment should be of similar size as most of the other MLPA amplification products. The four DQ fragments will hardly be visible when >100 ng of sample

DNA is used. They will be visible but with a much smaller peak size than the 94-bp amplification product when 50 ng of human DNA is used in the MLPA reaction. If the 64-, 70-, 76-, and 82-bp DQ amplification products have similar or larger peak sizes than the 94-bp fragment and the 130–472-bp MLPA probe amplification products, either the ligation reaction failed or the amount of sample DNA was <20 ng. Regardless, the results obtained may not be reliable.

12. The manual hot start as used in this protocol is recommended but not essential. Preparation of the complete PCR reaction at room temperature or on ice will usually result in almost identical results. Start the PCR as soon as possible after addition of the polymerase mixture! Some users prepare a mixture of water, PCR buffer, and polymerase mix; distribute this mix among all vials; and start the PCR reaction by addition of the ligation reaction.
13. Denaturation time of the PCR reaction can be reduced to 20 s in most thermal cyclers.
14. For samples with very low amounts of DNA, it is possible to increase the probe signals by increasing the number of PCR cycles up to 40.
15. PCR products can be stored in the dark at 4°C for at least 4 days.
16. Analysis of a limited number of samples or rapid quality inspection can be done by visual examination of the capillary electrophoresis peak profiles. For the analysis of large numbers of samples, export of peak heights or areas to Excel files/analysis programs is necessary. Software for data analysis is available at www.MLPA.com and www.chromosomelab.dk.

4.1. Trouble Shooting

4.1.1. No Bands Visible

1. Is the molecular weight marker pattern OK? If not, check capillaries and electrophoresis conditions.
2. Is the PCR primer peak present, and off-scale? If not, check whether the correct fluorescent label is used. Usually, 5-carboxyfluorescein label for Applied Biosystems apparatus, D4 label for Beckman CEQ apparatus, and IR800 label for LI-COR apparatus (LI-COR, Lincoln, NE).
3. Are the four control fragments visible? Each SALSA probe mix will generate four fragments of 64, 70, 76, and 82 bp even when ligation is omitted. These bands will be small or even not visible when >50 ng of human DNA is used for MLPA.

4.1.2. Primer Peak OK, but No Control Fragments and No Probe Amplification Products

PCR reaction failed. Repeat the PCR reaction using the same ligation reactions, e.g., using a different thermal cycler. Reduce all volumes by 50% to save reagents.

4.1.3. Primer Peak OK, Control Fragments Clearly Visible, but No Probe Amplification Products

No DNA was present in your sample, or the ligation reaction failed. The MLPA reagents are quite stable. After storage of a complete MLPA kit for 12 weeks at room temperature, including probes, ligase, and polymerase, good results were obtained. However, the ligase cofactor NAD, present in ligase buffer A might be inactivated by >15 freeze-thaw cycles. Enzymes might be inactivated by repeated freeze-thaw cycles in freezers operating at –25/–30°C instead of –20°C.

4.1.4. Bands Visible, but Weak Signals

The tubes of the PCR reaction can be placed in the thermal cycler for another four cycles, or a new PCR using the same ligation reactions but with 36–40 cycles might help. Reduce all volumes by 50% to save reagents. Increasing the number of PCR cycles to 40 will not influence results!

4.1.5. Signal Strength OK, but Large Differences in Relative Peak Areas between Different Samples

Peak area differences might be due to impurities in the DNA preparations. The use of lower amounts of sample often solves the problem. Another possibility is overloading of capillaries, resulting in saturation of the fluorescence detection device. Broad peaks, especially of longer fragments, can be caused by incomplete denaturation during the electrophoresis, old capillaries, or deterioration of the gel. In this case, the markers also have broad peaks.

4.1.6. Extra Peaks

Incomplete denaturation during the electrophoresis can result in extra peaks and broad peaks, especially of longer fragments. An increase in the run temperature might solve this problem. In some apparatuses, such as ABI3700, overloading of capillaries might result in signals in neighboring capillaries.

4.1.7. Low Peak Area of Longer Probes

This problem can have many causes. The volume of the DNA sample + probe mix + MLPA buffer is 8 µl. Evaporation during the 16-h hybridization period is low due to the heated lid. In general, >6 µl of fluid remains at the bottom of the tube after this 16-h incubation. If the volume is <5 µl, perhaps due to bad closure of the tubes, or incorrect heated lid temperature, "fading" of the peak heights with increased length of amplification product can occur.

Impurities in DNA samples also affect the relative peak sizes, and they can, in particular, reduce the peak heights of longer probes and those probes that already have a lower than average peak area. These impurities are often small ionic contaminants that can be removed by ethanol precipitation. The electrokinetic injection procedure of some capillary electrophoresis seems to select for injection of smaller fragments. This selection seems to be more pronounced with older types of ABI capillary sequencers. Finally, the use of deteriorated capillaries (too long in use) and deteriorated gel are a major source of problems. Gel for capillary electrophoresis should be stored at 4°C between runs.

References

1. Schouten, J. P., McElgunn, C. J., Waaijer, R., Zwijnenburg, D., Diepvens, F., and Pals, G. (2002) Relative quantification of 40 nucleic acid sequences by multiplex ligation-dependent probe amplification. *Nucleic Acids Res.* **30,** e57.
2. Slater, H. R., Bruno, D. L., Ren, H., Pertile, M., Schouten, J. P., and Choo, K. H. (2003) Rapid, high throughput prenatal detection of aneuploidy using a novel quantitative method (MLPA). *J. Med . Genet .* **40,** 907–912.
3. Gerdes, T., Kirchhoff, M., Lind, A. M., Larsen, G. V., Schwartz, M., and Lundsteen, C. (2005) Computer-assisted prenatal aneuploidy screening for chromosome 13, 18, 21, X and Y based on multiplex ligation-dependent probe amplification (MLPA). *Eur. J . Hum . Genet .* **13,** 171–175.
4. Nygren, A. O., Ameziane, N., Duarte, H. M., et al. (2005) Methylation-specific MLPA (MS-MLPA): simultaneous detection of CpG methylation and copy number changes of up to 40 sequences. *Nucleic Acids Res.* **33,** e128.
5. Hochstenbach, R., Meijer, J., van den Brug, J., et al. (2005) Rapid detection of chromosomal aneuploidies in uncultured amniocytes by multiplex ligation-dependent probe amplification (MLPA). *Prenat. Diagnos .* **25,** 1032–1039.

9

MALDI-TOF Mass Spectrometry for Trisomy Detection

Dorothy J. Huang, Matthew R. Nelson, and Wolfgang Holzgreve

Summary

Matrix-associated laser desorption/ionization time-of-flight (MALDI-TOF) mass spectrometry is a tool currently under investigation for use in prenatal detection of abnormalities in chromosome number, such as trisomy 21. Because of its ability to detect extremely small differences in mass, even to the level of a single nucleotide difference, this method can be applied to the detection of single-nucleotide polymorphisms (SNPs), which when present in a heterozygous state, can yield quantitative information regarding chromosome status from diagnostic specimens such as amniotic fluid or chorionic villus samples. MALDI-TOF mass spectrometry has several potential advantages over traditional karyotyping methods, including its amenability to high-throughput analyses and its nonreliance on prior cell culture. The method described here is based on the MassEXTEND® protocol developed by Sequenom, Inc., although any mass spectrometry platform sensitive enough to detect the small difference in mass between SNPs could be applied for this purpose.

Key Words: Matrix-associated laser desorption/ionization time-of-flight (MALDI-TOF); mass spectrometry; prenatal diagnosis.

1. Introduction

Chromosomal abnormalities have been associated with up to 50% of all clinically recognized pregnancy losses, and those involving chromosomes 13, 18, 21, X, and Y are compatible with life and are of special concern *(1)*. Diagnostic methods most commonly involve obtaining cells of fetal origin either by amniocentesis or chorionic villus sampling (CVS), followed by karyotyping. Karyotyping is a reliable and time-tested method, but it requires prior culturing of cells, and the results may take up to two weeks or more to obtain. This delay can

From: *Methods in Molecular Biology, vol. 444: Prenatal Diagnosis*
Edited by: S. Hahn and L. G. Jackson © Humana Press, Totowa, NJ

cause significant parental anxiety, and failure of cell culture results in the need to repeat amniocentesis or CVS, or reliance on other methods of diagnosis.

For these and other reasons, alternative methods to karyotyping, many of which are described in this book, have been examined ***(2)***. These methods often make use of frequently occurring genetic markers such as short tandem repeats or single-nucleotide polymorphisms (SNPs) ***(3–7)***. SNPs are frequently found in the human genome, which has approximately one SNP per kilobase ***(8)***.

Mass spectrometric analysis of SNPs has been recently evaluated for the detection of trisomy 21 ***(9,10)***. In brief, mass spectrometry allows for measurement of the mass of individual molecules, which have been first converted into ions ***(11)***. The ions are sorted in the mass analyzer according to their mass-to-charge (*m/z*) ratios, and the data are converted into a mass spectrum, with various peaks corresponding to molecules of differing masses. In matrix-associated laser desorption/ionization time-of-flight (MALDI-TOF) mass spectrometry, samples to be analyzed are first bound to a crystalline matrix, after which laser bombardment evaporates the molecules from the matrix and ionizes them. The time-of-flight analyzer then measures the time required for the ions to travel through an electric field, known as a flight tube, before reaching a detector. This measurement reflects the *m/z* ratio. Thus, when comparing molecules of equal charge, larger molecules travel more slowly through the field and reach the detector at a later time than smaller molecules. Tsui et al. ***(9)*** recently described the use of MALDI-TOF mass spectrometry for the detection of trisomy 21 in a study of 13 trisomic placental and nine trisomic CVS samples. In collaboration with Sequenom, Inc., our lab also has demonstrated the reliability and reproducibility of this method for the detection of trisomy 21 in a large study of 350 DNA samples from uncultured amniotic fluid, amniocyte culture, CVS and amniotic fluid supernatant ***(10)***. The basic concept involves first selecting SNPs on chromosome 21, which are frequently present in the heterozygous state in the population of interest (i.e., minor allele frequency of close to 0.5). SNPs can only be "informative" in the heterozygous state; that is, two allele peaks must be present to allow for comparison and relative quantification of chromosome number. Amplification is then performed on genomic DNA from the sample of interest, followed by a primer extension (homogenous MassEXTEND®, or hME) reaction, in which a primer annealing just adjacent to the SNP site "extends" the amplicon by one or a few bases only. The difference in mass between the two products is detected on mass spectrometry and results in the generation of two separate peaks on the mass spectrum. The ratio of the two peak areas can then be calculated and chromosome number determined, because euploid samples would be expected to have an allelic ratio of 1:1, and trisomic samples either a ratio of 1:2 or 2:1.

Before samples can be analyzed for chromosome number, a set of control data generated from known euploid samples must be established for comparison purposes. This involves first performing MALDI-TOF mass spectrometric analysis on at least 10–20 known euploid samples by using the method described in this protocol, followed by generation of a "skew correction factor," which corrects for allele-specific preferences during the amplification, extension, and mass spectrometric steps. Establishment of the skew correction factor is described at the end of the methods section.

The protocol described here assumes that a technical representative from Sequenom, Inc. has assembled and calibrated all the necessary equipment for use. Therefore, no extensive details are described in this protocol regarding particulars about technical setup, but rather only the main considerations for this particular analysis. In addition, details regarding defining, creating and editing assays, defining and editing samples to be analyzed, creation of plates, reviewing the data and generating reports using the MassARRAY® Typer software (version 3.4) are not described in detail here. For this information, a Sequenom technical representative, the MassARRAY® Typer User's Guide, or both, should be consulted. Finally, the notes section of this chapter should be carefully reviewed before using this protocol.

2. Materials

2.1. Specialized Equipment and Hardware

The software corresponding to each instrument should be installed and tested by a technical representative from Sequenom, Inc.

1. MassARRAY® Compact Analyzer (Bruker, Newark, DE).
2. MassARRAY® Nanodispenser (Samsung, Irvine, CA).
3. Multimek 96 Automated 96-Channel Pipettor (Beckman Coulter, Fullerton, MA).
4. "Skirted" Thermo-Fast 96-well PCR plates (ABgene, Epsom, Surrey, UK, cat. no. AB-0800).
5. Microtest 96-well plates (Vee plate, Sarstedt, Nümbrecht, Germany, cat. no. 82.1583).
6. 96-well dimple plate (Sequenom, Inc., Hamburg, Germany).
7. Adhesive PCR film (ABgene, cat. no. AB-0558).

2.2. Polymerase Chain Reaction (PCR)

1. DNA for amplification (minimum of 2 ng/reaction recommended).
2. Forward and reverse oligo primers (*see* **Notes 1** and **2**), each 12.5 μM concentration, high-performance liquid chromatography purified, MALDI-TOF mass spectrometry checked, if possible.
3. AmpliTaq Gold polymerase (5 U/μl) (Applied Biosystems, Foster City, CA).

4. GeneAmp 10× PCR Gold buffer (Applied Biosystems).
5. 25 mM $MgCl_2$ solution (Applied Biosystems).
6. 100 mM dNTP set, PCR grade.

2.3. Shrimp Alkaline Phosphatase (SAP) Treatment

1. SAP (1 U/µl) (Sequenom, Inc., item no. 10002).
2. hME buffer (Sequenom, Inc., item no. 10055).

2.4. MassARRAY® hME Primer Extension

1. MassEXTEND® Mix Tri (termination mix chosen based on assays; *see* **Note 3**) (Sequenom, Inc., various item nos.).
2. hME primer (50 µM each).
3. Thermo Sequenase enzyme (32 U/µl) (Sequenom, Inc., item no. 10052).

2.5. Clean-Up of Extension Products

1. Clean Resin (Sequenom, Inc., item no. 10053).

2.6. reparation of Chip for Mass Spectrometry

1. SpectroCHIP G384, 10 pack (Sequenom, Inc., item no. 00637).
2. 3-Oligo Calibrant Mix (Sequenom, Inc., item no. 00335).

3. Methods

3.1. PCR

1. Prepare (*see* **Note 4**) the mix (omitting DNA) according to **Table 1** in a 2.0-ml Eppendorf tube. Multiply the volumes for one reaction (*see* **Note 5**) by at least one more reaction than needed to compensate for pipetting inaccuracies (e.g. for 96 samples, make enough mix for 97 or 98 samples).
2. Aliquot the necessary amount of PCR mixture to each well of a skirted 96-well plate (e.g., 23-µl mixture if using 2 µl of DNA for a reaction volume of 25 µl).
3. Add the necessary amount of template DNA to each well.
4. Seal the plate and briefly centrifuge at 800*g* for 1 min at 4°C.
5. Perform PCR under the following reaction conditions:

   a. 10 min at 95°C.
   b. 45 cycles of 20 s at 95°C; 30 s at appropriate primer annealing temperature; 1 min at 72°C.
   c. 5 min at 72°C.
   d. Hold at 4°C.

Table 1
Polymerase chain reaction. Concentrations and volumes of the components shown are for a 25 μl reaction volume

	Original conc.	Final conc.	Volume for one reaction (μl)
Sterile deionized water			Up to 25 μl final volume
10× PCR Gold Buffer	10×	1×	2.50
$MgCl_2$	25 mM	3.5 mM	3.50
dNTP mixture	25 mM each	250 μM	0.25
Primer mixture (forward + reverse)	12.5 μM each	100 nM	0.20
AmpliTaq Gold polymerase	5 U/μ	1 U	0.20
Template DNA	As measured	2 ng (minimum)	As needed

3.2. SAP Treatment

1. Make SAP mixture according to **Table 2**. For a full 96-well plate and use with the 96-channel automated pipettor (*see* **Note 6**), use the volumes indicated in the last column of the table.
2. Transfer 10 μl of each PCR reaction (*see* **Note 7**) to a new skirted 96-well plate. If using the Beckman automated 96-channel pipettor, place plates in the indicated positions in the instrument and run program "96 hME 10 μl PCR aliquot."
3. Add 4 μl of SAP mixture to each 10 μl of PCR reaction.

Table 2
Shrimp alkaline phosphatase treatment. Amounts of each reagent calculated for one sample and 96 samples

	Volume for one reaction (μl)	Volume for 96 reactions (μl)
Sterile deionized water	3.06	528.70
hME buffer (10× conc.)	0.34	58.70
SAP	0.60	103.70
Final volume	4.00	691.00

4. If using the mix for a full 96-well plate and the Beckman automated pipettor (as indicated table 2) skip **step 3** and instead aliquot 52 µl of the SAP mixture to each well in the first row of a 96-well Vee plate. Using a 12-channel pipettor, aliquot 6 µl to each remaining well of the plate, then place plates in the appropriate positions in the Beckman automated pipettor and run program "96 hME SAP Add" to transfer 4 µl of SAP mixture to each well of the PCR plate.
5. Seal the plate and thermocycle under the following conditions:

    a. 20 min at 37°C.
    b. 5 min at 85°C.
    c. Hold at 4°C.

3.3. Primer Extension and Cleanup of Extension Products

1. Prepare (*see* **Note 4**) the hME reaction cocktail solution as described in **Table 3**. For use with a full 96-well plate and the Beckman automated pipettor, use the volumes indicated in the last column of the table.
2. Transfer 4 µl of the reaction cocktail solution to each well of the plate containing SAP-treated PCR products.
3. If using the mix for 96 reactions and the Beckman automated pipettor, skip **step 2** and instead aliquot 52 µl of hME reaction cocktail solution into the first row of a 96-well Vee plate. Using a 12-channel pipettor, transfer 6 µl of hME reaction cocktail solution into each well of the Vee-well plate, then transfer 4 µl of the hME reaction cocktail solution into each well of the skirted PCR plate containing SAP-treated PCR product by using the Beckman automated 96-channel pipettor (program hME Cocktail Mix addition).

Table 3
Mix for primer extension reaction. Amounts of each reagent calculated for one sample and 96 samples

	Original conc.	Final conc.	Volume for one reaction (µl)	Volume for 96 reactions (µl)
Sterile deionized water			3.34	529.06
hME termination mix	10×	1×	0.40	63.36
hME primer	50 µM	600 nM	0.22	34.85
Thermo Sequenase	32 U/µl	4 U	0.04	6.34
Total volume			4.00	633.61

4. Seal the plate and thermocycle using the following program:

   a. 2 min at 94°C.
   b. 55 cycles of 5 s at 94°C; 5 s at appropriate primer annealing temperature; 5 s at 72°C.
   c. Hold at 4°C.

5. While the extension reaction is running, prepare the cleanup of the hME reaction products by transferring 6 mg of CLEAN resin to each well of a 96-well dimple plate (spoon out resin onto the left side of the plate and then use the flat plastic spreader to spread the resin into the wells from left to right).
6. Also, while the extension reaction is running, wash and condition the pin heads of the MassARRAY® nanodispenser.
7. Fill a tip box lid with enough sterile deionized water to cover the bottom of the lid, and place it in the appropriate position in the automated 96-channel pipettor.
8. Place the PCR plate containing the extension products into the instrument and transfer 32 µl of H_2O into each well by running the program: "96 hME Water Add [32 µl to 18 µl]."
9. Slowly invert the PCR plate onto the resin-containing dimple plate and then invert the plates together and gently knock the resin into the PCR wells.
10. Seal the PCR plate and then slowly and continually rotate the plate for 5–10 min to mix.
11. Centrifuge the plate for 5 min at 500*g* to bring down the resin.

3.4. MALDI-TOF Mass Spectrometry

1. Carefully place the plate in the nanodispenser and fix it down using the plastic affixers.
2. Add 70 µl of calibrant to the calibrant reservoir.
3. Place an unused SpectroCHIP in the holder, making sure it is in the correct position as indicated on the screen.
4. Transfer the samples to the chip, making sure that the Transfer Definition File is set to "96 to 384 (x4D)" and that the Load/Dispense Settings are set to

   Load time: 3 s.
   Load offset: 1 mm.
   Load speed: 80 mm/s.
   Disp. time: 0 s.
   Disp. offset: 1 mm.
   Disp. speed: 110 mm/s.

5. Load the chip into the MassARRAY® compact analyzer.
6. On the corresponding computer, make sure the programs Flex Control, Server Control, and MassARRAY® Caller are open and running.
7. Open the program "Chip Linker."

a. Under "Customer," highlight the project to be analyzed (previously created; refer to MassARRAY® Typer User's Guide).
b. Select process method "Allelotype."
c. Set dispenser setting to "Nanodispenser S-96(4) to 384."
d. Enter an experiment name.
e. Give the chip an identifying barcode number.
f. Click "Add"–the experiment and chip details will appear in the box below.
g. If all the information is correct, click "Create."

8. Open the program "SpectroAcquire."

a. Under the "Auto Run Setup" tab, enter the chip barcode in the box corresponding to the chip placement on the holder for the Bruker Compact Analyzer.
b. Select "Barcode Report"– this should read "Found."
c. We suggest setting the acquisition parameters to the following:

Shots [n]: 10.
Maximum acquisitions: 9.
Minimum good spectra: 5.
Maximum good spectra: 5.

d. Under the "Auto Run" tab, click "Start Auto Run" to begin acquiring mass spectra.

9. Retrieve data using the MassARRAY® Typer software.

3.5. Generation and Use of Skew Correction Factors

1. In the MassARRAY® Typer Analyzer, create a "Genotype Area Report" for the run(s) involving only known euploid samples (*see* **Note 8**). This will generate a skew correction file.
2. Give the skew correction file a name, and then save the file.
3. When analyzing data for sample plates, choose "Allelotype Correction Report," enter the skew correction file name when prompted. Save the file under a new name.
4. The skew-corrected peak area values, along with other data, will appear automatically in a Microsoft Excel file.

4. Notes

1. We recommend using the MassARRAY® Designer software from Sequenom, Inc. for designing assays, especially when designing multiplex reactions. This program designs both the PCR and hME primers for each SNP to be investigated, and it minimizes the chances of overlapping peaks in the mass spectra when designing multiplex reactions. It also minimizes potential intra- and interprimer interactions. For details and additional information on designing assays using this software, the MassARRAY® Assay Design Software User's Guide should be reviewed.

2. If not using the MassARRAY® Designer software, the following guidelines, in addition to the usual considerations used in designing PCR primers, should be considered in the design of the assays:

   a. Amplification primers optimally should be 80 to 120 bp.
   b. 5′ end tags are useful for increasing the mass of the primer, such that it falls out of the range of the mass spectrum. The 3′ end of the extension primer should be immediately adjacent to the SNP site.
   c. Extension primers should be between 16 to 25 nucleotides long, or approximately 4,800–7,500 Da.
   d. Consider both DNA strands when designing the extension primer.
   e. Check that the mass of the amplification primer is different from the extension primer and its products to avoid confusion in the mass spectrum.
   f. Choose an appropriate termination mix, which ideally extends the primer at the polymorphic site by one nucleotide for one allele, and soon thereafter for the other allele (usually yielding an extended primer of two nucleotides, including terminator).
   g. The mass of the extended primers should be within the range of 5,000 to 8,500 Da.

3. When designing multiplex reactions, group reactions by termination mix and make sure that all extension primers and products differ by at least 50 Da for the clearest separation of peaks on mass spectrum.
4. Set up PCR and primer extension reactions away from areas where DNA is handled to avoid potential contamination.
5. The protocol described here is for single reactions; however, the various reaction mixes can be adjusted as needed for multiplexing purposes.
6. For optimum consistency in pipetting and reduction of time between and during steps, we recommend using a 12-channel manual pipettor whenever possible and an automated 96-channel pipettor such as the Beckman Multimek 96 automated 96-channel pipettor described in this protocol.
7. Remove seals from PCR plates as carefully as possible to avoid potential contamination into other wells and onto work areas.
8. When performing the analyses on control samples to establish the skew correction files, we recommend running at least two analyses and combining the data before generating the skew correction factors to minimize the effect of any specific run conditions.

References

1. Gardner, R. J. M. and Sutherland, G. R. (1996) Pregnancy loss and infertility. In: Chromosome abnormalities and genetic counseling, vol 29. Oxford University Press, New York, 311–321.
2. Dudarewicz, L., Holzgreve, W., Jeziorowska, A., Jahnbowski, L., and Zimmermann, B. (2005) Molecular methods for rapid detection of aneuploidy. *J. Appl. Genet.* **46,** 207–215.

3. Verma, L., Macdonald, F., Leedham, P., Marton, T., Quirke, P., and Papp, Z. (1998) Rapid and simple prenatal DNA diagnosis of Down's syndrome. *Lancet* **352,** 9–12.
4. Findlay, I., Toth, T., Matthews, P., Marton, T., Quirke, P., and Papp. Z. (1998) Rapid trisomy diagnosis (21, 18, and 13) using fluorescent PCR and short tandem repeats: applications for prenatal diagnosis and preimplantation genetic diagnosis. *J. Assist. Reprod. Genet.* **15,** 266–275.
5. Levett, L. J., Liddle, S., and Meredith, R. (2001) A large-scale evaluation of amnio-PCR for the rapid prenatal diagnosis of fetal trisomy. *Ultrasound. Obstet. Gynecol.* **17,** 115–118.
6. Mansfield, E. S. (1993) Diagnosis of Down syndrome and other aneuploidies using quantitative polymerase chain reaction and small tandem repeat polymorphisms. *Hum. Mol. Genet.* **2,** 43–50.
7. Pertl, B., Kopp, S., Kroisel, P. M., et al. (1997) Quantitative fluorescence polymerase chain reaction for the rapid prenatal detection of common aneuoploidies and fetal sex. *Am. J. Obstet. Gynecol.* **117,** 899–906.
8. Wang, D. G., Fan, J. B., Siao, C. J., et al. (1998) Large-scale identification, mapping, and genotyping of single-nucleotide polymorphisms in the human genome. *Science* **280,** 1077–1082.
9. Tsui, N. B. Y., Chiu, R. W. K., Ding, C., et al. (2005) Detection of trisomy 21 by quantitative mass spectrometric analysis of single-nucleotide polymorphisms. *Clin. Chem.* **51,** 2358–2362.
10. Huang D. J., et al. (2006) Reliable detection of trisomy 21 using MALDI-TOF mass spectrometry. *Genet. Med.* **8,** 728–784.
11. Chiu, C. M. and Muddiman, D. C. (2001) What is Mass Spectrometry? (www.asms.org). American Society for Mass Spectrometry.

10

Rapid Detection of Fetal Mendelian Disorders: *Thalassemia and Sickle Cell Syndromes*

Joanne Traeger-Synodinos, Christina Vrettou, and Emmanuel Kanavakis

Summary

The inherited disorders of hemoglobin synthesis constitute the most common monogenic diseases worldwide. The clinical severity of β-thalassemia major and the sickle cell syndromes targets them as priority genetic diseases for prevention programs, which incorporates population screening to identify heterozygotes, with the option of prenatal diagnosis for carrier couples. Rapid genotype characterization is fundamental in the diagnostic laboratory, especially when offering prenatal diagnosis. The application of real-time polymerase chain reaction (PCR) provides a means for rapid and potentially high-throughput assays, without compromising accuracy. It has several advantages over endpoint PCR analysis, including the elimination of post-PCR processing steps and a wide dynamic range of detection with a high degree of sensitivity. Although there are >180 mutations associated with the β-thalassemia and sickle cell syndromes, the relatively small size of the β-globin gene (<2,000 base pairs) and the proximity of most mutations facilitates the design of a minimal number of real-time PCR assays by using the LightCycler system (Roche Diagnostics [Hellas] A.E., Athens, Greece), which are capable of detecting the majority of most common β-gene mutations worldwide. These assays are highly appropriate for rapid genotyping of parental and fetal DNA samples with respect to β-thalassemia and sickle cell syndromes.

Key Words: Prenatal diagnosis; β-thalassemia and sickle cell syndromes; real-time polymerase chain reaction (PCR).

1. Introduction

Prenatal diagnosis (PND) aims to provide an accurate, rapid result as early in pregnancy as possible. A prerequisite involves obtaining fetal material promptly and safely. In addition, for monogenic diseases the parental mutation(s) have to

From: *Methods in Molecular Biology, vol. 444: Prenatal Diagnosis*
Edited by: S. Hahn and L. G. Jackson © Humana Press, Totowa, NJ

be characterized before analysis of the fetal sample. The majority of methods currently used for genotyping parental samples and for performing prenatal diagnosis are based on the polymerase chain reaction (PCR) (e.g., allele-specific oligonucleotide hybridization analysis of PCR amplicons, amplification refractory mutation system [ARMS]-PCR, restriction endonuclease analysis of PCR amplicons, and direct DNA sequencing). Most of these techniques are relatively simple, fairly quick, and inexpensive (with the possible exception of sequencing), but none are very "high throughput," which can be a disadvantage with respect to identifying the parental mutations if the disease-associated gene has a wide spectrum of potential mutations. The application of real-time PCR offers a means for rapid and potentially high-throughput assays, without compromising accuracy, making it an ideal approach for genotyping parental and fetal DNA samples in the context of prenatal diagnosis.

1.1. Introduction to Real-Time PCR for Genotyping Applications

Real-time PCR integrates microvolume rapid-cycle PCR with fluorometry, allowing real-time fluorescent monitoring of the amplification reaction for quantitative PCR, qualitative characterization of PCR products, or both. The latter application provides a means for rapid genotyping, precluding any post-PCR sample manipulation. Several real-time PCR instruments are available on the market; in addition, there are several detection chemistries, some of which can be used on any instrument and others that are instrument specific. Real-time PCR machines can be classified as either "flexible" or high-throughput. The flexible instruments, usually faster and with a wider choice of running parameters, are more suitable for smaller batches of samples, whereas the high-throughput instruments are more appropriate for running large batches of samples requiring a smaller repertoire of assays. Flexible instruments are probably more appropriate for PND applications, because the processing of samples in small batches precludes potential errors, e.g., through the occurrence of a tube-switch. However, high-throughput instruments may be advantageous when screening parental samples to characterize mutations before performing the prenatal diagnosis.

There are several detection chemistries suitable for genotyping, but those more commonly described for use in genotyping monogenic diseases include hybridization probes, Taqman probes, and molecular beacons *(1)*. The protocol that we describe here uses of a flexible instrument (the LightCycler 1.0 or 1.5, Roche Diagnostics [Hellas] A.E., Athens, Greece) with "hybridization probes."

Hybridization probes *(2)* involve a dual-probe system. The two fluorescently labeled probes hybridize to adjacent sequences within the amplified target DNA, one of which covers the region expected to contain the mutation(s).

Proximity of annealed probes facilitates fluorescence resonance energy transfer (FRET) between them, and a fluorescent signal is only generated when both of the probes are hybridized to the target amplicon (*see* **Fig. 1A**). The two probes of the pair are designed to have different melting temperatures (Tm), whereby the probe with the lower Tm lies over the mutation site. Monitoring the emitted fluorescent signal under conditions of increasing temperature will detect a loss of fluorescence as the lower Tm probe melts off the template. A single base mismatch under this probe results in a Tm shift of 5–10°C, allowing easy distinction between wild-type and mutant alleles (*see* **Fig. 1B**). The ability to detect any nucleotide mismatches under the low Tm probe (mutation detection probe) and the use of different-colored probes (according to the properties of the real-time PCR instrument in use) can allow more than one mutation to be assayed in a single PCR reaction (*see* **Fig. 2**).

1.2. Molecular Basis of β-Hemoglobinopathies and Design of Real-Time PCR Mutation Detection Assays

The β-globin gene (*HBB:* OMIM 141900) is a relatively small gene (<2,000 bp) located in the short arm of chromosome 11 (11p15.5, GenBank NM_000518). Although >180 causative mutations have been identified for β-thalassemia syndromes (http://www.globin.cse.psu.edu/), the spectrum of mutations and their frequency in most populations usually include a limited number of common mutations (e.g., up to approximately six) plus a slightly larger number of rare mutations (e.g., approx 10). In most populations worldwide, the majority of the most common mutations tend to cluster within small distances within the β-globin gene. The probes sets described here were designed based on those mutations most commonly found in the Mediterranean populations, although they are also suitable for detecting several of the common mutations found in other world populations (e.g., Asian Indian, South East Asian, and Chinese) *(3,4)*. Overall, they facilitate mutation detection with the use of a small number of probe sets within a minimal number of amplicons.

All the allele-specific (mutation-detection) probes described here were designed to be complementary to the wild-type sequence of the β-globin gene, rather than complementary to each specific mutation. This potentially allows the distinction of normal alleles from any allele with a nucleotide variation under the length of the probe, minimizing the number of assays that need to be performed, and potentially reducing costs and time required to identify the mutations in parental samples, and subsequently perform the PND. Although most mutations have a distinct melting profile and can be implicated by comparison with controls, the definitive characterization of each mutation should be achieved

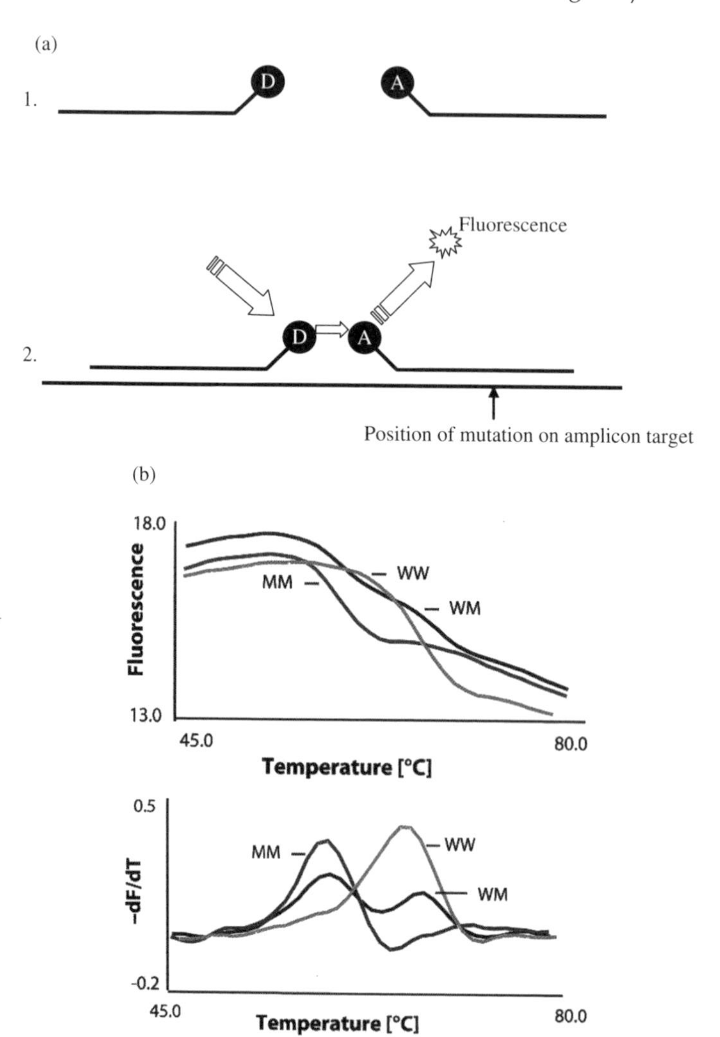

Fig. 1. Principle of hybridization probes for allele discrimination. **(A)** 1. One probe is labeled with a donor fluorophor (*D*) and the other with an acceptor fluorophore (*A*). The probe with the acceptor fluorophore is complimentary to the region including the expected mutation(s). 2. Proximity of annealed probes facilitates FRET and the emission of a fluorescent signal. **(B)** Plots of fluorescence (*F*) versus temperature (*T*) from melting curve analysis by using hybridization probes. *Top plot* shows the raw melting peak data with fluorescence versus temperature. *Bottom plot* shows the melting peaks displayed when the computer software automatically converts and displays the first negative derivative of fluorescence to temperature vs temperature (–dF/dT vs T). The latter facilitates easy discrimination between wild-type and mutant alleles.

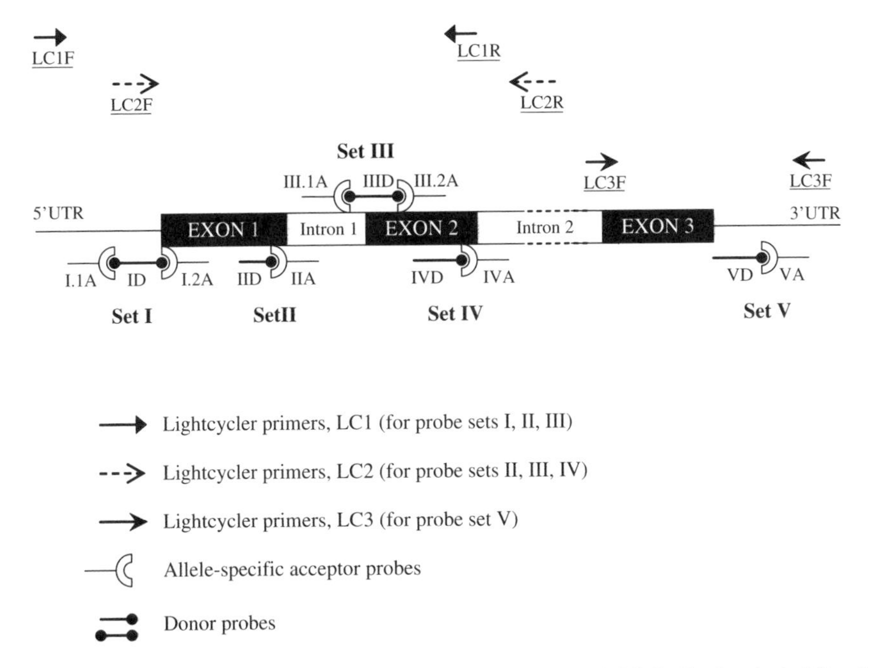

Fig. 2. The position of the β-globin gene primers and LightCycler hybridization probe sets appropriate for prenatal diagnosis and preimplantation genetic diagnosis protocols. Three PCR primer sets can be used (either LC1 or LC2, which can both be multiplexed with LC3 if required). F, forward primer; R, reverse primer. Probe sets I–V are used according to the mutations under investigation (*see* **Table 1**).

by a second method such as an ARMS-PCR assay *(5,6)*. (As recommended for best practice, any laboratory performing DNA diagnostics and PND should have more than one available mutation detection method *(7,8)*).

The LightCycler PCR primers (sets LC1, LC2, and LC3) were designed with the aid of computer software (Amplify version 2.0, Bill Engels, 1992–1995). LC1 and LC2 amplify two overlapping regions of the β-globin gene surrounding the majority of the most common β-thalassemia mutations in all world populations, along with the HbS mutation, and when multiplexing mutation detection they should be selected according to the mutations under investigation (*see* **Table 1; Fig. 2**). LC3 amplifies a 3' region of the β-globin gene, which contains some rare β-thalassemia mutations, and. if necessary. It can be multiplexed with either set LC1 or LC2.

Design of LightCycler mutation detection probe sets took into account secondary structure properties (e.g., hairpin-loops) and potential primer–dimer formation between the probes themselves and the PCR primers, evaluated by computation before synthesis (TIB MOLBIOL, Berlin, Germany). In addition,

Table 1
Lightcycler mutation detection probe sets for the most common β-thalassemias mutations worldwide (and HbS mutation)

Probe set	Acceptor probe name and sequence	Donor probe name and sequence	Beta-gene mutation(s) detected
Set I	I.1A 5′-ttc tga cac aac tgt gtt cac tag ca-3′ LC Red**	ID FITC 5′-cct caa aca gac acc atg gtg cac c-3′ FITC	CAP +20 (C>T)* CAP+22 (G>A)
	I.2A LC Red** 5′-gac tcc tga gga gaa gtc tgc-3′ P		HbS (Cd 6 A>T) Cd5 (–CT) Cd6 (–A) Cd8 (–AA) Cd 8/9 (+G)
Set II	IIA LC Red** 5′-tgt aac ctt gat acc aac ctg ccc a-3′ P	IID 5′-tgc cca gtt tct att ggt ctc ctt aaa cct gtc-3′ FITC	IVSI-1 (G>A) IVSI-1 (G>T) IVSI-2 (T>G) IVSI-2 (T>C) IVSI-2 (T>A) IVSI-5 (G>A) IVSI-5 (G>C) IVSI-5 (G>T) IVSI-6 (T>C)

Set III	III.1A 5′-tct gcc tat tgg tct att ttc cc-3′ LC Red**	IIID FITC 5′-ccc tta ggc tgc tgg tgg tc-3′ FITC	IVSI-110 (G>A) IVSI-116 (T>G)
	III.2A LC Red** 5′-acc ctt gga ccc aga ggt tct t-3′ P		Cd39 (C>T) Cd37 (TGG>TGA) Cd41/42 (delTTCT)
Set IV	IVA LC Red** 5′-tct cag gat cca cgt gca gct tg-3′P	VID 5′-gtc cca tag act cac cct gaa g-3′ FITC	IVSII-1(G>A)
SetV	VA LC Red** 5′gct caa ggc cct ttc ata ata tcc cc	VD 5′ ttt ttc att agg cag aat cca ga-3′ FITC	polyA signal mutation AATAAA>AACAAA AATAAA>AATGAA AATAAA>AATAGA AATAAA>AATAAG AATAAA>AA- - AA AATAAA>A- - - - -

The β-globin gene-specific PCR primers include (*see* Fig. 10.2): For probes sets I, II, III: LC1F: 5′-GCT GTC ATC ACT TAG ACC TCA-3′; LC1R 5′-CAC AGT GCA GCT CAC TCA G-3′. For probes sets II, III,IV: LC2F 5′-CAA CTG TGT TCA CTA GCA AC-3′; LC2R 5′-AAA CGA TCC TGA GAC TTC CA-3′. For probe set V: LC3F 5′-ATT TCT GAG TCC AAG CTG GGC -3′; LC3R 5′-AAA TGC ACT GAC CTC CCA-3′.

FITC, fluorescein ; P, phosphorylated.

*, Polymorphism linked with the IVSII-745 (C>G) mutation; **, LC Red: The fluorescent label used for each probe will depend upon the relative frequency of mutations in the population under study and the potential requirement of multiplexed assays.

design of mutation detection probes avoided the regions of the β-globin gene known to contain common polymorphic point mutations such as codon 2 (CAC>CAT). The LightCycler systems 1.0 and 1.5 can detect two fluorescent labels (LightCycler Red 640 [LC Red 640] and LightCycler Red 705 [LC Red 705]) and SYBR Green. The choice of fluorescent label used for each probe depends upon the relative frequency of mutations in the population under study and on the potential requirement for multiplexed assays when more than one mutation is investigated within a single sample. For example, in the Greek population, IVSI-110 G>A is the most common mutation, so the probes for most other mutations encountered in Greece are labeled with the opposite fluorescent marker to that used for IVSI-110 G>A.

More specifically, two of the probe combinations (named set I and set III) include two acceptor (mutation detection) probes with one central donor probe (*see* **Fig. 2**), foreseeing the use of one or both acceptor probes of the set according to the needs of any genotyping assay. Each of the acceptor probes in sets I and III are labeled with different acceptor fluorophores, and the central donor probe, designed to span the distance between the two acceptor probes, is labeled with a fluorescein molecule at both 5′ and 3′ ends. Sets II and V were designed to screen for several neighboring mutations with use of a single mutation detection (acceptor) probe. Set IV was designed to detect a single mutation each and involved a donor probe that was labeled with fluorescein only at the end adjacent to the acceptor probe (*see* **Table 1; Fig. 2**). In all sets, the mutation-screening (acceptor) probes were designed to have a lower Tm relative to the donor probes, thereby ensuring that the fluorescent signal generated during the melting curve is determined only by the specificity of mutation-screening probe.

2. Materials

2.1. Equipment

The method described is to be used with a LightCycler instrument version 1.0 or 1.5. Besides the LightCycler instrument, essential equipment includes the following:

1. Aluminum cooling block, which holds 32 LightCycler centrifuge adapters (cat. no. 1 909 312, Roche Diagnostics [Hellas] A.E., Athens, Greece), in which the real-time PCR reactions are set up.
2. LightCycler glass capillary tubes (20 μl) (cat. no. 11 909 339 001, Roche Diagnostics [Hellas] A.E., Athens, Greece) in which the real-time PCR reactions are run in the LightCycler instrument.

3. A bench centrifuge with a well depth of approx 4.5 cm, for centrifugation (maximum 3,000*g*) to pull the reaction volume (20 µl) to the base of the glass capillary, before loading in the LightCycler instrument.

2.2. Reagents

1. QIAamp DNA Mini kit (cat. no. 51104, QIAGEN, Hilden, Germany) for extracting DNA from chorionic villi samples or amniocytes.
2. Pair(s) of PCR primers selected according to mutations under study (either LC1F/LC1R or LC2F/LC2R, if necessary, with LC3F/LC3R, as shown in **Table 1** and **Fig. 2**).
3. Fluorescently labeled mutation detection probe sets, appropriate for mutations under study (*see* **Table 1; Fig. 2**).
4. LightCycler DNA Master hybridization probes kit (cat. no. 2 015 102, Roche Diagnostics [Hellas] A.E., Athens, Greece), which also includes 25 mM $MgCl_2$ and PCR-grade water.

2.3. Handling and Storage of PCR Reagents

1. All PCR primers to be used on the LightCycler are diluted as stock solutions of 100 µM, divided into aliquots of convenient volume (e.g., 25 µl), and stored at –20°C. For primer working solutions, the stock solutions are diluted to 10 µM and can be subsequently stored at 4°C for up to 3 months.
2. The LightCycler hybridization probes are diluted to 3 µM and stored in aliquots of relatively small volume (e.g., 20 µl) at –20°C. A thawed aliquot should not be refrozen, but it can be used up to 1 month when stored at 4°C.
3. The "Master Mix" from the LightCycler-DNA Master hybridization probes kit should not be refrozen once thawed, but it can be used for up to 1 month when stored at 4°C.

3. Methods

3.1. DNA Extraction from Chorionic Villi Samples or Amniocytes

The real-time PCR genotyping method described below assumes that assays are performed on good-quality genomic DNA samples (parental or fetal) with a concentration of 30–50 ng/µl. The kit that we use in the laboratory in Athens is the QIAamp DNA Mini kit (QIAGEN), and we follow the Blood and Body Fluid Spin Protocol outlined in the manual supplied with the kit.

3.2. Real-Time PCR Reaction Setup

1. In an Eppendorf tube, make a premix for the amplification reactions for a total reaction volume of 20 µl/sample. Each reaction should contain the ready-to-use reaction mix provided by the manufacturer (LightCycler DNA Master

Table 2
A typical PCR reaction for single color detection for one sample

	Stock Conc	Final Conc	µl/sample
H_2O (PCR grade)			9.6
$MgCl_2$	25 mM	4 mM	2.4
β-Globin forward primer LC1(F) or LC2(F) or LC3(F)	10 µM	0.5 µM	1
β-Globin reverse primer LC1(R) or LC2(R) or LC3(R)	10 µM	0.5 µM	1
LC Red allele-specific probe (640 or 705)	3 µM	0.15 µM	1
FITC donor	3 µM	0.15 µM	1
Master Mix		2	
Premix volume			18 µl
DNA sample volume			2 µl
Total reaction volume			20 µl

The volume of water is always adjusted to give final reaction volume of 20 µl/sample even when more than one primer or probe set is included in the reaction. For example, a PCR reaction with dual-color detection by using two allele-specific LC probes (one labeled with Red 640 and the other with Red 705) and a common (central) doubly labeled FITC probe, or even two sets of LC donor-acceptor probes (i.e., four probes).

hybridization probes) plus $MgCl_2$, a β-globin gene PCR primer pair (i.e., LC1, LC2, and LCR3), and LightCycler fluorescent probe sets for the relevant mutations. A typical PCR reaction for single color detection for one sample is shown in **Table 2**.

2. When calculating the premix volume, make premix enough for the number of samples being genotyped, a PCR premix blank plus controls for the mutation(s) under investigation. The controls should include a homozygous wild-type sample (N/N), a sample heterozygous for the mutation (M/N), and a sample homozygous for the mutation (M/M).
3. Place the appropriate number of LightCycler glass capillary tubes in the centrifuge adapters in an aluminum-cooling block.
4. Distribute accurately 18 µl of premix in all the capillaries.
5. Add 2 µl of genomic DNA (approx 50 ng) per sample and controls and 2 µl of double-distilled water to the PCR blank.
6. Once the PCR reactions have been set up in the capillaries at 4°C, place the caps carefully on each capillary, without pressing down yet.
7. Remove the aluminium centrifuge adaptors containing the capillaries from the cooling block and place in a bench centrifuge with well deep enough to hold the aluminium centrifuge adaptors (approx 4.5 cm).

8. Spin at a maximum of 3,000*g* for 10 s to pull the 20-µl reaction volume to the base of the glass capillary.
9. Place each glass capillary carefully into the LightCycler carousel by letting it simply "slip" in place. Then, press gently the cap fully in to the capillary and simultaneously the glass capillary fully down into position in the LightCycler carousel.
10. Put the loaded carousel into the LightCycler and initiate the PCR cycles and melting curve protocols by using the LightCycler software version 3.5.1, according to the manufacturer's specifications.

3.3. Amplification and Melting Curve Analysis

1. Preprogram the LightCycler software for the following amplification steps: a first denaturation step of 30 s at 95°C followed by 40 cycles of 95°C for 3 s, 58°C for 5 s, and 72°C for 20 s with a temperature ramp of 20°C/s. During the PCR, emitted fluorescence can be measured at the end of the annealing step of each amplification cycle to monitor amplification.
2. Immediately after the amplification step, the LightCycler is programmed to perform melting curve analysis to determine the genotypes. This involves a momentary rise of temperature to 95°C, cooling to 45°C for 2 min to achieve maximum probe hybridization, and then heating to 85°C with a rate of 0.4°C/s during which time the melting curve is recorded.
3. Emitted fluorescence is measured continuously (by both channels F2 [640 nm] and F3 [705 nm], if necessary) to monitor the dissociation of the fluorophore-labeled detection probes from the complementary single-stranded DNA (F/T) (F, fluorescence emitted; T, temperature). The computer software automatically converts and displays the first negative derivative of fluorescence to temperature vs temperature (–dF/dT vs T), and the resulting melting peak allows easy discrimination between wild-type and mutant alleles (*see* **Fig. 1B** and **Note 1**).

4. Notes

Besides the recommendations for handling and storing PCR primers, probes, and reagents (*see* **Subheading 2.3.**), the set up of reactions on the LightCycler is very straightforward. The only pitfalls that may arise are related to the melting curve analysis performed to identify mutations within PCR amplicons, and solutions to some of these problems are outlined below.

1. Trouble with melting curve analysis usually occurs when a relatively high number of samples is included in a single run. Under these circumstances, wide melting peaks are observed with only minor differences in the central Tm peak of the melting curve for all genotypes, e.g., M/M, N/N, and M/N. Furthermore, heterozygous DNAs will not give a double peak after melting curve analysis, but instead a single peak, which lies between that of the N/N and M/M controls/samples.

The LightCycler system 1.0 and 1.5 (software version 3.5) is designed to run a maximum of 32 samples in a single run, but our experience indicates that analysis of >20 samples produces wide and "flattened" melting curves. The problem is probably due to the way that the Lightcycler performs the melting curve analysis. During the LightCycler melting curve analysis, the temperature increases from 45°C to 90°C, which takes ≈150 s when using the "continuous acquisition" mode and a temperature increment of 0.3°C/s. In continuous acquisition mode, the temperature increases continuously, and it does not take into account that the measurement of fluorescence between one capillary and the next takes a certain time. If a small number of samples is analyzed, the fluorescence of each sample will be read more often during the 150 s of melting curve data acquisition compared with a run with more samples. In the latter situation, there are too few measurement points to calculate a detailed melting curve.

There are three possible solutions to this problem:

a) For high sample numbers, it is recommended that the temperature increment used for the melting curve is decreased, e.g., to 0.1°C/s when using continuous acquisition mode or to 0.4°C/s when using "stepwise acquisition" mode.
b) If the melting curve still fails to give a satisfactory result, we have found that additional melting curves can be performed using other temperature increments, although it must be noted that the quality of the melting curves is reduced each time an analysis is performed (and this is not recommended by the manufacturer).
c) In cases when the melting curve still fails to give satisfactory results when analyzing >20 samples, pause the LightCycler program after the amplification step and perform melting curve analyses in batches (including the appropriate controls with each melting curve analysis).

References

1. Wilhelm, J. and Pingoud, A. (2003) Real-time polymerase chain reaction. *Chembiochem* **4,** 1120–1128.
2. Lyon, E. (2001) Mutation detection using fluorescent hybridization probes and melting curve analysis. *Expert Rev. Mol. Diagn.* **1,** 92–101.
3. Vrettou, C., Traeger-Synodinos, J., Tzetis, M., Malamis, G., and Kanavakis, E. (2003) Rapid screening of multiple β-globin gene mutations by real time PCR (LightCycler™): application to carrier screening and prenatal diagnosis for thalassemia syndromes. *Clin. Chem.* **49,** 769–776.
4. Vrettou, C., Traeger-Synodinos, J., Tzetis, M., Palmer, G., Sofocleous, C., and Kanavakis, E. (2004) PCR for single cell genotyping in sickle cell and thalassemia syndromes as a rapid, accurate, reliable and widely applicable protocol for preimplantation genetic diagnosis. *Hum. Mutat.* **23,** 513–521.

5. Old, J. M., Varawalla, N. Y., and Weatherall, D. J. (1990) Rapid detection and prenatal diagnosis of β-thalassaemia: studies in Indian and Cypriot populations in the UK. *Lancet* **336,** 834–837.
6. Kanavakis, E., Traeger-Synodinos, J., Vrettou, C., Maragoudaki, E., Tzetis, M., and Kattamis, C. (1997) Prenatal diagnosis of the thalassemia syndromes by rapid DNA analytical methods. *Mol. Hum. Reprod.* **3,** 523–528.
7. Old, J., Petrou, M., Varnavides, L., Layton, M., and Modell, B. (2000) Accuracy of prenatal diagnosis for haemoglobin disorders in the UK: 25 years' experience. *Prenat. Diagn.* **20,** 986–991.
8. Traeger-Synodinos, J., Old, J. M., Petrou, M., and Galanello, R. (2002) Best Practice Guidelines for carrier identification and prenatal diagnosis of haemoglobinopathies. European Molecular Genetics Quality Network (EMQN). CMGC and EMQN, http://www.emqn.org.

11

Rapid Detection of Fetal Mendelian Disorders: *Tay-Sachs Disease*

Esther Guetta and Leah Peleg

Summary

Tay-Sachs disease is an autosomal recessive storage disease caused by the impaired activity of the lysosomal enzyme hexosaminidase A. In this fatal disease, the sphingolipid GM2 ganglioside accumulates in the neurons. Due to high carrier rates and the severity of the disease, population screening and prenatal diagnosis of Tay-Sachs disease are routinely carried out in Israel. Laboratory diagnosis of Tay-Sachs is carried out with biochemical and DNA-based methods in peripheral and umbilical cord blood, amniotic fluid, and chorionic villi samples. The assay of hexosaminidase A (Hex A) activity is carried out with synthetic substrates, 4-methylumbelliferyl-6-sulfo-*N*-acetyl-β-glucosaminide (4-MUGS) and 4-methylumbelliferil-*N*-acetyl-β-glucosamine (4-MUG), and the DNA-based analysis involves testing for the presence of specific known mutations in the α-subunit gene of Hex A. Prenatal diagnosis of Tay-Sachs disease is accomplished within 24–48 h from sampling. The preferred strategy is to simultaneously carry out enzymatic analysis in the amniotic fluid supernatant or in chorionic villi and molecular DNA-based testing in an amniotic fluid cell-pellet or in chorionic villi.

Key Words: Tay-Sachs disease; hexosaminidase; 4-methylumbelliferil-*N*-acetyl-β-glucosamine (4-MUG); 4-methylumbelliferyl-6-sulfo-*N*-acetyl-β-glucosaminide (4-MUGS); prenatal diagnosis; chorionic villi sampling; amniocentesis; HEXA gene; α-subunit.

1. Introduction

Tay-Sachs disease (TSD) is an autosomal recessive disease with a carrier frequency of 1:30 among the Jewish Ashkenazi population. Due to the relatively high frequency of carriers and severity of the disease, leading to the death of

From: *Methods in Molecular Biology, vol. 444: Prenatal Diagnosis*
Edited by: S. Hahn and L. G. Jackson © Humana Press, Totowa, NJ

affected individuals in infancy up to 4 years of age, a national screening program was established in Israel to identify couples at risk for affected offspring. The Danek Gertner Institute of Human Genetics at Chaim Sheba Medical Center is a national reference center for Tay-Sachs genetic counseling and diagnosis.

TSD is a lysosomal storage disease. The material stored is a sphingolipid, GM2 ganglioside, and its massive accumulation in the neurons is ultimately fatal. The degradation of gangliosides occurs in the lysosomes where hydrolases remove the sugar molecules by sequential enzymatic steps. The hydrolytic steps of GM2 ganglioside degradation are carried out by the enzyme hexosaminidase A (Hex A; EC 3.2.1.52), which catalyses the cleavage of the terminal β-bond of the amino-sugar *N*-acetyl-β-D-galactosamine ***(1)***.

The family of lysosomal hexosaminidases comprises several isoenzymes, the most prominent of which are Hex A and Hex B ***(1)***. Hex A is composed of two nonidentical peptides, an α-subunit (an acidic protein) and a β-subunit (a basic subunit). Hex B is composed of two β-subunits ***(2)***. They are nonspecific enzymes capable of catalyzing a group of high- and low-molecular-weight molecules (e.g., glycoproteins, glycolipids, glycosaminoglycans) ***(3)***. GM2-ganglioside is cleaved only by Hex A; thus, its absence will cause GM2 storage and accumulation. Because Hex A is composed of two different subunits, the patients were classified as type B (TSD), those having Hex B activity but no Hex A activity due to a defect in the α-subunit; or type 0 (Sandhoff disease), those with a defect in the β-subunit with no activity of either Hex. ***(4,5)***. In the cell, an additional protein is essential for the cleavage of GM2 ganglioside, which is the GM2-activator. The activator binds to a molecule of the membranous ganglioside, lifts it out of the membranous environment, and the whole complex (the real substrate of Hex A in vivo) is recognized by the enzyme ***(6)***. A deficiency of the GM2 activator also leads to the storage of GM2-ganglioside, and to the third variant of patients, the AB type (with active Hex A and Hex B) ***(6)***.

GM2 ganglioside is highly insoluble making it impractical to use this substrate in routine laboratories. Because the enzyme is able to catalyze the cleavage of a variety of terminal β-bonds of amino-sugars, artificial (soluble) colorimetric or fluorogenic substrates were successfully assayed. The most sensitive and commonly used synthetic substrate is 4-methylumbelliferil-*N*-acetyl-β-glucosamine (4-MUG), which does not require an activator. Prenatal diagnosis of TSD was first accomplished in 1969 ***(7)***, and it provided the rationale for prospective population screening for carriers ***(8,9)***. An important difference between the two isoenzymes was observed in 1981; Hex A but not Hex B releases *N*-acetylglucosamine-6-sulfate from keratin sulfate ***(10)***. This finding led to the synthesis of the sulfated substrates, which are hydrolyzed almost exclusively by Hex A. The fluorogenic substrate 4-methyl umbelliferyl-6-sulfo-*N*-acetyl-β-glucosaminide (4-MUGS) is efficiently used to assay Hex A with no need for differentiation between Hex A and Hex B ***(11)***.

The enzyme in the sample hydrolyzes the methylumbelliferyl (MU) residue of the synthetic substrates. After it has been removed, the MU's fluorescence can be measured, and the degree of fluorescence is directly proportional to the enzyme activity rate.

For rapid prenatal diagnosis, assays with both MUG and MUGS are carried out for chorionic villus sampling (CVS), whereas only MUGS is used in amniotic fluid samples. The enzymatic assays are carried out on the day of the procedure (amniocentesis or CVS), and the results are reported to the patient in the late afternoon.

In our laboratory, the initial rapid diagnosis is routinely verified by carrying out the analysis in cultured cells. However, if results of the rapid procedure indicate that a fetus is affected, the result is reported and the couple is referred to immediate genetic counseling.

1.1. Biochemical Assay with 4-MUG

The two isoenzymes Hex A and Hex B react with 4-MUG, and they have a similar K_m/V_{max} value and similar pH optimum of 4.4 ***(12)***. The active site for the cleavage of 4-MUG is located on the β-subunit of both enzymes ***(12)***, which renders some complexity because it is crucial to assay the Hex A exclusively, which is the sole enzyme able to catalyze GM2 in vivo. However, differences in the two isoenzymes in several of their biochemical qualities, for example, sensitivity to temperature, are largely applied to distinguish between them. Hex A is thermosensitive and Hex B is thermostable. Usually, the sample is divided: half is heated and then the enzymatic assay is carried out on both halves. The unheated part demonstrates the total Hex activity (Hex A + Hex B), in the heated half, only Hex B is measured (Hex A was inactivated). The percentage of activity of Hex A is deduced from the difference between the two tubes.

1.2. Biochemical Assay with 4-MUGS

The active site for the cleavage of 4-MUGS is located on the α-subunit; therefore, Hex A but not Hex B is capable of hydrolyzing it. Thus, Hex A has two different active sites on its two subunits ***(13–16)***. The active sites have different properties concerning, e.g., substrate specificity, kinetic values, optimal pH, activation energy, sensitivity to temperature, and ionic strength ***(12)***. The enzymatic activity with 4-MUGS is expressed as nanomoles of MU released per milligram of protein or by ratios of the activities against both substrates ***(17)***.

Because heat inactivation of amniotic fluid and the 4-MUG assay results in overlap between affected and nonaffected embryos, the rapid diagnosis in amniotic fluid is carried out with 4-MUGS only.

1.3. Molecular Genetics of TSD

In the mid-1980s, the cDNA of the α- and β-subunits was cloned ***(18–20)*** and mapped to chromosome 15q23-24 and 5q13, respectively ***(21,22)***. These two genes are similar in size, 35 and 40 kb, with 13 introns and 14 exons. The polypeptides are identical in >50% of their amino acids ***(20)***. Mutations in the α-subunit gene (HEXA) result in TSD, whereas Sandhoff disease (clinically identical) is caused by mutations in the β-subunit gene (HEXB). Clones encoding the GM2 activator also were isolated, the GM2A gene is approx. 16 kb, contains four exons, and was mapped to 5q31.1-31.3 (GM2A has a pseudogene on chromosome 3) ***(12)***. More than 130 mutations were reported ***(12)*** in the Hex A gene, three mutations are predominant and specific to the Ashkenazi Jews and two mutations to French Canadians. In these populations, the disease is relatively frequent, with an incidence of 1:4,000. Specific mutations also were found among Moroccan Jews and none-Jews of Celtic background. The underlying mutations differ in each group ***(12)***. Other mutations and combinations of the above-mentioned mutations were reported in population isolates (such as the Cajuns in Louisiana, Japan, and Portugal); however, most of the mutations are private ***(12)***. In the Ashkenazi population, two mutations are associated with the infantile form of the disease: a four-base pair insertion (5′TATC3′) in exon 11 and a G-to-C alteration in a conserved splice junction of the 5′ boundary of intron 12 ***(23)***. Another mutation responsible for the adult onset of the disease is a G-to-A change in nucleotide 805 of exon 7 (G269S) ***(24)***. Here, we described step-by-step procedures for the diagnosis of the two infantile mutations.

2. Materials

2.1. Enzymatic Assays

1. 0.1% bovine serum albumin (BSA), 100 mM citrate buffer, pH 5. Dissolve 31.5 g of citric acid in 1,500 ml of double-distilled water (DDW) (solution 1). Dissolve 73.5 g of sodium citrate in 2,500 ml of DDW (solution 2). Mix 1,025 mL of solution 1 with 1,775 ml of solution 2. Correct the pH by adding solutions 1 (acidic) or 2 (basic) and add 5 g of BSA. Add water up to a final volume of 5,000 ml.
2. 100 mM citrate buffer, pH 4.5.

   a. Dissolve 73.5 g of sodium citrate in 2,500 ml of DDW (solution 1).
   b. Dissolve 52.5 g citric acid in 2500 ml of DDW (solution 2).

c. Mix 1,495 ml of solution 1 with 1,390 ml of solution 2. Correct the pH by adding solutions 1 or 2 and add water up to a final volume of 5,000 ml.

3. 100 mM citrate buffer, pH 4.2: Dissolve 29.4 g of sodium citrate in 1,000 ml of DDW (solution 1). Dissolve 21 g citric acid in 1,000 ml of DDW (solution 2). Mix 460 ml of solution 1 with 540 ml of solution 2. Correct the pH by adding solutions 1 or 2.
4. 4-MUG: Dissolve 0.125 mg/ml 4-MUG (Melford Laboratories Limited, Chelsworth, Ipswich Suffolk, UK) in citrate buffer, pH 4.5. The substrate solution should be freshly prepared on the day of analysis.
5. 4-MUGS.

   a. Dissolve 0.25 mg/ml 4-MUGS (Melford Laboratories Limited) in citrate buffer, pH 4.2.
   b. The substrate can be stored frozen in –20°C.

6. MU: 15 mM in DDW (Sigma-Aldrich, St. Louis, MO).
7. 200 mM glycine buffer, pH 10.5: Dissolve 37.5 g of glycine and 18.2 g of NaOH in 5 liters of DDW. Correct pH with a solution of 50% NaOH.

2.2. Materials for DNA-Based Molecular Diagnosis

1. Taq DNA polymerase: 1–1.5 units from the stock for each reaction in the buffer supplemented by the manufacturer. (The enzyme from any supplier should be stored at –20°C.)
2. dNTPs: a stock mixture of 2 mM of all four deoxynucleotides (=10×). The stock should be divided into aliquots and stored at –20°C. Final concentration is adjusted to 0.2 mM for each reaction.
3. Specific primers.

   a. Adjust primer concentration to 100 pmol/µl.
   b. Prepare a 1:10 working dilution.
   c. The final concentration in the reaction mixture is 0.5–1pmol/µl.
   d. Primers should be kept frozen in small aliquots to avoid repeated thawing and freezing.
   e. Primers for PCR on the splice mutation: primer F 5′-AGTTA-CCCCACCATCACCAGACTG-3′ and primer R 5′-TTGGGTCTCTAAG G-GAGAACTCCT-3′. Primers for PCR on the insertion mutation: primer F 5′-CCAGGAATCTCCTCAGCTTTGTGT-3′ and primer R 5′-AAGCCTCCTTTGGTTAGCAAGG-3.

4. Dimethyl sulfoxide (DMSO) at a final concentration of 8% in the reaction mixture.
5. Tris borate-EDTA (TBE): Dissolve 10.8 g of Tris-base (final concentration 90 mM) and 5.5 g of boric acid (final concentration 90 mM). Add 4 ml of 0.5 M EDTA (in DDW, pH 8.0) (final concentration 2 mM). Add DDW up to a final volume of 1,000 ml.

6. Agarose gel: Use ready-to-use gels: 4%E-Gel (Invitrogen, Paisley, UK) for the insertion mutation or prepare 3% agarose in TBE for the splice site mutation.
7. Bromphenol blue loading dye: Concentration of bromphenol blue is 0.25% in 50% sucrose solution in DDW. Dissolve 8 g of bromphenol blue powder and 20 g of sucrose into a final volume of 40 mL of DDW. The dye solution is added to the sample before loading on the gel.
8. Ethidium bromide: Final concentration of ethidium bromide is 30–50 μg/ml. Add to the gel (binds to DNA molecules, which then fluoresce under UV illumination).
9. DdeI restriction enzyme (buffer is supplied by the manufacturer).

3. Methods

3.1. Enzymatic Assay Procedures

3.1.1. Controls and Standards

In each run, normal and positive controls should be used and a blank control and a standard. The controls are pools of samples from previously diagnosed normal and affected embryos. Small aliquots of the pools are kept frozen (–20°C). The standard is a commercial MU (Sigma-Aldrich), stock solution concentration of 15 nmol/ml. Blanks contain the entire mix excluding the sample, or MU.

3.1.2. Analysis of Hexosaminidase Activity in Chorionic Villi

3.1.2.1. Preparation of Chorionic Villi (CV) Sample

1. Homogenize 5–10 mg of the CV tissue (manually in a 1-ml homogenizer) in 0.3–0.5 ml of DDW. The tissue should be blotted very gently on a filter paper before weighing.
2. Sonicate the disrupted tissue for 5 s (20 kHz), wait 10 s, and sonicate for 5 additional seconds. It is essential to keep the tubes on ice throughout the entire sonication process.
3. Centrifuge the sonicated samples (in microcentrifuge tubes) for 7 min at 17,000*g*.
4. The supernatant is ready for enzymatic analysis (can be kept frozen at –20°C for retesting).

3.1.2.2. Enzymatic Assay of CV Samples with 4-MUGS Substrate

1. Dilute 20 μl of the CV sonicate in 80 μl of DDW. Use the same dilution for the normal and positive controls. Prepare duplicates of each sample, including two blank tubes containing 20 μl of water.
2. Add 0.2 ml of 4-MUGS to the tube and incubate in a shaker bath for 60 min at 37°C (*see* **Note 1**) to each sample.
3. Transfer the samples to an ice bath and add 0.8 ml of glycine buffer, pH 10.5 (*see* **Note 2**), to all samples except for the standard MU samples to which 1 ml of glycine buffer should be added.

4. Read the fluorescence of all samples at 355-nm emission and 442-nm excitation.
5. Calculate the activity: nanomoles of substrate hydrolyzed in the sample, according to the fluorescence reading of the standard samples (MU 15 nmol/ml).
6. Measure protein concentration by method of preference (e.g., the Lowry or Bradford methods).
7. Calculate the final enzymatic activity: nanomoles of substrate hydrolyzed divided by protein concentration (**step 5**).

3.1.2.3. Enzymatic Assay of CV Samples with 4-MUG Substrate

1. Use 20–50 µl of sample and add citrate buffer, pH 5.0, containing 0.1% BSA up to 0.6 ml.
2. Place 0.1 ml of each sample into four tubes. Heat 2 tubes of each sample at 50°C for 3 h and keep the unheated sample on ice. The tubes should be properly closed to avoid evaporation of the samples during heating.
3. Cool the heated samples on ice for a few minutes and add 0.2 ml of substrate (4-MUG) to all samples (heated and unheated) except for the MU standard and incubate in a shaker bath for 30 min at 37°C (*see* **Note 1**).
4. Add 0.8 ml of glycine buffer pH 10.5 (*see* **Note 2**) to all samples except for the standard MU to which 0.9 ml should be added.
5. Read fluorescence of all samples at 355-nm emission and 442-nm excitation.
6. Calculate the average activity of the total Hex activity (unheated samples) and the Hex B activity (the residual heat-stable activity).
7. Calculate the percent of Hex B from total activity. Hex A (the heat labile activity) is calculated by subtracting the percent activity of Hex B from 100%.

3.1.3. Analysis of Hexosaminidase in Amniotic Fluid

1. Centrifuge the amniotic fluid for 10 min at 290*g* (minimum volume of 0.5 ml).
2. Place 100 µl of the sample supernatant, positive and negative amniotic fluid controls (all in duplicate), and two blanks in test tubes.
3. Add 0.2 ml of 4-MUGS to each sample and incubate in a shaker bath for 60 min at 37°C (*see* **Note 1**).
4. Transfer the samples to an ice bath and add 0.8 ml of glycine buffer, pH 10.5 (*see* **Note 2**), to all samples except for the standard MU samples to which 0.9 ml of glycine buffer should be added.
5. Read the fluorescence of all samples at 355-nm emission and 442-nm excitation.
6. Calculate the activity: nanomoles of substrate hydrolyzed in the sample in 1 ml of amniotic fluid, according to the fluorescence reading of the standard samples.

3.1.4. Analysis of Biochemical Results

Establishment of normal and affected ranges is essential in the case of enzyme-assay-based diagnosis. An embryo can be a normal homozygote, mutant homozygote, or heterozygote. It is imperative to be sure that there is no overlap between heterozygotes (which are healthy in the case of TS) and the affected. In our laboratory, a CVS activity of <40 nmol/mg protein points to an affected embryo, and >60 nmol/mg protein is the activity found in healthy individuals (carriers and noncarriers).

For amniotic fluid, the ranges are <2.5 nmol/ml for affected and >5 nmol/mL for nonaffected. The ranges for the reaction with MUG are <12% for affected and >18% for healthy (*see* **Note 3**). When the values fall in the overlapping zones (this happens mostly in the case of amniotic fluid), the reactions should be repeated, and other methods should be used (such as DNA analysis).

3.2. DNA-Based Molecular Diagnosis of TSD

For molecular rapid prenatal diagnosis, it is necessary to know the specific mutations carried by the couple at risk. The diagnosis is based on PCR of a segment inclusive of the region of the mutation and detection of the lesion by different techniques.

If a restriction site is altered by the mutation (a new site is created or a known site is eliminated), treatment with a restriction enzyme will distinguish between normal and mutant alleles. The size of the fragments after enzyme treatment can be estimated by electrophoresis in a gel containing a fluorescence DNA-linker (such as ethidium bromide) and according to a size ladder run with the samples. The sizes also can be measured accurately by using fluorescent primers or fluorescent dNTPs and analysis with computerized systems (e.g., Applied Biosystem 3100 AVANT genotyper in our laboratory). Insertions and deletions can be detected directly after the amplification assay without the restriction enzyme step *(23)*.

If the common mutations are not found, the issue of pseudodeficiency mutation should be kept in mind *(25)*. There are two mutations that were proven to be harmless, even though the enzymatic activity in vitro is diminished. The laboratory should be able to diagnose these mutations in order not to terminate a pregnancy in which the fetus harbors such a mutation in one of its alleles.

3.2.1. Detection of the Two Major Mutations in Ashkenazi Jews

These two mutations will serve as examples for DNA-based diagnosis of TS. The diagnosis of most other mutations is similar.

In addition to the methods routinely employed in our laboratory and described below, we also tested two kits that were found suitable for TS mutation detection: Elucigene Ashplex 1 and Pronto diagnostics. The first kit is based on the amplification refractory mutation system (ARMS) strategy; the second kit is based on single nucleotide primer extension.

After DNA isolation, all the techniques mentioned can be completed within 24–48 h.

3.2.2. DNA Isolation

DNA can be isolated from fetal samples with any method available in the laboratory, manual or automated. The Gentra system (Puregene DNA purification system) is routinely used in our laboratory. The procedure takes ≈ 2.5–3 h. DNA isolation is carried out on several milligrams of CV or a cell pellet from 5–10 ml of amniotic fluid.

3.2.3. Detailed Procedure for Diagnosis of Insertion Mutation

1. Prepare reaction mixture with Taq DNA polymerase (*see* **Subheading 2.2, item 1**), dNTPs (*see* **Subheading 2.2, item 2**), and specific primers (*see* **Subheading 2.2, item 3**) (*see* **Note 4**) and add DMSO to the mixture. The volume of each reagent should be calculated according to the number of samples to be tested. Add DDW up to a volume of 0.023 ml per sample (*see* **Note 5**).
2. Add 2 µl of sample DNA (*see* **Note 6**): fetal samples, control samples (normal and mutation carriers), and blank (no DNA). Place the tubes in the PCR apparatus and start the following program: 94°C for 2 min followed by 30 cycles of 94°C for 30 s, 56°C for 30 s, and 72°C for 45 s. After the 30th cycle, start the following program: 72°C for 5 min and the process is ended at 10°C. The PCR product of the normal allele is 276 bases and the mutant 280 bases.
3. Volumes of the reaction should not exceed 25 µl. If it is necessary to add extra reagents or DNA, accordingly less water should be added.
4. The diagnosis of the insertion mutation (+1278TATC ins) can be carried out directly without a restriction enzyme reaction step. Place 5–8 µl of the PCR product of each embryo sample into a tube, add the same volume of a known normal DNA, and heat the mixture at 95°C for 5 min. Allow cooling to room temperature.
5. Place 5–8 µl of each PCR product (nonmixed samples, mixed heated samples, control, and no DNA samples) into new tubes. Add 1 µl of bromphenol blue dye solution to each tube, and DDW to 23 µl. Load the entire volume onto the 4% E-Gel gel and separate by electrophoresis at 57 V for 30–50 min. Run a DNA size ladder parallel to the samples.
6. Inspect the gel showing the PCR products under UV light (*see* **Note 7**).

3.2.3.1. Analysis of Results

After the PCR products are cooled, they form double-stranded fragments. The hybrid fragments (those made up of normal and mutant strands) run slower in the gel due to secondary structures created by the annealing of two fragments of different sizes (the mutant is larger by four nucleotides). The mixing of embryonic samples with normal control DNA facilitates diagnosis of affected embryos: a hybrid is formed with reduced electrophoretic mobility compared with PCR products of homozygotes. Homozygous normal embryos will not have the slower band (after mixing) and heterozygotes will have the two bands before the mixing. In homozygous samples (normal and mutant), a single band is observed before mixing. All samples can be run in the same gel.

An alternative method is to carry out fluorescent PCR under the same amplification conditions with one fluorescent primer (for example tagged with VIC) in which the two alleles appear as two distinct peaks with a four-base difference in the 3100 AVANT genotyper. In this four-capillary apparatus, analysis of four samples is completed in 40 min.

3.2.4. Detailed Procedure for Diagnosis of Splice Site Mutation

1. Prepare the reaction mixture with Taq DNA polymerase (*see* **Subheading 2.2., item 1**), dNTPs (*see* **Subheading 2.2., item 2**), and specific primers (*see* **Subheading 2.2., item 3**) (*see* **Note 4**). The volume of each reagent should be calculated according to the number of samples to be tested. Add DDW to a volume of 0.048 ml per sample (*see* **Note 5**).
2. Add 2 μl of DNA samples (*see* **Note 6**): fetal samples, control samples (normal and mutation carriers), and blank (no DNA). Place the tubes in the PCR apparatus and start the following program: 94°C for 2 min followed by 30 cycles of 94°C for 30 s, 56°C for 30 s, and 72°C for 45 s. After the 30th cycle, start the following program:72°C for 5 min and the process is ended at 10°C. Reaction volume should not exceed 50 μl. If it is necessary to add extra reagents or DNA, less water should be added accordingly.
3. For diagnosis of the splice site mutation (IVS12+1G_C), a restriction reaction is needed. Prepare Dde1 enzyme mixture (5 units of enzyme per tube in buffer) according to the number of tubes to be tested (samples, controls, and blanks). Add 12 μl of PCR product to each tube and incubate in 37°C for 3 h (can also incubate overnight).
4. After incubation, add 2 μl of bromphenol blue dye solution to each sample and load the entire reaction mixture on a 50-ml 3% agarose gel. Separate by electrophoresis at 90 V for 30–40 min. (Run controls, blanks, and a DNA size ladder in parallel to samples.)
5. Inspect the bands under UV light (*see* **Note 7**).

3.2.4.1. Analysis of Results

The 203-bp PCR product has three natural restriction sites for Dde1; thus, in the normal allele, one band of 120 bp is observed and the other two fragments are too small to be detected in the electrophoresis conditions. The mutation creates an additional site and the PCR product of the mutant allele shows a smaller band of 85 bp (the 35-bp fragment is not detected in the gel). In practice, one large band of 120 bp indicates a normal homozygote sample, one small band (85 bp) indicates an affected embryo, and the presence of two bands indicates a heterozygote. An alternative method is to perform fluorescent PCR under the same conditions with two fluorescent primers (e.g., tagged with VIC). The analysis of the fragments is carried with a genetic analyser (3100 AVANT genotyper, Applied Biosystems) in which after Dde1 treatment the two alleles appear as two distinct peaks of 120 and 85 bases. In the four-capillary apparatus, analysis of four samples is completed in 40 min.

4. Notes

1. Inspect the bath temperature during the reaction; it should not exceed 39°C.
2. Glycine buffer must be at least pH 10 or there will be no fluorescence readings.
3. If there is a considerable discrepancy between the reaction with the two substrates (in CVS diagnosis), for example, the reaction with MUGS shows an affected fetus and with MUG a nonaffected fetus, the possibility should be considered that a mutation has occurred in the β-subunit gene, causing the Hex B isoenzyme to be heat labile and to appear as Hex A activity *(17)*.
4. For each mutation, specific primers should be add to the reaction mixture.
5. Prepare the mixture in a clean spot station by using filtered tips.
6. DNA samples should be added outside the clean spot, with filtered tips.
7. If the controls show the correct fragment size and there is no amplification in the blank (no DNA) sample, then the diagnosis can be continued.

References

1. Robinson, D. and Stirling, J. L. (1968) N-Acetyl-β-glucosaminidase in human spleen. *Biochem. J.* **107,** 321–327.
2. Srivastava, S. and Beutler, E. (1973) Hexosaminidase A and hexosaminidase B: studies in Tay-Sachs disease and Sandhoff's disease. *Nature* **241,** 463.
3. Mahuran, D. J., Tsui, F., Gravel, R. A., and Lowden, J. A. (1982) Evidence for two dissimilar polypeptide chains in the beta subunits of hexosaminidase. *Proc. Natl. Acad. Sci. USA* **79,** 1602–1605.
4. Okada, S. and O'Brien, J. S. (1969) Tay Sachs disease: generalized absence of a β-D-*N*-acetylhexosaminidase component. *Science* **165,** 698–700.
5. Geiger, B. and Arnon, R. (1976) Chemical characterization and subunit structure of human *N*-acetylhexosaminidase A and B. *Biochemistry* **15,** 3484–3493.
6. Conzelmann, E. and Sandhoff, K. (1978) AB variant of infantile GM2-gangliosidosis: deficiency of a factor necessary for stimulation of hexosaminidase A-catalyzed degradation of ganglioside GM2 and glycolipid GA2. *Proc. Natl. Acad. Sci. USA* **75,** 3979–3983.
7. O'Brien, J., Okada, S., Fillerup, D., et al. (1971) Tay-Sachs disease–prenatal diagnosis. *Science* **172,** 61–64.

8. Kaback, M. M., and Zeigler, R. S. (1972) Heterozygote detection in Tay-Sachs disease: a prototype community screening program for the prevention of recessive genetic disorders. In: Volk, B.W. and Aronson, S.M. (eds.), Sphingolipids, sphingolipidosis and allied disorders: advances in experimental medicine and biology. Plenum, New York, pp 613–632.
9. Kaback, M., Bailin, G., Hirsch, P., and Roy, C. (1977) Automated thermal fractionation of serum hexosaminidase: effects of alteration in reaction variables and implications for Tay-Sachs disease heterozygote screening and prevention. In: Kaback, M., Rimoin, D., and O'Brien, J. (eds.), Tay-Sachs disease: screening and Prevention. Alan R. Liss , New York, p 197.
10. Kresse, H., Fuchs, W., Glossl, J., Holtfrerich, D., and Gilberg, W. (1981) Liberation of N-acetylglucosamine-6-sulfate by human beta-N-acetylhexosaminidase. *J. Biol. Chem.* **256,** 12926–12932.
11. Bayleran, J., Hetchman, P., and Saray, W. (1984) Synthesis of 4-methylumbelliferyl-β-D-*N*-acetylglucoseamine-6-sulfate and its use in classification of GM2 gangliosidosis genotypes. *Clin. Chim. Acta* **143,** 73–89.
12. Gravel, R. A., Clarke, J. T. R., Kaback, M. M., Mahuran, D., Sandhoff, K., and Suzuki, K. (1995) The GM_2 gangliosidoses. In: Scriver, C. R., Beaudet, A. L., Sly, W. S., et al. (eds.), The metabolic and molecular basis of inherited disease, 7th edn. McGraw-Hill, New York, pp 2839–2879.
13. Kytzia, H. J. and Sandhoff, K. (1985) Evidence for two different active sites on human beta-hexosaminidase A. Interaction of GM2 activator protein with beta-hexosaminidase A. *J. Biol. Chem.* **260,** 7568–7572.
14. Hou, Y., Tse, R., and Mahuran, D. J. (1996) Direct determination of the substrate specificity of the α active site in heterodimeric β-hexosaminidase A. *Biochemistry* **35,** 3963–3969,
15. Tse, R., Vavougios, G., Hou, Y., and Mahuran, D. J. (1996) Identification of an active acidic residue in the catalytic site of β-hexosaminidase. *Biochemistry* **35,** 7599–7607.
16. Sharma, R., Deny, H., Leung, A., and Mahuran, D. (2001) Identification of 6-sulfate binding unique to α-subunit containing isozymes of human β-hexosaminidase. *Biochemistry* **40,** 5440–5446.
17. Peleg, L. and Goldman, B. (1994) Detection of Tay-Sachs disease carriers among individuals with thermolabile hexosaminidase B. *Eur. J. Clin. Chem. Clin. Biochem.* **32,** 65–69.
18. Myerovitz, R., Piekarz, R., Neufeld, E. F., Shows, T. B., and Suzuki, K. (1985) Human β-hexosaminidase α-chain: coding sequence and homology with the β chain. *Proc. Natl. Acad. Sci. USA* **82,** 7830–7834.
19. Proia, R. L. and Soravia, E. (1987) Organization of the gene encoding the human β-hexosaminidase β-chain. *J. Biol. Chem.* **262,** 5677–5681.
20. Proia, R. L. (1988) Gene encoding the human human β-hexosaminidase β-chain: extensive homology of intron placement in the α- and β-chains gene. *Proc. Natl. Acad. Sci. USA* **85,** 1883–1887.
21. Takeda, K., Nakai, H., Hagiwara, H., et al. (1990) Fine assignment of beta-hexosaminidase A alpha subunit on 15q23–24 by high resolution in situ hybridization. *Tohoku J. Exo. Med.* **160,** 203–211.

22. Fox, M. F., DoToit, D. K., Warnich, L., and Retief, A. E. (1984) Regional localization of alpha-galactosidase to Xpter-q22, hexosaminidase B (HEXB) to 5q13-qter and arylsulfatase B (ARSB) to 5pter-q13. *Cytogenet. Cell. Genet.* **38,** 45–49.
23. Triggs-Raine, B. L., Feigenbaum, A. S., Natowitcz, M., et al. (1990) Screening for carriers of Tay Sachs disease among Ashkenazi Jews. *N. Engl. J. Med.* **323**, 6–12.
24. Navon, R. and Proia, R. L. (1989) The mutations in Ashkenazi Jews with adult GM2 gangliosidosis, the adult form of Tay Sachs disease. *Science* **243,** 1471–1474.
25. Triggs-Raine, B. L., Mules, E. H., et al. (1992) A pseudodeficiency allele common in non-Jewish Tay-Sachs carriers: Implication for carrier screening. *Am. J. Hum. Genet.* **51,** 793–801.

12

Arrayed Primer Extension Reaction for Genotyping on Oligonucleotide Microarray

Janne Pullat and Andres Metspalu

Summary

Arrayed primer extension reaction (APEX) is a straightforward and robust enzymatic genotyping method in which hundreds to thousands of variations in the genome are simultaneously analyzed in a single multiplexed reaction. It differs from allele-specific hybridization in that the genotype information in APEX is obtained by single base extension, performed by a specific DNA polymerase, together with four different dye terminators. This approach ensures highly specific discrimination without allele-specific hybridization, because the primer to be extended anneals just adjacent to the DNA base that needs to be identified. Selection of primers for specific sites or their consecutive placement in tiled format, shifting them by one base, permits single-nucleotide polymorphism analysis, mutation detection, or resequencing of the DNA template. Careful primer design also permits the analysis of insertions, deletions, splice variants, gene copy numbers, or CpG islands within the genome for gene methylation studies, by performing additional bisulfite reactions.

Key Words: Arrayed primer extension reaction (APEX); genotyping; minisequencing; molecular diagnostics; mutation analysis; oligonucleotide microarray; resequencing.

1. Introduction

Arrayed Primer Extension reaction (APEX) occurs by a two-step reaction mechanism: (1) targeting of DNA hybridization to the complementary oligoprimers and (2) single base extension of these primers with appropriate dye-labeled dideoxynucleotides that match the nucleotide on polymorphic site by DNA polymerase (*see* **Fig. 1**). Thus, these dye-labeled dideoxynucleotides are

From: *Methods in Molecular Biology, vol. 444: Prenatal Diagnosis*
Edited by: S. Hahn and L. G. Jackson © Humana Press, Totowa, NJ

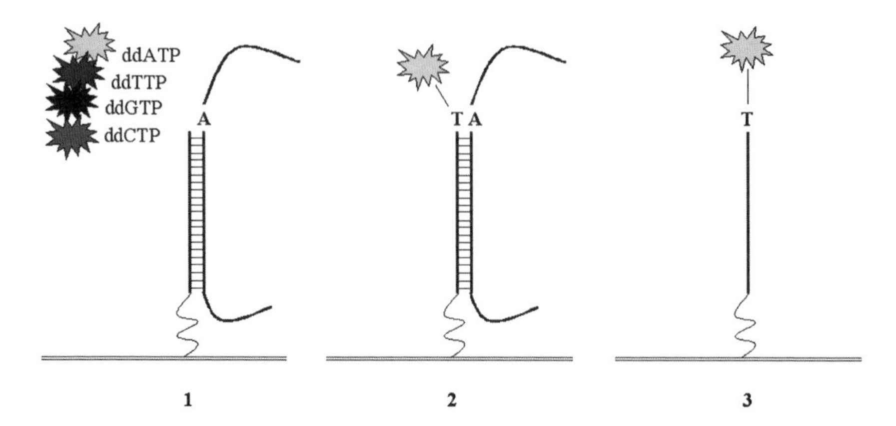

Fig. 1. The APEX approach. (**1**) Known oligonucleotides are attached to the surface of two-dimensional arrays at their amino-modified 3′ends. (**2**) The oligoprobe hybridizes the target DNA molecule, and it is extended enzymatically using dye terminators. (**3**) Signal detection.

used to terminate the extension reaction directly at their incorporation site, complementarily representing the DNA base in question *(1,2)*. In the current protocol, APEX is performed on two-dimensional spotted oligomicroarray glass slides.

The design of oligonucleotide probes for either SNP or point mutation typing is presented on **Fig. 2A**. For unknown mutational analysis, resequencing of the gene region is suggested **Fig. 2B**. Allele-specific oligoprimers (ASO) could be considered for identification of insertions/deletions.

Each base is identified by two unique 25-mer oligonucleotides, one for the sense and one for the antisense strand, with their 3′ ends one base upstream of the base to be identified. Probe selection, using specific oligonucleotide selection and design programs (http://bioinfo.ebc.ee/apex2/), is recommended to achieve the best discrimination. In the current protocol, an automatic laser imaging system, Genorama QuattroImager (Asper Biotech, Tartu, Estonia, www.asperbio.com), is used to read the APEX slide for result extraction *(3)*. Fluorescent labels (fluorescein isothiocyanate, cyanine [Cy]3, Cy5, and Texas Red) for the four dideoxynulceotides are read at emission wavelengths of 488, 532, 594, and 635 nm. The signal-to-noise ratio in APEX was typically between 30:1 to 100:1.

Photolithographic *in situ* synthesis, such as performed by Affymetrix (Santa Clara, CA), has been proposed for oligoarray fabrication *(4)*. This photolithographic synthesis of the entire oligonucleotide matrix is performed from photoprotected phosphoramidites by light irradiation.

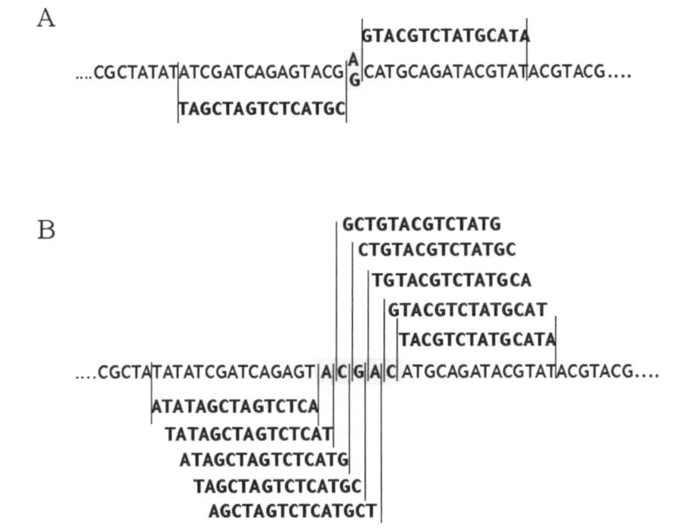

Fig. 2. Oligoprobe design for (**A**) point-variation detection and (**B**) resequencing.

Alternative approaches use virtual masks generated on a computer that are then relayed to a micromirror array *(5,6)*. The synthesis is thus digitally controlled, which enables a flexible chip production without substantial masks, and with suitable nucleotide precursors in the 5′-3′ or 3′-5′ direction. All designed oligoprobes can be verified at once (*see* **Fig. 3**) and redesigned when necessary. This *in situ* oligo microarray synthesis method decreases oligoarray developmental time.

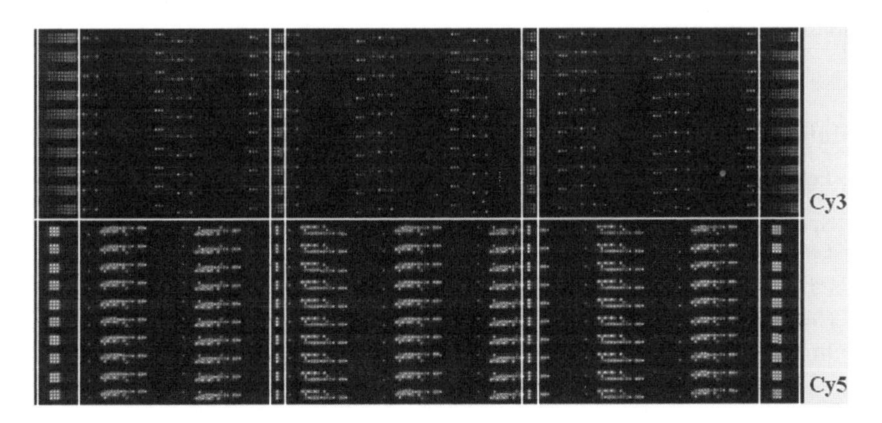

Fig. 3. APEX performed on an in situ-synthesized oligonucleotide microarray. Oligos were synthesized within biochip microchannels. Signals surrounded with white frames correspond to self-extending control oligos.

In order to spot oligoprimers on aminosilane microslides, primers have to include amino modification so they can be immobilized onto the glass surface at their 5′ ends. The DNA polymerase extends the DNA strand from the 3′-end; consequently, the single-stranded oligonucleotide probe on microarray must be immobilized from the 5′ end. Thus, classical Affymetrix arrays cannot be used in APEX reactions. To efficiently attach the oligoprobe onto the slide, the slide surface has to be chemically modified. Glass slides are silanized with 3-aminopropyl-trimethoxysilane *(2)* and linked with a spacer molecule (1,4-phenylene diisothiocyanate) to increase the accessibility of target hybridization to the oligoprimer *(7)*. To spot oligoprimers on aminosilane microslides, primers have to include amino modification first have to undergo amino modification, so they can be immobilized onto the glass surface at their 5′ ends.

In summary, the six-step microarray APEX process was developed as follows: (1) Design of APEX and polymerase chain reaction (PCR) primers. (2) *In silico* control with bioinformatic tools for possible secondary structures and cross-reaction effects and uniqueness of oligoprimers (http://bioinfo.ebc.ee/apex2/). (3) Spotting of oligonucleotide probes onto glass slides by using a pipetting robot. (4) PCR amplification of DNA fragments spanning variation of interest. (5) Validation of APEX reaction and optional primer redesign to identify oligos that performed unsatisfactorily (*see* **Notes 1** and **2**). (6) Imaging, genotype recording, and data analysis.

The primer extension reaction was performed from 48 to 56°C for 15 min. To achieve better target hybridization efficiency, PCR products were fragmented. The optimal length of the product for the APEX reaction was approximately 100 bp. Fragmentation was achieved by replacing 15–30% of dTTP in the PCR Master Mix with dUTP followed by purification of the PCR product and incubation with uracil *N*-glycosylase (UNG) digestion *(8)*.

2. Materials

2.1. Equipment

1. Spotted microarray slides.
2. Lifter Slip slides 22 × 25 mm^2 (Erie Scientific Company, Portsmouth, NH).
3. PTC 100 programmable thermal controller plate with humidified reaction chamber (MJ Research, Watertown, MA).
4. Imaging and signal recording by QuattroImager (Asper Biotech).

2.2. Preprocessing and Fragmentation of Target DNA

1. QIAquick PCR purification kit (QIAGEN, Hilden, Germany).
2. 1 U/μl UNG (Epicenter Technologies, Madison, WI).

3. 10× UNG buffer: 500 mM Tris-HCl, pH 9.0, 200 mM $(NH_4)_2SO_4$. Store at –20°C.
4. 1 U/μl shrimp alkaline phosphatase (SAP) (Sigma Chemie, Deisenhofen, Germany).

2.3. APEX Components

1. Purified and fragmented PCR products.
2. Thermo Sequenase DNA polymerase (GE Healthcare, Chalfont St. Giles, U.K.) supplemented with Thermo Sequenase reaction buffer (concentrated) and Thermo Sequenase dilution buffer. All stocks must be stored at –20°C. For the working solution of Thermo Sequenase, dilute the stock enzyme solution in the dilution buffer to a concentration of 3 U/μl.
3. Cy3-ddUTP (2.5 nM at 0.1 mM), Cy5-ddCTP (2.5 nM at 0.1 mM), Texas Red-5-ddATP (1.0 M) and fluorescein-12-ddGTP (1.0 mM) (PerkinElmer Life and Analytical Sciences, Boston, MA).
4. SloFade Light Antifade reagent (Invitrogen, Carlsbad, CA).
5. Alcanox detergent solution (Alcanox Inc., New York, NY).

3. Methods

3.1. PCR

1. Amplifiy template DNA fragments spanning the variation of interest by PCR. To achieve fragmentation of the PCR product, 20% of the dTTP nucleotides must be replaced with dUTP nucleotides in the PCR amplification mixture (*see* **Note 3**).
2. Pool approximately 200 ng of each amplified product during the purification step, following the manufacturer's instructions (QIAquik PCR purification kit), and elute with 15 μl of Milli-Q water (Millipore Corporation, Billerica, MA).

3.2. Fragmentation and Inactivation of dNTPs

1. To inactivate unused dNTPs from the PCR reaction, combine 15 μl of the purified PCR product (*see* **Subheading 3.1., step 2**) with 3 μl of 10× UNG buffer, 1 μl of SAP, 2 μl of UNG, and add 9 μl of double distilled H_2O (ddH_2O) to total reaction volume of 30 μl.
2. Incubate for 1 h at 37°C.
3. Use a 2.5% agarose gel in Tris-acetate-EDTA buffer to analyze fragmentation efficiency. Denature aliquots of 2 μl of each sample for 10 min at 95°C and snap-cool before applying on the gel. Use nondenatured samples as controls.

3.3. APEX

1. Wash prespotted and processed glass slides twice for 2 min each in Milli-Q H_2O at 95°C and gently remove the slides from the water bath with forceps. The

surface of the slides remains dry as the upper layer slide surface chemistry is hydrophobic. Place slides on a prewarmed thermoplate by using a programmable thermal controller.

2. Mix 28 µl of the amplified DNA sample (*see* **Subheading 3.2.**) with 4 µl of 10× Thermo Sequenase reaction buffer and 2 µl of ddH_2O, and denature for 10 min at 95°C.
3. Meanwhile, carefully mix 1 µl of fluorescein-ddGTP, 1 l of Cy3-ddCTP, 1 µl of Cy5-ddUTP, 1 l of Texas Red-ddAT, and 2 l of diluted Thermo Sequenase.
4. Briefly centrifuge the denatured reaction mix from **step 2** and immediately put on ice.
5. Add the reagent mix from **step 3** (6 l), mix carefully, and apply the entire mixture onto the oligonucleotide microarray grid slide.
6. Carefully cover the solution on the slide with a LifterSlip (Nalge Nunc, International, Rochester, NY) and take care to avoid air bubbles and evaporation.
7. Incubate the array for 15 min at 58°C.
8. Remove the LifterSlip and wash the slide for 2 min with Milli-Q H_2O at 95°C for 3 min with 0.3% Alcanox detergent solution at room temperature and for 2 min with Milli-Q H_2O at 95°C.
9. Remove the slide from the water and apply SlowFade®Light Antifade reagent to minimize bleaching.

3.4. Imaging and Signal Recording by QuattroImager

Imaging and data processing utilized *Genorama*™ 4.2.9. software (www.Asperbiotech.com).

4. Notes

1. To validate the designed primer set on the microarray, up to 50 APEX reactions should be performed with different individual DNA templates. In case of poor performance, missing, false positive, or poor signals, the APEX primer should be redesigned; the experimenter should modify reaction conditions, such as temperature, incubation time, DNA sample concentration, enzyme concentration, and different fluorescent dyes; or a combination. However, primer selection is of primary importance and for SNP typing, we suggest using the next SNP within the same LD block. Additional time is necessary to develop a well-performing assay for the analysis of specific mutations that cannot be changed; for some sites, this may be very difficult.
2. Additionally, APEX without template DNA (negative control) identifies possible secondary oligoprimer structures by producing signals due to self-extension or dimer formation between primers.
3. Note that when using the APEX method, the dUTP incorporation rate is highly dependent on the DNA sequence. In case of poor fragmentation, the amount of dUTP in PCR Master Mix can be modified (10–30%) so as to obtain the best dUTP to dTTP ratio.

Reference

1. Kurg, A., Tonisson, N., Georgiou, I., Shumaker, J., Tollett, J., and Metspalu, A. (2000) Arrayed primer extension: solid-phase four-color DNA resequencing and mutation detection technology. *Genet. Test.* **4,** 1–7.
2. Shumaker, J. M., Metspalu, A., and Caskey, C. T. (1996) Mutation detection by solid phase primer extension. *Hum. Mutat.* **7,** 346–354.
3. Tonisson, N., Zernant, J., Kurg, A., et al. (2002) Evaluating the arrayed primer extension resequencing assay of TP53 tumor suppressor gene. *Proc. Natl. Acad. Sci. USA* **99,** 5503–5508.
4. Fodor, S. P., Rava, R. P., Huang, X. C., Pease, A. C., Holmes, C. P., and Adams, C. L. (1993) Multiplexed biochemical assays with biological chips. *Nature* **364,** 555–556.
5. Beier, M. and Hoheisel, J. D. (2000) *Nucleic Acids Res.* **28,** E11.
6. Singh-Gasson, S., Green, R. D., Yue, Y., et al. (1999) Maskless fabrication of light-directed oligonucleotide microarrays using a digital micromirror array. *Nat. Biotechnol.* **17,** 974–978.
7. Southern, E., Mir, K., and Shchepinov, M. (1999) Molecular interactions on microarrays. *Nat. Genet.* **21,** 5–9.
8. Cronin, M. T., Fucini, R. V., Kim, S. M., Masino, R. S., Wespi, R. M., and Miyada, C. G. (1996) Cystic fibrosis mutation detection by hybridization to light-generated DNA probe arrays. *Hum. Mutat.* **7,** 244–255.

13

A Fast Microelectronic Array for Screening and Prenatal Diagnosis of β-Thalassemia

Barbara Foglieni, Silvia Galbiati, Maurizio Ferrari, and Laura Cremonesi

Summary

The electronic microchip is a recently developed technology for the fast and reliable detection of known single-nucleotide polymorphisms (SNPs) in the genome. The DNA fragment to be analyzed is directed electrophoretically into the chip, and then it is hybridized with fluorescent-tagged DNA probes specific for the mutant and wild-type sequences. The presence or absence of the mutation is detected by the fluorescence signal. Electronic stringency provides quality control for the hybridization process and ensures that any bound pairs of DNA are truly complementary; the microchip can be easily customized by the end user, allowing for assembly of specific probes onto the microchip to perform individualized analyses. Assays for 10 frequent mutations in the β-globin gene causing β-thalassemia and sickle cell anemia are presented that can be applied, in turn, to population screening or family study and prenatal diagnosis in single cases.

Key Words: β-Thalassemia; mutation detection; microarray; microelectronics, prenatal diagnosis.

1. Introduction

Hemoglobinopathies are caused by β-globin gene mutations that are spread at different prevalences in each geographical area affected by the disease *(1)*. This demands the use of an open, flexible platform with an assay format that may be individually designed based on each local situation.

The electronic microchip is a recently developed technology for the fast and reliable detection of known single-nucleotides polymorphisms (SNPs) in the genome. This approach was applied by Nanogen for the development

From: *Methods in Molecular Biology, vol. 444: Prenatal Diagnosis*
Edited by: S. Hahn and L. G. Jackson © Humana Press, Totowa, NJ

of an integrated system, which consists of a chip loader, a fluorescent detection/reader, a computer control interface, and a data display screen *(2–4)*. In this system, the biotinylated amplicon is electronically placed on the chip and electrophoresed to specific positively charged test sites on the cartridge, where it remains embedded through interaction with streptavidin in the permeation layer covering the microchip. After hybridization with fluorescently labeled wild-type and mutant reporter probes, a thermal stringency step specific for each mutation is performed, raising the temperature inside the chip to detach all reporters nonperfectly matched to template DNA. Hybridization is detected by fluorescence automated scanning by means of dedicated software. Alleles are assigned by evaluating the fluorescence ratio of the probes, allowing to assess the patient genotype.

The platform provides a completely automated device allowing the scanning of up to four chips (400 samples) in a single day, facilitating the screening of a large number of samples, thus representing a good example of low-medium density microarray with high-throughput analytical power.

The assay requires four components: the amplicon, the oligonucleotide stabilizer to open the secondary structure of the DNA template, and the two allele-specific oligonucleotide reporter probes for either the wild-type or the mutant sequence, labeled with specific fluorescent tags.

Two different reporter formats can be used: a traditional format with the fluorophores linked directly to specific probes; and a universal format, based on reporters containing a sequence-specific discriminator prolonged by a tail that hybridizes to either a wild-type or mutant universal-labeled probe *(5,6)*. The universal format displays the same or an even better efficiency than the traditional format, and it is also preferable, because it is cost-effective *(7)*.

Concerning stabilizers, sequence context can significantly influence the analysis. The presence of oligonucleotides to open the secondary structure of the DNA template generally produces higher fluorescent values and a better discrimination ratio. For some mutations, the use of a stabilizer is mandatory, and a minority of mutations demands the use of two stabilizers, probably due to the complexity of the secondary structure of the sequence. The other mutations can be analyzed even in the absence of stabilizers. Adjacent mutations can be analyzed by the use of the same "shared" stabilizer.

The system is able to correctly identify any kind of mutation, including deletions and conservative transversions. Under optimized conditions, no cross-hybridization is found, even for mutations affecting the same or adjacent nucleotide positions *(7)*.

Up to now, several applications of Nanogen technology for DNA variation identification have been described in different fields (for review, *see* **ref.** *(8)*).

The microchip can be easily adjusted to personalize diagnostic needs. Samples can be spotted and analyzed according to daily necessity, and the same cartridge can be used at different times by addressing and analyzing a new set of pads at each time, while kept in a refrigerator. This implies that different genes and mutations can be analyzed on the same cartridge. Several multiplexing formats can be performed, including multiplex polymerase chain reaction (PCR), multiple addressing of different amplicons to the same pad, and serial reporter hybridization, allowing correct genotyping until the end of the analysis *(7)*. All these features decrease costs and the time per analysis, which is also important in a diagnostic setting.

Here, a microelectronic microchip-based strategy is described for the detection of the nine prevalent mutations causing β-thalassemia in the Mediterranean population (–87C>G, cd6delA, cd8delAA, IVSI-1G>A, IVSI-6T>C, IVSI-110G>A, cd39C>T, IVSII-1G>A, and IVSII-745C>G) and HbS (cd6A>T) *(9)*. The assay format can be easily scaled up and down according to the necessity, from large-scale screening applications to individual family studies for genotyping of couples at risk for the disease and prenatal diagnosis of specific mutations.

2. Materials

1. NMW 1000 NanoChip Molecular Biology Workstation. Composed of a loader, in which PCR fragments are addressed to the microchip, and a reader, which processes and scans the microchip.
2. NanoChip Cartridge. Store at 4°C; use within 1 month.
3. High-performance liquid chromatography-purified oligonucleotides: PCR primers [5′-biotinylated and unmodified]. Reporter and oligonucleotide stabilizer. Cyanine (Cy)3-, Cy5-, Alexa532-, and Alexa647-labeled universal probes.
4. PCR reagents and thermal cycler.
5. 96-well format desalting plates (Multiscreen-PCR plates MANU 030, Millipore Corporation, Billerica, MA) and Vacuum manifold (Multiscreen Separation System, Millipore Corporation).
6. Conductivity meter.
7. Nunc 96-/384-well plates (Nalge Nunc, Rochester, NY).
8. 50 mM and 100 mM L-histidine: Prepare in a sterile recipient. Filter using a 0.2-μm filter. Check the conductivity (<100 μS/cm for the 50 mM, <100–200 μS/cm for the 100 mM). Store at 2–8°C in the dark. Prepare fresh every week the 50 mM histidine. Store approximately 5-ml aliquots at –20°C for up to 2 months the 100 mM histidine.
9. 0.1 N NaOH. Prepare fresh every week.
10. High salt buffer (HSB): 500 mM sodium chloride and 50 mM sodium phosphate, pH 7.4. (prepared from a 1 M sodium phosphate dibasic/monobasic [4:1 stock solution]). Filter using a 0.2-μm filter. Store at room temperature.

11. Low salt buffer (LSB): 50 mM sodium phosphate, pH 7 (prepared from a 1 M sodium phosphate dibasic/monobasic [7:3 stock solution]). Filter using a 0.2-µm filter. Store at room temperature.

3. Methods

3.1. DNA Extraction

Genomic DNA is isolated from peripheral blood by standard methods (phenol/chloroform, salting-out protocol or by commercial kits). For prenatal diagnosis, genomic DNA is extracted from maternal amniocytes or chorionic villi by standard methods.

3.2. PCR Amplification Conditions

The β-globin gene DNA regions (promoter, exons 1 and 2, and introns 1 and 2) containing the 10 mutations of interest are amplified using β-specific sets of primers (*see* **Table 1**) (*see* **Note 1**). **Figure 1** shows the localization of each single amplicon. For each set, one of the primers was 5' biotinylated. Base composition and primer length were designed to allow the same thermal cycling conditions for all fragments.

The SNP assay should include a wild-type, a heterozygous, and, if available, a homozygous DNA control sample specific for each mutation to be analyzed. Control samples must be processed in parallel and at the same conditions as unknown samples.

Concerning the PCR protocol, different conditions must be applied depending on the aim of the study. For the screening of unknown samples

Table 1
PCR primer sequences

Amplicon	PCR primer[a]
1	1 For 5′-GTTACAAGACAGGTTTAAGGAGACC-3′
	1 Rev 5′-Biotin-GCACTTTCTTGCCATGAGCCTTC-3′
2	2 For 5′-Biotin-ACGTGGATGAAGTTGGTGGT-3′
	2 Rev 5′-CCCCTTCCTATGACATGAACTTAACC-3′
3	3 For 5′-ATCATGCCTCTTTGCACCATTCT-3′
	3 Rev 5′-Biotin-CAGGAGCTGTGGGAGGAAGAT-3′
4	4 For 5′-Biotin-GACAGGTACGGCTGTCATCACT-3′
	4 Rev 5′-CTTCATCCACGTTCACCTTGCC-3′

[a]For, forward; Rev, reverse primer sequences.

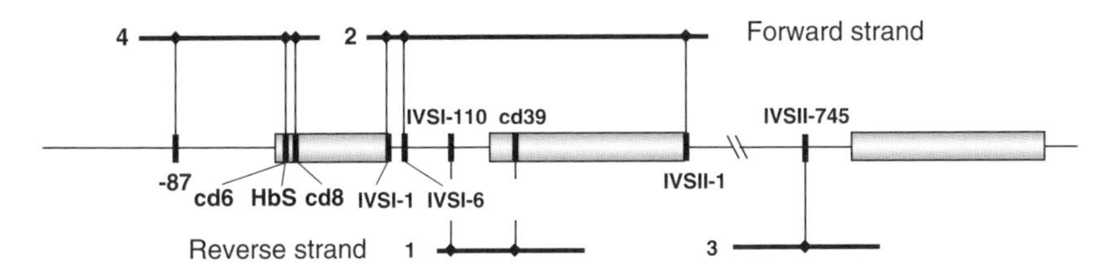

Fig. 1. β-Globin gene amplification scheme. DNA regions including the nine β-thalassemia and HbS mutations are amplified by the use of four sets of primers, according to the following scheme: amplicon 1 includes IVSI-110G>A and cd39C>T; amplicon 2 includes IVSI-1G>A, IVSI-6T>C, and IVSII-1G>A; amplicon 3 includes IVSII-745C>G; and amplicon 4 includes –87C>G, cd6delA, HbS, and cd8delAA. Amplicons 1 and 3 are analyzed by using the reverse strand as target for hybridization, amplicons 2 and 4 are analyzed by using the forward strand.

for both large-population or individual family studies, the complete panel of the 10 β-globin mutations should be evaluated by the multiplex PCR approach (*see* **Note 2**). For the analysis of specific mutations, as for prenatal diagnosis in previously characterized families at risk for the disease, only the target mutations should be tested.

Regarding the multiplex PCR reaction, due to partial overlapping of amplicons 1 and 2, and to competition of primer sets between the two adjacent amplicons 1 and 4 (**Fig. 1**), PCR multiplexing is performed by independent processing of the two duplexes, one duplex including amplicons 1 and 3 and the other duplex including amplicons 2 and 4.

PCR is performed in 50-µl reactions containing 100 ng of DNA, 200 µM dNTP, 10 mM Tris-HCl, pH 8.3, 50 mM KCl, 1.5 mM $MgCl_2$, 1.3 units of thermostable DNA polymerase (AmpliTaq Gold, Applied Biosystems, Foster City, CA), and 20 pmol of each primer, except for fragment 2 for which 40 pmol are used. Cycling conditions are as follows: 95°C for 10 min (one cycle); 95°C for 30 s, 57°C for 30 s, 72°C for 30 s (35 cycles); and 72°C for 10 min (one cycle).

If the parental mutations have already been identified, prenatal diagnosis requires a single or duplex PCR to amplify only fragments carrying mutations of interest. PCR cycling conditions for the single-fragment amplification are the same as described above.

3.3. PCR Product Purification

At the end of the amplification process, each single or duplex PCR is purified and desalted by the use of a 96-well format plate (Multiscreen-PCR

Plates MANU 030) coupled with the Multiscreen Separation System (Millipore Corporation), according to the following protocol:

1. For the multiplex amplification, the two duplex PCR are pooled into the same well of the plate. This allows the mixture of the four amplicons to be applied to the same pad for further processing as a unique sample (*see* **Note 2**). For the single amplification, each PCR is loaded in a single plate well. To have good purification results, it is important to purify the same volume of different PCR reactions on the same plate.
2. Add 100 µl of distilled water to each loaded well of the plate.
3. Apply vacuum until all liquid is pulled through the filter (approx 10 min at 10 inHg).
4. Add 100 µl of distilled water to each well, and repeat **step 3**.
5. Add 70 µl of distilled water. This final volume is enough for two loading processes (*see* **Note 3**).
6. Place the plate on an orbital shaker for 10–15 min, mix each sample, and transfer it into a clean PCR tube. Store samples at 4°C.
7. Check sample conductivity with the conductivity meter supplied with the Nanogen instrumentation (optimal range <100 µS/cm). If the conductivity is out of range, an additional washing with 100 µl of distilled water can be performed in new wells of the desalting plate.

Sample preparation, including amplification and purification, takes approximately 3.5 h for 96 samples.

3.4. Microchip β-Thalassemia Universal Probe Design

All reporters for this protocol are designed in a universal format, containing a tail sequence that hybridizes to either a wild-type or mutant complementary labeled probe (universal probes) *(5)* (*see* **Table 2**). Reporters are designed either to have the base variation located in the middle in a dot-blot format (e.g., cd39C>T) or located either at their 3′ or 5′ end in a base-stacking format (e.g., IVSII-1G>A, –87C>G), depending on the sequence context surrounding the DNA variation of interest *(8)*.

With the base-stacking format, depending on the position of the base variation in the reporter, a Cy5/Cy3 or Alexa532/Alexa647 labeling should be located at either the free 5′ or 3′ end of the wild-type/mutant universal probes, respectively.

Reporter, stabilizers and primer oligonucleotides are designed with the help of Web-free programs (dnamfold, http://www.bioinfo.rpi.edu/ezukerm/rna; and oligoanalyzer 3.0, http://www.idtdna.com), following specific guidelines *(8)*.

Table 2
Reporter, stabilizer, and universal probe sequences[a]

Mutation	Probe, stabilizer	Temp (°C)
cd39C>T	Wt 5′-TGGACC**C**AGAGGTctgagtccgaacattgag-3′ Mut 5′-GGACC**T**AGAGGTgcagtatatcgcttgaca-3′ Stab 5′-CTTTGAGTCCTTTGGGGATCTGTCCACTCCTG-3′	26
IVSI-110G>A	Wt 5′-CCTATT**G**GTCTATTctgagtccgaacattgag-3′ Mut 5′-GCCTATT**A**GTCTATTgcagtatatcgcttgaca-3′ Stab 5′-AGACTCTTGGGTTTCTGATAGGCACTGACT-3′	28
IVSI-1G>A	Wt 5′-CCAA**C**CTGCCctgagtccgaacattgag-3’ Mut 5’-CCAA**T**CTGCCCgcagtatatcgcttgaca-3′ Stab 5′-AGGGCCTCACCACCAACTTCATCCACGTTCACC-3′	34
IVSI-6T>C	Wt 5′-CCTTGAT**A**CCAActgagtccgaacattgag-3′ Mut 5′-CTTGAT**G**CCAAgcagtatatcgcttgaca-3′ Stab 5′-AGGGCCTCACCACCAACTTCATCCACGTTCACC-3′	26
IVSII-1G>A	Wt 5′-ctgagtccgaacattgagCCATAGACTCA**C-**3′ Mut 5′-gcagtatatcgcttgacaCCATAGACTCA**T-**3′ Stab 5′CCTGAAGTTCTCAGGATCCACGTGCAGCTT-3′	30
IVSII-745C>G	Wt 5′-ctgagtccgaacattgagCTACAATCCAG**C-**3′ Mut 5′-gcagtatatcgcttgacaCTACAATCCAG**G-**3′ Stab 5′-TACCATTCTGCTTTTATTTTATGGTTGGGATAAGGC-3′	35
–87C>G	Wt 5′-**G**GTGTGGCTCctgagtccgaacattgag-3′ Mut 5′-**C**GTGTGGCTCgcagtatatcgcttgaca-3′ Stab1 5′-CCCTGGCTCCTGCCCTCCCTGCTCC-3′ Stab2 5′-TGCTCCTGGGAGTAGATTGGCCAACCCTAG-3′	30

(Continued)

Table 2
(Continued)

Mutation	Probe, stabilizer	Temp (°C)
cd6delA	Wt 5′-ctgagtccgaacattgagAGACTTCTCC**T**C-3′ Mut 5′-agcagtatatcgcttgacaGACTTCTCCCA-3′ Stab 5′-GGAGTCAGGTGCACCATGGTGTCTGTTTGAGGT-3′	27
cd8delAA	Wt 5′-CAGAC**TT**CTCCctgagtccgaacattgag-3′ Mut 5′-CAGACCTCCTgcagtatatcgcttgaca-3′ Stab 5′-CAGGAGTCAGGTGCA-3′	32
HbS	Wt 5′-ctgagtccgaacattgagAGACTTCTCC**T**CA-3′ Mut 5′-gcagtatatcgcttgacaAGACTTCTCC**A**CA-5′ Stab 5′-GGAGTCAGGTGCACCATGGTGTCTGTTTGAGGT-3′	31
5′-labeled universal probes	Wt 5′-Cy3-ctcaatgttcggactcag-3′ Mut 5′-Cy5-tgtcaagcgatatactgc-3′	
3′-labeled universal probes	Wt 5′-ctcaatgttcggactcag-Alexa532-3′ Mut 5′-tgtcaagcgatatactgc-Alexa647-3′	

[a]Lowercase letters indicate the universal tail. The base on the sequence that matches the position of the mutation is in bold. Universal wild-type and mutant reporters labeled with Alexa532 and Alexa647 are used for the IVSII-1G>A, IVSII-745C>G, cd6delA, and HbS mutations. For all the remaining mutations, the universal wild-type and mutant reporters labeled with Cy3 and Cy5 are used. The last column indicates temperatures for thermal stringency after hybridization. Wt, wild-type; Mut, mutant; stab, stabilizer.

3.5. Microarray Analysis

Details of the components of the Nanogen Workstation were already reported ***(2,4,10)***.

3.5.1. Microchip Addressing

The addressing process consists of an electronic loading of biotinylated PCR products to streptavidin-containing gel pads (*see* **Note 4**). Each PCR-purified sample is mixed together with a loading solution and electronically addressed to a specific test site of the microarray.

The SNP assay includes a negative control (only 50 mM histidine) for background signal subtraction along with a wild-type, heterozygous, and, if present, homozygous control specific for each mutation to be analyzed processed with the same procedure.

1. Loading Guidelines.
   a. Mix 30 μl of each PCR with 30 μl of 100 mM histidine (50 mM final concentration) and load it in a Nunc 96- or 384-well plate.
   b. Add 60 μl of 0.1 N NaOH into a plate-well.
   c. Load the plate and a new cartridge into the loader.
2. Create a new Loader Map File by designating the addressing site on the cartridge for each sample in the loader. The sodium hydroxide is addressed to the cartridge after all samples and controls, to denature PCR products, removing the nonbiotinylated single-stranded DNA. In this way, the biotinylated strand is available for subsequent hybridization steps.
3. Start the Loading Process
4. As control step, the loader performs a conductivity test to check pad activation before loading and during each addressing process. The entire loading process of 100 DNA samples is completely automated and takes approximately 5 h; thus, it would be advisable to perform it overnight.

3.5.2. Cartridge Hybridization

The cartridge is manually incubated with the mixture specific for each mutation to be analyzed. If samples are to be screened for all the 10 β-globin mutations, the mutation order reported in **Table 2** should be followed.

Details on the cartridge hybridization protocol are as follows:

1. Incubate for 3 min at room temperature the cartridge with 100 μl of a mixture containing 1 μM stabilizer in HSB.
2. Wash twice with HSB and incubate for 3 min the cartridge with 100 μl of a mixture containing specific wild-type and mutant reporters and the pair of universal probes (1μM each), in HSB.

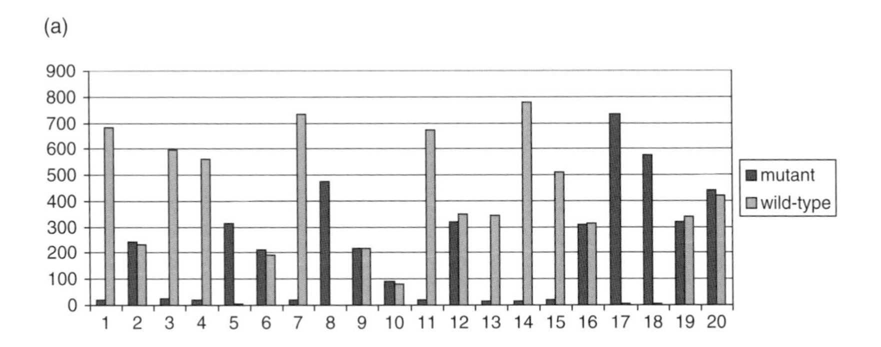

B

Results for cd39 (scan 26)					
Sample	**Black**	**Gray**	**Pads**	**Ratio (B:G)**	**Probe Designation**
1	18	685	1	1:38.1	C/C
2	245	231	1	1:1.1	C/T
3	23	598	1	1:26	C/C
4	18	561	1	1:31.2	C/C
5	315	3	1	105:1	T/T
6	210	193	1	1.1:1	C/T
7	18	733	1	1:40.7	C/C
8	474	2	1	237:1	T/T
9	218	218	1	1:1	C/T
10	91	83	1	1.1:1	C/T
11	20	674	1	1:33.7	C/C
12	318	349	1	1:1.1	C/T
13	16	342	1	1:21.4	C/C
14	16	780	1	1:48.8	C/C
15	18	512	1	28.4	C/C
16	310	315	1	1:1	C/T
17	733	5	1	146.6:1	T/T
18	574	3	1	191.3:1	T/T
19	320	340	1	1:1.1	C/T
20	440	420	1	1.1:1	C/T

Fig. 2. Microchip analysis for the cd39C>T β-globin gene mutation. **(A)** The graph shows fluorescence signals for 20 samples analyzed at 26°C stringency temperature. Sample 9 was used as heterozygous control for normalization of gray (wild type) and black (mutant) signals. **(B)** Fluorescence values and genotype assignment. For each sample fluorescence wild-type and mutant probe signals, number of pads addressed with the same sample (Pads), black/grey (B:G) ratio, and genotype (probe designation) are indicated.

3. Analyze the chip in the reader.
4. If amplicons should be analyzed for more than one mutation, previous reporter probes and stabilizers need to be stripped, by incubating the cartridge for 5 min with 0.1 N NaOH and washed five times with 150 μl of H_2O and five times with 150 μl of HSB (*see* **Note 5**).
5. Repeat **steps 1–4** for each mutations to be analyzed. The entire process for one mutation for the whole chip takes approximately 10 min.

3.5.3. Fluorescence Analysis

To detach fluorescent probes not perfectly matched to target DNA, a thermal stringency step is performed by raising the temperature inside the cartridge. Hybridization is detected by fluorescence automated scanning (*see* **Note 6**).

1. Place the cartridge into the reader.
2. Start the specific Reader Protocol for each mutation, which includes heating at the specific discrimination temperature (*see* **Table 2**), washing, and scanning steps.
3. The addressed cartridge can be stored at 4°C up to 1 month (*see* **Note 7**). The automated fluorescence scanning for the whole chip takes approximately 30 min.

3.5.4. Interpretation of Results

Quantitative analysis of the hybridization results is performed by a dedicated software accessible through the Web, according to the following steps:

1. Background control value subtraction: only samples displaying a signal to noise ratio greater than five-fold are considered.

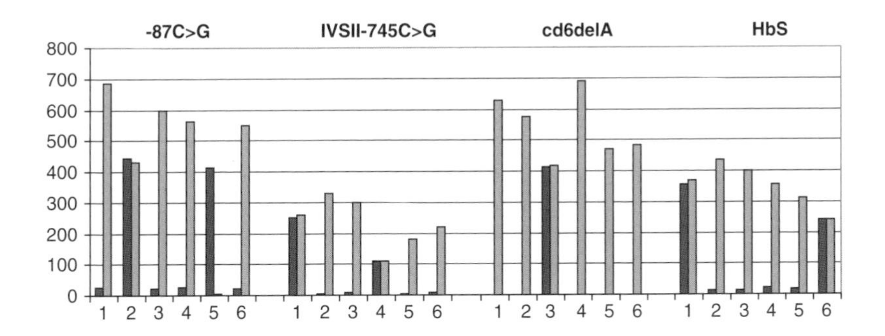

Fig. 3. Examples of microchip results for multiple addressing of the two duplex PCR reactions to the same pad and serial probe hybridization with all 10 sets of wild-type and mutant universal reporters. Analysis of the –87C>G, IVSII745C>G, cd6delA, HbS mutations on six patients is reported. Genotypes were as follows: sample 1, IVSII745/HbS; sample 2, –87/WT; sample 3, cd6/WT; sample 4, IVSII-745/WT; sample 5, –87/-87; and sample 6, HbS/WT. Analysis of the remaining mutations has already been reported *(9)*.

2. Signals normalization according to wild-type-to-mutant ratio of heterozygous control sample.
3. Genotyping is based on the following thresholds. Samples displaying wild-type-to-mutant signal ratio of between 1:1 and 1:2 are assigned as heterozygotes; samples with wild-type-to-mutant signal ratio of >5:1 or >1:5 are assigned as wild-type homozygotes or mutant homozygotes, respectively; samples with a wild-type-to-mutant signal ratio of between 1:2 and 1:5 are not scored (no call) (*see* **Notes 8** and **9 Figs. 2** and **3**).

Row data are recorded and archived in specific files on the workstation and can be exported through a zip disk. Processed data are available in a useful format that includes genotype designation.

4. Notes

1. Most of the DNA regions of the β-globin gene interested by β-thalassemia mutations are already included into the amplicons designed for the present work. This implies that additional mutations can be easily analyzed in the multiplex session by only adding their specific probes, thereby allowing expansion of the panel of the mutations to be identified.
2. The combination of multiplex PCR with multiple addressing and serial hybridization makes it possible to analyze all 10 mutations for the same sample on one test site of the chip. The recently developed new version of the chip (NanoChip 400; www.nanogen.com) includes 400 pads per cartridge, further expanding and simplifying mutational screening for the same or different disease genes on large-population samples.
3. For scarce fetal DNA samples extracted from either chorionic villi or amniocites, after the PCR reaction, the vacuum-purified sample can be directly resuspended into 60 μl of 50 mM histidine, which are all addressed to the pad to increase its concentration. Store samples at –20°C if not immediately processed.
4. In the protocol described here, a biotinylated amplicon is addressed to the pad. This "amplicon-down" format is particularly suitable for the analysis of a panel of several mutations on the same target sample. Alternatively, when a mutation is prevalent in a given population, a "capture-down" format can be used, where a biotinylated capture oligonucleotide is immobilized on the pad and binds to target-amplified DNA, which is then denatured and hybridized with specific reporter probes *(11)*.
5. Serial reporter hybridization formats have already been extensively reported *(6, 12,13)*. More than 20 different mutations were subsequently analyzed on the same amplified fragments obtaining reliable identification of mutations and signal-to-noise fluorescence ratio >5 until the end of the analysis *(7)*.
6. Cross-hybridization is a problem that is often overlooked. Under optimized conditions no cross-hybridization has been observed, even for mutations affecting the same or adjacent nucleotide positions *(7)*.
7. In case the chip is not completely loaded, it can be kept in a refrigerator at 4°C until subsequent addressing of a new set of samples. Before loading a new set of

samples, it is advisable to perform a stripping procedure with NaOH as described above to eliminate any possible fluorescence background. This procedure can be repeated until complete loading of the chip, within 1 month.

8. An overall 1% no-call rate was obtained because of either a lack or insufficient amplification or sample loss during purification as revealed by subsequent agarose gel electrophoresis ***(9)***.
9. The β-thalassemia assay was blindly validated on 250 individuals randomly chosen from wild-type or mutant homozygotes, heterozygotes, or compound heterozygotes for one of the nine mutations causing β-thalassemia, with complete concordance of results ***(9)***. Overall, extensive blind validation of known control samples (>10,000 patients) carrying different mutations in a variety of genes related with relevant genetic diseases resulted in 100% accuracy of the microelectronic microchip system compared with the sequencing reference procedure ***(6,12,14–16)***. Methods for several clinically relevant assays have been published and optimized ***(6,9,10,12,14–18)***, and validated analyte-specific reagents tests are currently available for hereditary hemocromatosis; factor V, factor II, and prothrombin; nonsyndromic deafness; β-thalassemia; cystic fibrosis; and apolipoprotien E mutations.

Acknowledgments

This work was supported by Telethon GGP04016 (to L.C.), Fondo per gli Investimenti della Ricerca di Base grant RBNE01SLRJ (to M.F.), and European Commission for the Special Non-invasive Advances in Fetal and Neonatal Evaluation (SAFE) Network of Excellence (LSHB-CT-2004-503243) (to L.C.).

References

1. Higgs, D. R., Thein, S. L., and Woods, W. G. (2001) The molecular pathology of the thalassaemias. In: Weatherall, D. J. and Clegg, B. (eds.). The thalassaemia syndromes, 4th ed. Blackwell Science, Oxford, England, pp 133–191.
2. Heller, M. J., Forster, A. H., and Tu, E. (2000) Active microeletronic chip devices which utilize controlled electrophoretic fields for multiplex DNA hybridization and other genomic applications. *Electrophoresis* **21,** 157–164.
3. Edman, C. F., Raymond, D. E., Wu, D. J., et al. (1997) Electric field directed nucleic acid hybridization on microchips. *Nucleic Acids Res.* **25,** 4907–4914.
4. Gilles, P. N., Wu, D. J., Foster, C. B., Dillon, P. J., and Chanock, S. J. (1999) Single nucleotide polymorphic discrimination by an electronic dot blot assay on semiconductor microchips. *Nat. Biotechnol.* **17,** 365–370.
5. Cooper, K. L. and Goering, R. V. (2003) Development of a universal probe for electronic microarray and its application in characterization of the *Staphylococcus aureus* polC gene. *J. Mol. Diagn.* **5,** 28–33.

6. Moutereau, S., Narwa, R., Matheron, C., Vongmany, N., Simon, E., and Goossens, M. (2004) An improved electronic microarray-based diagnostic assay for identification of MEFV mutations. *Hum. Mutat.* **23,** 621–628.
7. Ferrari, F., Foglieni, B., Arosio, P. et al. (2006) Microelectronic DNA chip for hereditary hyperferritinemia cataract syndrome, a model for large-scale analysis of disorders of iron metabolism. *Hum. Mutat.* **27,** 201–208.
8. Ferrari, M., Cremonesi, L., Bonini, P., Foglieni, B., and Stenirri, S. (2005) Single-nucleotide polymorphism and mutation identification by the nanogen microelectronic chip technology. *Methods Mol. Med.* **114,** 93–106.
9. Foglieni, B., Cremonesi, L., Travi, M., et al. (2004) beta-Thalassemia microelectronic chip: a fast and accurate method for mutation detection. *Clin. Chem.* **50,** 73–79.
10. Santacroce, R., Ratti, A., Caroli, F., et al. (2002) Analysis of clinically relevant single-nucleotide polymorphisms by use of microelectronic array technology. *Clin. Chem.* **48,** 2124–2130.
11. Sosnowski, R., Heller, M. J., Tu, E., Forster, A. H., and Radtkey, R. (2002) Active microelectronic array system for DNA hybridization, genotyping and pharmacogenomic applications. *Psychiatr. Genet.* **12,** 181–192.
12. Thistlethwaite, W. A., Moses, L. M., Hoffbuhr, K. C., Devaney, J. M., and Hoffman, E. P. (2003) Rapid genotyping of common MeCP2 mutations with an electronic DNA microchip using serial differential hybridization. *J. Mol. Diagn.* **5,** 121–126.
13. Frusconi, S., Giusti, B., Rossi, L., et al. (2004) Improvement of low-density microelectronic array technology to characterize 14 mutations/single-nucleotide polymorphisms from several human genes on a large scale. *Clin. Chem.* **50,** 775–777.
14. Erali, M., Schmidt, B., Lyon, E., and Wittwer, C. (2003) Evaluation of electronic microarrays for genotyping factor V, factor II, and MTHFR. *Clin. Chem.* **49,** 732–739.
15. Evans, J. G., and Lee-Tataseo, C. (2002) Determination of the factor V Leiden single-nucleotide polymorphism in a commercial clinical laboratory by use of NanoChip microelectronic array technology. *Clin. Chem.* **48,** 1406–1411.
16. Sethi, A. A., Tybjaerg-Hansen, A., Andersen, R. V., and Nordestgaard, B. G. (2004) Nanogen microelectronic chip for large-scale genotyping. *Clin. Chem.* **50,** 443–446.
17. Sohni, Y. R., Cerhan, J. R., and O'Kane, D. (2003) Microarray and microfluidic methodology for genotyping cytokine gene polymorphisms. *Hum. Immunol.* **64,** 990–997.
18. Nagan, N. and O'Kane, D. J. (2001) Validation of a single nucleotide polymorphism genotyping assay for the human serum paraoxonase gene using electronically active customized microarrays. *Clin. Biochem.* **34,** 589–592.

II

Noninvasive Approaches

14

RHD Genotyping from Maternal Plasma: *Guidelines and Technical Challenges*

Neil D. Avent

Summary

Rhesus D (*RhD*) blood group incompatibility between mother and fetus can occasionally result in maternal alloimmunization where the resultant anti-D can cross the placenta and attack the fetal red cells, which in worse case scenarios can cause fetal anemia and ultimately death. Fetal *RHD* genotyping was introduced in the mid-1990s after the molecular characterization of the RH genes as an aid to the clinical management of these cases. Initially, these tests used fetal DNA extracted invasively from chorionic villus and amniocyte samples.

RHD genotyping of fetuses carried by *RhD*-negative women has become the first large-scale application of noninvasive prenatal diagnosis (NIPD). Initially the real-time polymerase chain reaction (PCR)-based tests were devised to characterize free fetal DNA in maternal plasma and serum, and *RHD* genotyping was a convenient assay to develop this exciting new technology, because the accuracy of tests could easily be confirmed after the simple *RhD* phenotyping of fetal cord blood cells after birth. “First generation” *RHD* genotyping tests were based on the incorrect concept that all D-negative phenotypes were caused by a complete *RHD* gene deletion. Thus, it was a relatively simple task to develop diagnostic PCR strategies based on the detection of *RHD* where D-negative genomes will completely lack *RHD*. Subsequent research into the molecular basis of D-negative phenotypes revealed that a significant number of D-negative genomes possess fragments of, or mutated *RHD* genes, the most notable of which is the *RHD* pseudogene found in Africans. Thus, more comprehensive *RHD* genotyping tests have evolved to differentiate these alleles, and are more appropriate in the diagnosis of multi-ethnic population groups such as those found in Europe and North America.

Many European Union countries have suggested the mass application of *RHD* NIPD for all fetuses carried by D-negative women. This is of clear benefit, because most *RhD* prophylaxis programs have switched to antenatal administration. This will help conserve anti-D stocks and it will prevent unnecessary administration of a human-derived blood

From: *Methods in Molecular Biology, vol. 444: Prenatal Diagnosis*
Edited by: S. Hahn and L. G. Jackson © Humana Press, Totowa, NJ

product to a vulnerable patient group. Although anti-D stocks are inherently safe, there is a moral obligation to eliminate unnecessary administration of it because there have been instances of hepatitis C infection due to contamination. Furthermore, as yet undescribed viruses may be contaminants of blood products. Mass-scale *RHD* NIPD will shortly be implemented in several countries in the European Community as a consequence.

Key Words: *RHD* genotyping; fetal genotyping; hemolytic disease of the newborn and fetus.

1. Introduction

Despite the huge success of the prophylactic anti-D program of the 1960s, there is an ongoing desire to further improve and eventually eliminate maternal alloimmunization to the *RhD* antigen completely. The low incidence of hemolytic disease of the newborn and fetus (HDNF) is largely due to a comprehensive screening program to identify at risk mothers during pregnancy ***(1)***. Such screening includes the monitoring of levels of maternal anti-blood group-specific antibodies (especially anti-D and anti-c and to a lesser extent anti-K). It is worthwhile noting here that anti-c and anti-E can cause hemolytic disease of the newborn, and testing for paternal inheritance of these antigens should be considered as part of the arsenal of tests available to the diagnostic immunohematology laboratory involved in HDNF investigations. For that matter, anti-G (an antigen expressed only on D-positive and C-positive red cells) has been known to cause HDNF ***(2)***. Thus, D-negative fetuses that carry the Cde (r') (i.e., that are C,G-positive) gene complex also should be considered. The frequency of the Cde complex in Caucasian populations is 1%; thus, within a D-negative cohort being investigated, its frequency will be much higher.

The detection of elevated levels of anti-D in D-negative mothers automatically triggers a series of events whereby further assessment of potential fetal anemia can be monitored (e.g., ultrasound for middle cerebral artery blood flow velocity). If significant anemia is detected (e.g., detection of hydrops), then fetal transfusion will be performed using antigen-negative blood by the process of cordocentesis. Rhesus D (*RhD*) is by far the major cause of HDNF, and it is the most immunogenic protein-based blood group antigen in humans. This is because *RhD*-negative persons have mutations or completely lack the *RHD* gene that encodes the *RhD* protein, but almost all individuals have on the surface of their erythrocytes the polymorphic RhCcEe protein (for reviews, *see* **refs. *(3–8)***). RhCcEe and *RhD* proteins have similar structures, but they differ by 35 to 36 amino acid exchanges (dependent on Rh C/c and E/e blood group status). Only nine of these changes are predicted to lie in extracellular positions, notably lining the inside of an aqueous (dubbed *antigenic*) extracellular-localized vestibule *(3,9)*. As with all immune responses to antigen, it has an absolute requirement for T cell responses, and T cell epitopes, of the D antigen have been mapped as a prelude to predict tolaragens to prevent alloimmunization in women of child bearing age *(10,11)*. These tolaragenic peptides may be fed to prepubescent D-negative females to induce tolerance to the D antigen. Until such treatment regimes are implemented, procedures for screening for potential alloimmunization of fetuses will be the frontline diagnosis that *RhD*-negative mothers will rely on. *RhD*-negative women have their sera tested for the presence of allo-anti-D, and if it is detected and reaches significant levels, this will lead to the consideration of invasive procedures, including in worse-case scenarios cordocentesis. One of the first clinical applications after the cloning of RhCcEe *(12,13)* and *RhD (14,15)* proteins was the implementation of *RHD* genotyping based on structural differences between the *RHCE* and *RHD* genes. Where fetuses have been identified by genotyping as *RhD* negative, then no further intervention is necessary.

RHD genotyping has for over a decade changed the clinical management of *RhD* alloimmunization *(16–18)*. Rapid definition of fetal *RHD* status can permit early intervention by in utero transfusion by cordocentesis in extreme cases, or it may inform if the mother requires the administration of prophylactic anti-D during pregnancy. It is of great personal satisfaction that mass-scale *RHD* genotyping is imminent and undoubtedly leads to further reductions in neonatal and perinatal deaths due to fetomaternal *RhD* incompatibility, because I was involved in the initial characterisation of the Rh protein species and their genes *(12,13,19–21)*. *RHD* genotyping was first implemented for the analysis of fetal DNA derived from invasively sampled material—extracted from amniocytes *(22)*. The procedure of amniocentesis was used routinely for Liley curve investigations of predicting the severity of HDNF; as a consequence, the cellular material containing fetal DNA was routinely discarded before OD measurement. From this, as soon as the cDNA encoding a *RhD* protein was characterized *(14,15)*, and its gene (*RHD*) was found to be absent from the majority of Caucasian D-negative genomes *(23)*, polymerase chain reaction (PCR)-based assays could be used to differentiate the presence or absence of *RHD (22)*.

2. Complexity of Rh and *RhD* Antigenicity

Rh antigen expression involves two genes, *RHD* and *RHCE*, located on chromosome 1p34-36 (*see* **Fig. 1**), and the essential coexpression of an Rh-associated glycoprotein (RhAG), whose coding gene lies on 6p11-21.1 *(24)*. *RHCE, RHD*, and *RHAG* are related gene families, with *RHCE* and *RHD* having 92% nucleotide sequence similarity and 96% amino acid sequence identity.

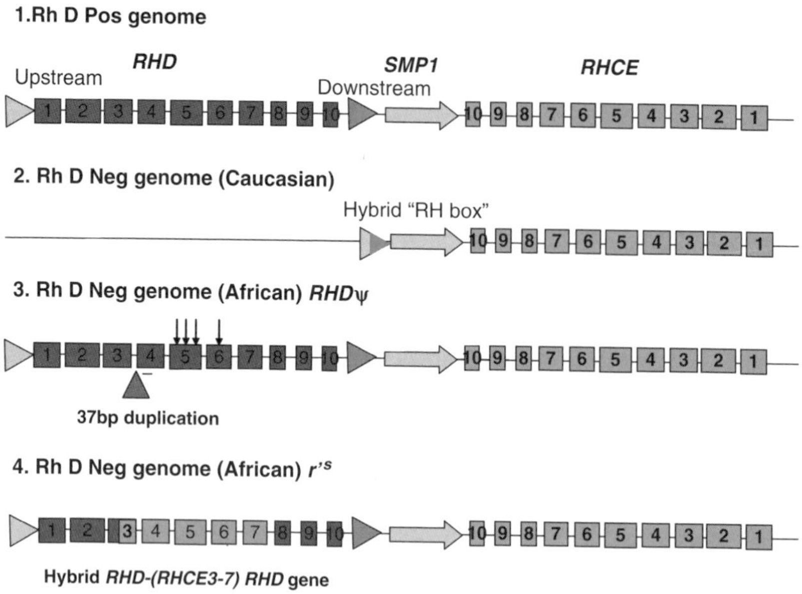

Fig. 1. RH locus structures of the common *RhD* phenotypes. Illustrated are the genomic structures of the RH genes (located at 1p34-36) in 1) normal D-positive individuals; 2) most Caucasian D negatives and a high proportion (approx 20%) of Europeans and North Americans of African descent; 3) *RHD* pseudogene found in individuals of African descent; 4) gene structure of the hybrid *RHD-RHCE-RHD* gene found in r'^{s} (Cde^s) phenotypes; and 5) *RHCE* gene (and derived in r'^{s}) exons (depicted as blue boxes) and *RHD* (red). Mutations in *RHD*Ψ (missense in exon 5, nonsense in exon 6) are depicted as arrows.

RhAG has 40% amino acid conservation to RhCE/*RhD* proteins. In the majority of Caucasian individuals, the D-negative phenotype is caused by a deletion of *RHD*, probably caused by unequal crossover between two homologous *Rhesus box* sequences that lie upstream and downstream of the *RHD* gene *(25)*. In D-negative phenotypes, a hybrid *Rhesus box*, comprising the 5′ segment of an upstream box and the 3′ segment of the downstream box is found (**Fig. 1**). It was initially hoped that detection of the deleted *RHD* gene could simplify *RHD* zygosity assignment (used to predict homozygosity or heterozygosity for *RHD* of fathers). Unfortunately, there are multiple breakpoints within the hybrid *Rhesus box* structures (named HRHB1-5), which makes the use of diagnostic assays in such cases complicated and not feasible for routine use *(26,27)*.

3. *RHD* Genotyping: Essential Due to Move to Use Antenatal Prophylactic Anti-D?

Many countries in the European Union (EU) have made a policy decision to implement the antenatal administration of prophylactic anti-D at 28 and 32 weeks gestation *(28)*. In the UK, in particular, the Royal College of Obstetricians and Gynaecologists and the National Institute of Clinical Excellence have published a series of guidelines for the implementation of this antenatal prophylaxis *(29)*. Antenatal prophylaxis helps prevent alloimmunization caused by fetomaternal hemorrhage during the third trimester of pregnancy. This is known to be a major cause of alloimmunization subsequent to the introduction of prophylactic anti-D in the 1960s. In The Netherlands, large-scale investigations with 2,359 maternal plasma samples have proven that *RHD* genotyping is viable as a diagnostic tool, with an accuracy of 99.4% *(30)*. Of these samples, 1,257 could have the PCR result directly compared with cord blood serology. Of the 2,359 samples, 15 samples were found to have *RHD* alleles, but the mothers were serologically D negative, 6 of these mothers carried the *RHD* pseudogene (*see* **Section 4**) These samples were not included in the cohort of 1,257 samples, because the PCR assay used (*RHD* exon 7) does not differentiate these alleles. Amongst this cohort, 1,249 results were concordant, with three false-negative and five false-positive results obtained. This Dutch study represents the largest cohort size to date, but similar findings have been observed in the UK (International Blood Group Reference Laboratory [IBGRL]) where 347 of 359 cases were correctly typed. Of the 12 not typed, there was insufficient fetal DNA present to enable reporting of results *(31)*. In France, a study involving 851 plasma samples from pregnant women has been reported *(32)*. Here, 649 of 654 (99%) fetuses were correctly diagnosed as being *RHD* positive, the three false-positive results not being defined except in one case where in vitro fertilization occurred, and the authors presumed that the false positive was a lost triplet. Of the 197 samples where a D-negative phenotype fetus was predicted by genotyping, four samples were false-negative results, due to low fetal DNA concentrations in the plasma samples. On repetition, all four samples were correctly typed. The most accurate of these recent studies (but also the smallest) is published by Gautier et al. *(33)*. Here, they described the analysis of 283 pregnancies to D-negative women; 170 were correctly assigned as D-positive fetuses and 113 as D-negative.

For mass-scale genotyping, false-negative results can be somewhat tolerated, because they will only lead to the unnecessary administration of anti-D to mothers. However, with the lack of a nonhuman blood product alternative for prophylactic anti-D, the potential for contamination of blood products with variant Creutzfeld Jacob Disease remains a remote possibility; so, measures should be considered to minimise false *RHD* genotyping results. Van der Schoot et al. *(30)* propose that if *RHD* DNA is detected in serologically *RhD*-negative

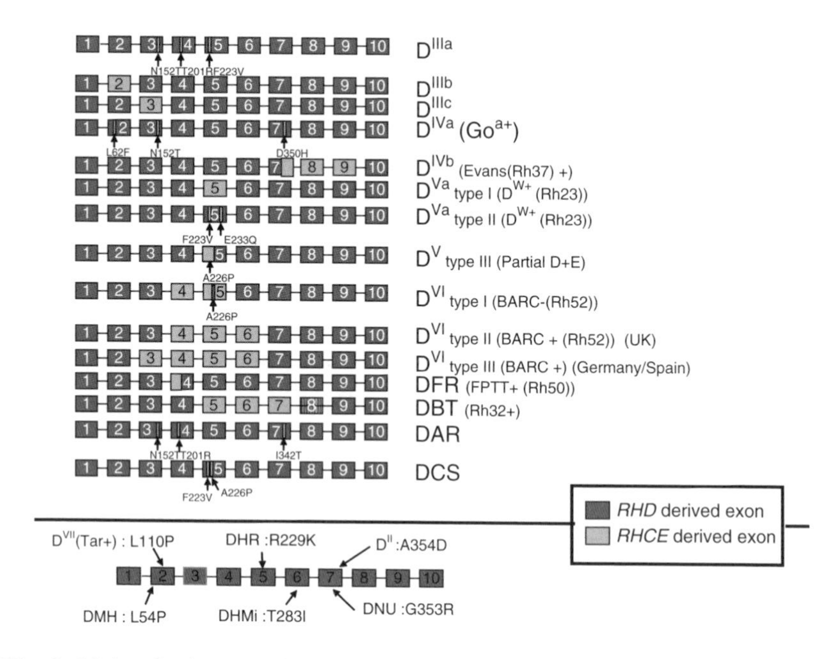

Fig. 2. Molecular bases underlying some partial D phenotypes. Depicted are some of the first described partial D phenotypes, generated to two different genetic mechanisms: hybrid *RHD-RHCE* genes (*top*) and by SNPs in the *RHD* gene (*bottom*). *RHD* exons are represented as red boxes and *RHCE* as yellow. The pertinent amino acid exchanges that result from the SNP are indicated by an arrow, as are the partial *RHCE*-to-*RHD* exchanges. The name of each partial D is indicated to the right of the panel and indicates which low-frequency Rh antigen the phenotype expresses. A full list of partial D alleles is available at http://www.uniulm.de/%7Efwagner/RH/RB/).

maternal blood samples at a concentration of >1,500pg/ml, then an assessment of the maternal genotype should be performed. If the mother is serologically *RhD* negative but genotypically *RHD* positive, then the basis of the *RHD* allele causing maternal D-negativity should be ascertained. Some of the *RHD* alleles that cause D-negative phenotypes and will produce false-positive results are illustrated in **Table 2**, and some partial D alleles that will cause false-negative results are shown in **Fig. 2**. It is, therefore, important that the variation in RH genetics that cause discrepant *RHD* genotyping are cataloged and understood. Sections 4 and 5 discuss what variation at the genetic level can cause false positive and negative *RHD* genotyping results and how laboratories performing such diagnoses should be alert to their existence.

Before such detailed discussions into the possible reasons for obtaining spurious results when analyzing *fetal* DNA, it is worth considering the possibility that the mother may be a variant Rh phenotype. In the UK, the *RhD* category VI (DVI) phenotype occurs at an approximate frequency of 1:4,000, and female DVI phenotype individuals are deliberately serotyped as D negative so that they receive prophylactic anti-D during and after their pregnancies ***(34)***. A mother with the DVI phenotype has been known to have a fetus with severe HDNF (actually a mortality) when carrying an infant with normal expression of the D antigen inherited from the father ***(35)***. It is, therefore, worth considering genotyping all mothers for potential expression of partial D alleles (or for that matter certain maternal weak D phenotypes that have been known to cause similar issues). The positive signals obtained for certain *RHD* exons in such cases would be explained by the variant *RHD* gene. A illustrative list of some partial D gene rearrangements is depicted in **Fig. 2**.

4. *RHD* Variation: The False-Positive Result

The genetic basis of individuals who are serologically *RHD* negative but have intact but dysfunctional *RHD* genes has been well characterized (for reviews, *see* **refs.** ***3,4,8,36,37,38,39***). They were first noted by Simsek et al. ***(40)*** and confirmed by others in the late 1990s ***(41–44)***. Extensive genomic analysis revealed that a large number of *RHD* alleles can cause D-negative phenotypes and that hybrid *RHCE-RHD* genes also cause D-negative phenotypes ***(45)***. In 2000, the second major genetic basis of the cause of a D-negative phenotype was described—the *RHD* pseudogene (*RHDψ*) found in individuals of African descent ***(46)***, furthermore an additional RH allele named r'^{s} (or Cde^s), which is a hybrid *RHD-RHCE-RHD* gene ***(47,48)*** is another cause of the D-negative phenotype in Africans (*see* **Fig. 1**). It is now considered critical that *RHD* genotyping assays are configured to detect wild-type *RHD* genes and do not detect *RHDψ*. For this reason, the "first generation" *RHD* genotyping assays, designed on the premise that the *RHD* gene was always deleted in D-negative genomes, are not effective in populations where African genes are likely to be present, namely, all of Northern Europe and America. Assays that avoid the detection of *RHDψ* have been described and used for the detection of D-positive infants carried by *RHDψ*-positive, serologically D-negative mothers ***(49)***. Further variations of *RHD* causing D-negative phenotypes are known; although they occur at a significantly low frequency, nevertheless they should not be discounted (*see* **Table 1**).

5. *RHD* Variation: The False-Negative Result

Although the disadvantages of incorrectly genotyping a fetus as *RHD* positive have been discussed, potentially disastrous consequences may befall a fetus incorrectly genotyped as *RHD* negative. The partial D phenotypes are the most

Table 1
Some known D-negative alleles described in the literature. For a full and current list, refer to the Rhesus site at http://www.uni-ulm.de/%7Efwagner/RH/RB/

D-negative allele	Nucleotide	Molecular basis	Reference
D negative		*RHD* gene deleted	***(23,42)***
*RHCE(1–9)-D(**10**)*	Hybrid gene	*RHCE* (1–9) *D **(10)***	***(45)***
RHD-CE(2–10)	Hybrid gene	*RHD **(1)** RHCE* (2–10)	***(68,69)***
RHD-CE(2–9)-D	Hybrid gene	*RHD **(1)** RHCE* (2–9) *RHD **(10)***	***(44,45)***
RHD-CE(2–8)-D	Hybrid gene	*RHD **(1)** RHCE* (2–8) *RHD* (9–10)	***(68)***
RHD-CE(2–7)-D	Hybrid gene	*RHD **(1)** RHCE* (2–7) *RHD* (8–10)	***(45)***
RHD-CE(3–7)-D	Hybrid gene	*RHD* (1–2)*RHCE* (3–7)*RHD* (8–10)	***(47,48,70)***
RHD-CE(4–7)-D	Hybrid gene	*RHD* (1–3) *RHCE* (4–7) *RHD* (8–10)	***(71)***
RHD-CE(4–8)-D	Hybrid gene	*RHD* (1–3) *RHCE* (4–8) *RHD* (9–10)	***(68)***
RHD(1–6)-CE (7–10)	Hybrid gene	*RHD* (1–6) *RHCE* (7–10)	
RHD-CE(8–9)-D	Hybrid gene	*RHD* (1–7) *RHCE* (8–9) *RHD **(10)***	***(72)***
RHD psi	G609A; G654C; T667G; C674T; T807G	37bp duplication at int3/ex 4 junction; stop codon in exon 6	***(46)***
RHD(W16X)	G48A	W16X	***(45)***
RHD(Q41X)	C121T; T643C; T646C; T988C	Q41X	***(42)***
RHD(W90X)	G270A	W90X	***(73)***
RHD(488del4)	Deletion of 4 bp from 488	M167X	***(43)***
RHD(del600)	Deletion 600G (CML patient)	L228X	***(74)***
RHD(G212V)	G635T	G212V	***(45)***
RHD(del711)	Deletion 711C	***(71)***	
RHD(906instggct, IVS6+1del4)	Insertion and deletion	IVS6+2T>A	***(73)***

(Continued)

Table 1
(Continued)

RHD(G314V)	G941T	G314V	***(75)***
RHD(Y330X)	C990G	Y330X	***(45)***
RHD(IVS8+1G>A)	IVS8+1G>A	***(45)***	
RHD(Y311X)	C933A	Y311X	***(76)***
RHD(IVS2+1G>A)	IVS2+1G>A	***(75)***	
RHD(W185X)	G554A	W185X	***(77)***
RHD(Y401X)	T1203A	Y401X	***(68)***

common cause of this type of *RHD* genotyping discrepancy, many of which are caused by hybrid *RHD-RHCE* genes (*see* **Fig. 2**). However, it also must be noted that a large number of *RHD* variants (weak D and partial D) are caused by single-nucleotide polymorphisms (SNPs) in the *RHD* gene, which could lead to false-negative results if the mutation was located within the PCR primer or probe binding sites. Fortunately, most weak D and partial D phenotypes are associated with significantly depressed levels of D antigen expression ***(50)***, and they are thus poorly immunogenic during pregnancy. However, some have high levels of D antigen sites; thus, they are potentially immunogenic—such known variants include the DVa and DVI type 3 ***(51)*** phenotypes. Fatal HDNF has been reported in at least one case of a DVa infant carried by a D-negative mother ***(52)***. In this instance, if an *RHD* genotyping assay were directed solely against exon 5 of the *RHD* gene, a false-negative result would be obtained, because the DVa gene is an *RHD-RHCE(5)-RHD* hybrid or has mutations in *RHD* exon 5 ***(4)***. HDNF also has been reported due to alloimmunization against low-frequency Rh antigens, including anti-Go^a (Gonzales) ***(53)*** and anti-Tar ***(54)***; thus, paternal inheritance of these antigens, although very infrequent, must not be ignored. Tar and Gonzales (Go^a) antigens are expressed by novel structures arising from the partial D genes DVII and DIVa, respectively. In addition, the extremely rare Rh and D—phenotypes can produce an antibody (anti-Rh 29), which also has been known to cause HDNF ***(55–57)***. Both these phenotypes have normal *RHD* genes, and they would have incorrect presumed phenotypes. It would be therefore prudent to test the maternal serum for the presence of anti-Rh29 if the mother was typed as *RhD* negative and Rhc negative. If anti-Rh29 is detected, then the potential rare maternal Rh null and D—phenotypes should be further investigated by specialised laboratories.

The false-negative result in *RHD* genotyping has always raised concerns as to whether fetal DNA is present the sample, because studies mentioned previously that obtained false-negative results had low levels of fetal DNA. For this reason, if in doubt, it is prudent to consider not reporting the result at all (as in the IBGRL), or to repeat the test. Almost all publications for *RHD* genotyping indicate that all samples are analyzed in triplicate and also include a control to monitor the levels of fetal DNA in the sample. The Y-chromosome gene *SRY* is used most commonly, and it is a better choice than the other commonly used Y-chromosome marker DYS14, because this target is multicopy ***(58)*** and would be detected more readily that the single copy *RHD* and SRY genes. In pregnancies where D-negative female fetuses were being carried, the male-specific gene targets are clearly inappropriate. The International Society of Blood Transfusion first workshop on molecular blood group genotyping concluded that in the absence of an internal control it is unwise to consider reporting such results as D negative ***(59)***. In such cases, maternal buffy coat DNA may be used to define potential paternal insertion/deletion alleles that may be analyzed to indicate the presence of fetal material in the plasma sample. Such insertion/deletions have been described and have previously been used in the assessment of minimal residual disease in hematological malignancy ***(60)***. Such tests are performed routinely for *diagnostic* analysis (i.e., in instances where there has been significant maternal alloimmunization), but not for screening strategies to identify D-negative mothers that require antenatal administered prophylactic anti-D. The "ultimate" controls to confirm the presence of fetal DNA may be hyper- and hypomethylated promoter sequences of fetal genes, so called epigenetic modifications of DNA. These sequences can be amplified after bisulfite conversion, and methylation-specific PCR amplification (methyl cytosines are converted to thymidine in this reaction). The *Maspin* gene promoter has been found to be hypomethylated in placental tissue, and it may be amenable as a maternal plasma internal control ***(61)***. However, significant developments must take place to automate and improve the bisulfite conversion process that requires significant amounts of maternal plasma to be functional, before routine implementation can be considered. Analysis of restriction modified DNA, specific for fetal methylated targets has recently proved to be a viable approach for assessment of the presence of fetal DNA within a given sample. For example the gene RASSF1A has been used Chang et al. ***(62)***.

6. Real-Time PCR-Based *RHD* Genotyping

RHD was used as a test-bed to prove the effectiveness of detecting a paternally inherited allele carried by circulating free fetal DNA in maternal plasma ***(63)***. The initial assay used the known difference between the *RHCE* and *RHD* genes in exon 10. Unfortunately, of all the published *RHD* assays, *RHD* exon 10 is the least accurate, because it will provide false-positive results in a number of different variant *RHD* genes, including *RHDψ*. As mentioned, at least two assays should be considered for routine *RHD* diagnosis to avoid false-negative results, especially from infants that carry *RHD* alleles encoding

Table 2
***RHD* PCR primers used in SAFE maternal plasma workshops. PCR primers and associated TaqMan probes used in the SAFE maternal plasma *RHD* genotyping workshops are displayed. The sequences are derived from Faas et al. *(78)* (superscript 1) and Finning et al. *(49)* (superscript 2). The report of the outcomes of the SAFE *RHD* workshops and recommendations for implementation of noninvasive *RHD* genotyping by using real-time PCR are described by Legler et al. (in preparation)**

Primer name	Sequence	Label	*RHD* region
RHD 940S[1]	GGGTGTTGTAACCGAGTGCTG	None	Exon 7
RHD 1064R[1]	CCGGCTCCGACGGTATC	None	Exon 7
RHD 968T	CCCACAGCTCCATCATGGGCTACAA	FAM-TAMRA	Exon 7
RHD ex5F[2]	CGCCCTCTTCTTGTGGATG	None	Exon 5
RHD ex5R[2]	GAACACGGCATTCTTCCTTTC	None	Exon 5
RHD ex5T	TCTGGCCAAGTTTCAACTCTGCTCTGCT	VIC-TAMRA	Exon 5

partial D antigens. To this end, the Special Advances in Fetal and Neonatal Evaluation (SAFE) European Community (EC)-funded network of excellence (www.safenoe.org/) have recommended the use of *RHD* exon 5 and exon 7 as routine screening assays ***(64)***. The *RHD* exon 5 assay is based on that of Finning et al. ***(49)***, and although not as sensitive as other *RHD* assays, it does not misinform giving false-positive results when *RHD*ψ is present from the maternal genome. A summary of the primers and probes recommended for use by the SAFE consortium is illustrated in **Table 2**.

7. Future *RHD* Genotyping Assays Using Maternal Plasma as Source of DNA?

Maternal plasma-based *RHD* genotyping has been successfully implemented in many European laboratories for use as a noninvasive prenatal test where alloimmunization has occurred. In all these organisations, real-time PCR has been the choice diagnostic test used, because of its ease of use, sensitivity, published assays, and ability to automate and thus avoid contamination issues. Many studies have shown the high degree of technical accuracy that can be achieved in these circumstances. Despite this widespread acceptance, there must always be the understanding that variant *RHD* alleles may confound the genotyping and that comprehensive *RHD* genotyping is not yet fully feasible technologically. Mass spectrometry (e.g., the technology provided by Sequenom, Inc., San Diego, CA) to detect SNPs may provide another alternative approach to prenatal *RHD* genotyping. Such assays meet the criteria of being high throughput. Moreover, they are sensitive enough to work from maternal plasma, and they can be multiplexed to detect multiple *RHD*

SNPs. Clearly, because Sequenom, Inc. has recently acquired intellectual property pertinent to maternal plasma diagnosis, they will have an interest in mass-scale *RHD* genotyping and associated technology. The multiplex ligation-dependent probe amplification system (MLPA) also has significant potential for application to maternal plasma-based typing. A prototype MLPA-based *RHD* genotyping assay has recently been described ***(65)***, and because the MLPA can be multiplexed may produce a viable alternative to real-time PCR with a higher degree of accuracy. Alternate approaches may be to use a specific *RHD* gene chip (some such chips have been described in the literature). The most comprehensive of these chips are produced by Progenika Biopharma (Derio, Vizcaya, Spain), which was the result of a 3-year EC-funded project BloodGen (www.bloodgen.com). Gene chips require relatively large amounts of target genomic DNA; therefore, they are not directly applicable to maternal plasma-based testing. Global DNA amplification protocols (e.g., AmpliPhi), that use DNA polymerase obtained from ϕ29, are not appropriate for analysis of maternal plasma, because it is highly fragmented ***(66,67)***. Thus, global DNA amplification protocols suitable for maternal plasma analysis need to be adapted. The high degree of accuracy possible for gene chip analysis targeting-known *RHD* alleles makes this a possibility, although fetal DNA-specific targets also should be included in such analysis.

In conclusion, *RHD* genotyping from maternal plasma should now be done exclusively in cases of maternal *RhD* alloimmunization. There is absolutely no case for still performing invasive diagnosis, indeed an international service is available at IBGRL to perform testing if the necessary infrastructure is unavailable. Invasive diagnosis, of course, exasperates maternal alloimmunization due to further exposure of the mother to fetal haemorrhage. It is also imperative that laboratories performing such *RHD* diagnosis include internal controls to confirm the presence of fetal DNA in every sample. Mass-scale diagnosis may not necessarily need such controls, because the high degree of accuracy of the testing may obviate such requirements according to legislative requirements. Nevertheless, efforts must be explored to improve the accuracy of such tests by producing assays that detect all known *RHD* alleles, especially those causing false-negative results. As mentioned in the penultimate paragraph this may not necessarily involve real-time PCR-based assays, currently the first choice diagnostic assay in immunohematology laboratories.

References

1. Urbaniak, S. J. and Greiss, M. A. (2000) RhD haemolytic disease of the fetus and the newborn. *Blood Rev.* **14,** 44–61.
2. Hadley, A. G., Poole, G. D., Poole, J., Anderson, N. A., and Robson, M. (1996) Haemolytic disease of the newborn due to anti-G. *Vox Sang.* **71,** 108–112.

3. Avent, N. D., Madgett, T. E., Lee, Z. E., Head, D. J., Maddocks, D. G., and Skinner, L. H. (2006) Molecular biology of Rh proteins and relevance to molecular medicine. *Expert Rev. Mol. Med.* **8,** 1–20.
4. Avent, N. D. and Reid, M. E. (2000) The Rh blood group system: a review. *Blood* **95,** 375–387.
5. Van Kim, C. L., Colin, Y., and Cartron, J. P. (2005) Rh proteins: key structural and functional components of the red cell membrane. *Blood Rev.* **20,** 93–110.
6. Huang, C. H. and Liu, P. Z. (2001) New insights into the Rh superfamily of genes and proteins in erythroid cells and nonerythroid tissues. *Blood Cells Mol. Dis.* **27,** 90–101.
7. Wagner, F. F. and Flegel, W. A. (2004) Review: the molecular basis of the Rh blood group phenotypes. *Immunohematology* **20,** 23–36.
8. Westhoff, C. M. (2004) The Rh blood group system in review: a new face for the next decade. *Transfusion* **44,** 1663–1673.
9. Conroy, M. J., Bullough, P. A., Merrick, M., and Avent, N. D. (2005) Modelling the human rhesus proteins: implications for structure and function. *Br. J. Haematol.* **131,** 543–551.
10. Stott, L. M., Barker, R. N., and Urbaniak, S. J. (2000) Identification of alloreactive T-cell epitopes on the Rhesus D protein. *Blood* **96,** 4011–4019.
11. Hall, A. M., Cairns, L. S., Altmann, D. M., Barker, R. N., and Urbaniak, S. J. (2005) Immune responses and tolerance to the RhD blood group protein in HLA-transgenic mice. *Blood* **105,** 2175–2179.
12. Cherif-Zahar, B., Bloy, C., Van Kim, C. L., et al. (1990) Molecular cloning and protein structure of a human blood group Rh polypeptide. *Proc. Natl. Acad. Sci. USA* **87,** 6243–6247.
13. Avent, N. D., Finning, K. M., Martin, P. G., and Soothill, P. W. (1990) cDNA cloning of a 30 kDa erythrocyte membrane protein associated with Rh (Rhesus)-blood-group-antigen expression. *Biochem. J.* **271,** 821–825.
14. Arce, M. A., Thompson, E. S., Wagner, S., Coyne, K. E., Ferdmann, B. A., and Lublin, D. M. (1993) Molecular cloning of RhD cDNA derived derived from a gene present in *RhD*-positive, but not *RhD*-negative individuals. *Blood* **82,** 651–655.
15. Van Kim, C. L., Mouro. I., Cherif-Zahar, B., et al. (1992) Molecular cloning and primary structure of the human blood group RhD polypeptide. *Proc. Natl. Acad. Sci. USA* **89,** 10925–10929.
16. Avent, N. D. (2002) Fetal genotyping. In: Hadley, A. G. and Soothill, P. W. (eds.). Alloimmune disorders of pregnancy, pp 121–139. Cambridge University Press, Cambridge, UK.
17. Avent, N. D., Finning, K. M., Martin, P. G., and Soothill, P. W. (2000) Prenatal determination of fetal blood group status. *Vox Sang.* **78,** 155–162.
18. Avent, N. D. (1998) Antenatal genotyping of the blood groups of the fetus. *Vox Sang.* **74,** 365–374.
19. Bloy, C., Blanchard, D., Dahr, W., Beyreuther, K., Salmon, C., and Cartron, J. P. (1988) Determination of the N-terminal sequence of human red cell Rh(D)

polypeptide and demonstration that the Rh(D), (c), and (E) antigens are carried by distinct polypeptide chains. *Blood* **72,** 661–666.

20. Avent, N. D., Ridgwell, K., Wanner, M. J., and Anstee, D. J. (1988) Protein-sequence studies on Rh-related polypeptides suggest the presence of at least two groups of proteins which associate in the human red-cell membrane. *Biochem. J.* **256,** 1043–1046.
21. Saboori, A. M., Smith, B. L., and Agre, P. (1988) Polymorphism in the Mr 32,000 Rh protein purified from Rh(D)-positive and -negative erythrocytes. *Proc. Natl. Acad. Sci. USA* **85,** 4042–4045.
22. Bennett, P. R., Le Van Kim, C., Colin, Y., Warwick, R. M., Cherif-Zahar, R., Fisk, N. and Cartron, J. P. (1993) Prenatal determination of fetal RhD type by DNA amplification. *N. Engl. J. Med.* **329,** 607–610.
23. Colin, Y., Cherif-Zahar, F., Le Van Kim, C., Raynal, V., Van Huffel, V., and Cartron, J. P. (1991) Genetic basis of the RhD-positive and RhD-positive and *RhD*-negative blood group polymorphism as analysis. *Blood* **78,** 2747–2752.
24. Ridgwell, K., Spurr, N. K., Laguda, B., MacGeoch, C., Avent, N. D. and Tanner, M. J. (1992) Isolation of cDNA clones for a 50 kDa glycoprotein of the human erythrocyte membrane associated with Rh (rhesus) blood-group antigen expression. *Biochem. J.* **287,** 223–338.
25. Wagner, F. F. and Flegel, W. A. (2000) RHD gene deletion occurred in the Rhesus box. *Blood* **95,** 3662–3668.
26. Wagner, F. F., Moulds, J. M., and Flegel, W. A. (2005) Genetic mechanisms of Rhesus box variation. *Transfusion* **45,** 338–344.
27. Matheson, K. A. and Denomme, G. A. (2002) Novel 3′ Rhesus box sequences confound RHD zygosity assignment. *Transfusion* **42,** 645–650.
28. Engelfriet, C. P., Reesink, H. W., Judd, W. J., et al. (2003) Current status of immunoprophylaxis with anti-D immunoglobin. *Vox Sang.* **85,** 328–337.
29. Urbaniak, S. J. (1998) Consensus conference on anti-D prophylaxis, April 7 & 8, 1997: final consensus statement. Royal College of Physicians of Edinburgh/Royal College of Obstetricians and Gynaecologists. *Transfusion.* **38,** 97–99.
30. Van der Schoot, C. E., Soussan, A. A., Koelewijn, J., Bonsel, G., Paget-Christiaens, L. G., and de Haas, M. (2006) Non-invasive antenatal RHD typing. *Transfus. Clin. Biol.* **13,** 53–57.
31. Finning, K., Martin, P. and Daniels, G. (2004) A clinical service in the UK to predict fetal Rh (Rhesus) D blood group using free fetal DNA in maternal plasma. *Ann. N Y Acad. Sci.* **1022,** 119–123.
32. Rouillac-Le Sciellour, C., Puillandre, P., Gillot, R., et al. (2004) Large-scale pre-diagnosis study of fetal RHD genotyping by PCR on plasma DNA from RhD-negative pregnant women. *Mol. Diagn.* **8,** 23–31.
33. Gautier, E., Benachi, A., Giovangrand, Y., et al., (2005) Fetal RhD genotyping by maternal serum analysis: a two-year experience. *Am. J. Obstet. Gynecol.* **192,** 666–669.
34. Lubenko, A., Contreras, M., and Habash, J. (1989) Should anti-Rh immunoglobulin be given D variant women? *Br. J. Hematol.* **72,** 429–433.

35. Lacey, P. A., Caskey, C. R., Werner, D. J., and Moulds, J. J. (1983) Fatal hemolytic disease of a newborn due to anti-D in an Rh-positive Du variant mother. *Transfusion* **23,** 91–94.
36. Flegel, W. A. (2006) Molecular genetics of RH and its clinical application Genetique moleculaire sur systeme RH et ses applications cliniques. *Transfus. Clin. Biol.* **13,** 4–12.
37. Avent, N. D. (2001) Molecular biology of the Rh blood group system. *J. Pediatr. Hematol. Oncol.* **23,** 394–402.
38. Avent, N. D. (1999) The rhesus blood group system: insights from recent advances in molecular biology. *Transfus. Med. Rev.* **13,** 245–266.
39. Denomme, G. A. (2004) The structure and function of the molecules that carry human red blood cell and platelet antigens. *Transfus. Med. Rev.* **18,** 203–231.
40. Simsek, S., Bleeker, P. M. and von dem Borne, A. E. (1994) Prenatal determination of fetal RhD type. *N. Engl. J. Med.* **330,** 795–796.
41. Carritt, B., Steers, F. J., and Avent, N. D. (1994) Prenatal determination of fetal RhD type. *Lancet* **344,** 205–206.
42. Avent, N. D., Martin, P. G., Armstrong-Fisher, S. S., *et al.* (1997) Evidence of genetic diversity underlying Rh D-, weak D (Du), and partial D phenotypes as determined by multiplex polymerase chain reaction analysis of the RHD gene. *Blood* **89,** 2568–2577.
43. Andrews, K. T., Wolter, L. C., Saul, A., and Hyland, C. A. (1998) The RhD– trait in a white patient with the RhCCee phenotype attributed to a four-nucleotide deletion in the RHD gene. *Blood* **92,** 1839–1840.
44. Huang, C. H. (1996) Alteration of RH gene structure and expression in human dCCee and DCW-red blood cells: phenotypic homozygosity versus genotypic heterozygosity. *Blood* **88,** 2326–2333.
45. Wagner, F. F., Frohmajer, A., and Flegel, W. A. (2001) RHD positive haplotypes in D negative Europeans. *BMC Genet.* **2,** 10.
46. Singleton, B. K., Green, C. A., Avent, N. D., et al. (2000) The presence of an RHD pseudogene containing a 37 base pair duplication and a nonsense mutation in Africans with the Rh D-negative blood group phenotype. *Blood* **95,** 12–18.
47. Daniels, G. L., Faas, B. H., Green, C. A., et al. (1998) The VS and V blood group polymorphisms in Africans: a serologic and molecular analysis. *Transfusion.* **38,** 951–958.
48. Faas, B. H., et al. (1997) Molecular background of VS and weak C expression in blacks. *Transfusion* **37,** 38–44.
49. Finning, K. M., Martin, P. G., Soothill, P. W., and Avent, N. D. (2002) Prediction of fetal D status from maternal plasma: introduction of a new noninvasive fetal RHD genotyping service. *Transfusion* **42,** 1079–1085.
50. Jones, J. W., Lloyd-Evans, P., and Kumpel, B. M. (1996) Quantitation of Rh D antigen sites on weak D and D variant red cells by flow cytometry. *Vox Sang.* **71,** 176–183.

51. Wagner, F. F., Gassner, C., Muller, T. H., Schonitzer, D., Schunter, F., and Flegel, W. A. (1998) Three molecular structures cause rhesus D category VI phenotypes with distinct immunohematologic features. *Blood* **91,** 2157–2168.
52. Mayne, K., Bowell, P., Woodward, T., Sibley, C., Lomas, C., and Tippett, P. (1990) Rh immunization by the partial D antigen of category DVa. *Br. J. Haematol.* **76,** 537–539.
53. Leschek, E., Pearlman, S. A., Boudreaux, I., and Meek, R. (1993) Severe hemolytic disease of the newborn caused by anti-Gonzales antibody. *Am. J. Perinatol.* **10,** 362–364.
54. Levene, C., Sela, R., Grunberg, L., Gale, R., Lomas, C., and Tippett, P. (1983) The Rh antigen Tar (Rh40) causing haemolytic disease of the newborn. *Clin. Lab. Haematol.* **5,** 303–305.
55. Gabra, G. S., Bruce, M., Watt, A., and Mitchell, R. (1987) Anti-Rh 29 in a primigravida with rhesus null syndrome resulting in haemolytic disease of the newborn. *Vox Sang.* **53,** 143–146.
56. Lenkiewicz, B. and Zupanska, B. (2000) Moderate hemolytic disease of the newborn due to anti-Hr0 in a mother with the D–/D– phenotype. *Immunohematology* **16,** 109–111.
57. Lubenko, A., Contreras, M., Portugal, C. L., et al. (1992) Severe haemolytic disease in an infant born to an Rh(null) proposita. *Vox Sang.* **63,** 43–47.
58. Zimmermann, B., El-Sheikhah, A., Nicolaides, K., Holzgreve, W., and Hahn, S. (2005) Optimized real-time quantitative PCR measurement of male fetal DNA in maternal plasma. *Clin. Chem.* **51,** 1598–1604.
59. Daniels, G., van der Schoot, C. E., and Olsson, M. L. (2005) Report of the First International Workshop on molecular blood group genotyping. *Vox Sang.* **88,** 136–142.
60. Alizadeh, M., Bernard, M., Danic, B., et al. (2002) Quantitative assessment of hematopoietic chimerism after bone marrow transplantation by real-time quantitative polymerase chain reaction. *Blood* **99,** 4618–4625.
61. Chim, S. S., Tong, Y. K., Chiu, R. W., et al. (2005) Detection of the placental epigenetic signature of the maspin gene in maternal plasma. *Proc. Natl. Acad. Sci. USA* **102,** 14753–14758.
62. Chang, et al., (2006) Hypermethylated RASSF1A in maternal plasma: A universal fetal DNA marker that improves the reliability of noninvasive prenatal diagnosis. *Clin Chem.* **52,** 2211–2218.
63. Lo, Y. M., Corbetta, N., Chamberlain, P. F., et al. (1997) Presence of fetal DNA in maternal plasma and serum. *Lancet* **350,** 485–487.
64. Legler, et al. (2007) Workshop report on the extraction of foetal DNA from maternal plasma. *Prenatal Diagnosis* **27,** 824–829.
65. Madgett, T. E. (2005) Rapid molecular genotyping for RHD using multiplex ligation-dependent probe amplification technology. *Transfusion* **45,** 129A.
66. Chan, K. C., Zhang, J., Hui, A. B., et al. (2004) Size distributions of maternal and fetal DNA in maternal plasma. *Clin. Chem.* **50,** 88–92.

67. Li, Y., Di Naro, E., Vitucci, A., Zimmermann, B., Holzgreve, W., and Hahn, S. (2005) Detection of paternally inherited fetal point mutations for beta-thalassemia using size-fractionated cell-free DNA in maternal plasma. *J. Am. Med. Assoc.* **293,** 843–849.
68. Gassner, C., Doescher, A., Drnovsek, T. D., et al. (2005) Presence of RHD in serologically D–, C/E+ individuals: a European multicenter study. *Transfusion.* **45,** 527–538.
69. Shao, C. P. and Xiong, W. (2004) A new hybrid RHD-positive, D antigen-negative allele. *Transfus. Med.* **14,** 185–186.
70. Blunt, T., Daniels, G., and Carritt, B. (1994) Serotype switching in a partially deleted RHD gene. *Vox Sang.* **67,** 397–401.
71. Faas, B. H., Beckers, E. A., Simsek, S., et al. (1996) Involvement of Ser103 of the Rh polypeptides in G epitope formation. *Transfusion* **36,** 506–11.
72. Gassner, C., Schmarda A., Kilga-Nogler, S., et al. (1997) RHD/CE typing by polymerase chain reaction using sequence-specific primers. *Transfusion* **37,** 1020–1026.
73. Shao, C. P., Maas, J. H., Su, Y. Q., Kohler, M., and Legler, T. J. (2002) Molecular background of Rh D-positive, D-negative, D(el), and weak D phenotypes in Chinese. *Vox Sang.* **83,** 156–161.
74. Cherif-Zahar, B., Bony, V., Steffensen, R., et al. (1998) Shift from Rh-positive to Rh-negative phenotype caused by a somatic mutation within the RHD gene in a patient with chronic myelocytic leukaemia. *Br. J. Haematol.* **102,** 1263–1270.
75. Okuda, H., Kawano, M., Iwamoto, S., et al. (1997) The RHD gene is highly detectable in RhD-negative Japanese donors. *J. Clin. Invest.* **100,** 373–379.
76. Qun, X., Grootkerk-Tax, M. G., Maaskant-van Wijk, P. A., and van der Schoot, C. E. (2005) Systemic analysis and zygosity determination of the RHD gene in a D-negative Chinese Han population reveals a novel D-negative RHD gene. *Vox Sang.* **88,** 35–40.
77. Kim, J. Y., Kim, S. Y., Kim, C. A., Yon, G. S., and Park, S. S. (2005) Molecular characterization of D– Korean persons: development of a diagnostic strategy. *Transfusion* **45,** 345–352.
78. Faas, B. H., Beuling, E. A., Christiaens, G. C., von dem Borne, A. E., and van der Schoot, C. E. (1998) Detection of fetal RHD-specific sequences in maternal plasma. *Lancet* **352,** 1196.

15

Isolation of Cell-Free DNA from Maternal Plasma Using Manual and Automated Systems

Dorothy J. Huang, Susanne Mergenthaler-Gatfield, Sinuhe Hahn, Wolfgang Holzgreve, and Xiao Yan Zhong

Summary

Cell-free fetal DNA present in the maternal circulation holds great potential for noninvasive prenatal diagnosis and analysis of fetal genetic traits. However, only approximately 3–6% of total DNA in the maternal plasma is of fetal origin. Because of its scarcity in the maternal circulation, various methods have been developed and tested to optimize the extraction of this rare material from plasma. Here, we first describe the commonly used protocol for separating plasma from whole blood samples. We also describe two commercially available methods for the extraction of cell-free DNA from maternal plasma, which we have found particularly straightforward and easy to use: a manual method using the High Pure PCR Template Preparation kit (Roche Diagnostics) and an automated system using the MagNA Pure LC instrument (Roche Diagnostics). Use of the methods described here will help to ensure maximum yield and purity of cell-free fetal DNA extracted from maternal plasma samples for downstream analyses.

Key Words: Cell-free DNA; maternal plasma; prenatal diagnosis; DNA extraction.

1. Introduction

Since the discovery of cell-free fetal DNA in the plasma of pregnant women in 1997 by Lo et al. *(1)*, several protocols for isolating this rare material from the maternal circulation have been evaluated and optimized. Of major importance with any of the techniques used is the avoidance of contamination by cellular material during separation of plasma from other blood components. This was demonstrated in a study by Chiu et al. *(2)* in 2001 in which different blood-processing protocols were shown to contribute to variations in the quantification

From: *Methods in Molecular Biology, vol. 444: Prenatal Diagnosis*
Edited by: S. Hahn and L. G. Jackson © Humana Press, Totowa, NJ

of total free DNA in maternal plasma, due mainly to the efficacy of different protocols in removing contaminating cells. The standard protocol for isolating plasma, which is described here, involves two centrifugation steps. One step is at high speed to ensure the removal of cellular material from the plasma layer. After separation of the plasma, several different commercial kits have been used to isolate the cell-free DNA from this layer. The technique described here, a modified version of the High Pure PCR Template Preparation kit (Roche Diagnostics, Rotkreuz, Switzerland), involves use of spin-column technology to isolate and purify DNA from plasma. We have had much experience with this technique and have found it to be an easy-to-use and efficient method. Another popular commercial kit, not described here, is the QIAmp DNA Blood Mini kit (QIAGEN, Hilden, Germany), which uses similar technology.

Methods are also available for the extraction of cell-free DNA from plasma that involve the use of an automated machine such as the MagNA Pure LC instrument (Roche Diagnostics). Many different automated systems are currently available, and they are particularly useful when many samples need to be processed. Once the samples and solutions have been loaded into the machine, the rest of the procedure is automated; thus, the time needed and potential for errors in processing are reduced. We found that in our own comparison of the High Pure PCR Template Preparation kit and the MagNA Pure LC instrument (using the MagNA Pure LC DNA Isolation kit–Large Volume), the yield and purity of cell-free fetal DNA were improved through use of the automated instrument ***(3)***. Here, we describe the use of this particular automated system for the isolation of plasma DNA. With this system, up to 32 samples can be processed simultaneously within approximately 2.5–3 h, including set-up time.

2. Materials

2.1. Separation of Plasma from Whole Blood

1. No special solutions are required.

*2.2. High Pure PCR Template Preparation kit (cat. no. 11 796 828 001, Roche Diagnostics)**

1. Binding buffer: 6 M guanidine-HCl, 10 mM urea, 10 mM Tris-HCl, 20% Triton X-100 (v/v), pH 4.4.
2. Inhibitor removal buffer: 5 M guanidine-HCl, 20 mM Tris-HCl, pH 6.6; final concentration after addition of 20 ml of absolute ethanol to 33-ml volume.
3. Wash buffer: 20 mM NaCl, 2 mM Tris-HCl, pH 7.5; final concentration after addition of 80 ml of absolute ethanol to 20-ml volume.
4. Elution buffer: 10 mM Tris-HCI, pH 8.5.

5. Proteinase K, recombinant, polymerase chin reaction (PCR) grade; reconstituted in double-distilled H_2O (ddH_2O).

* All materials except for proteinase K are included in the High Pure PCR Template Preparation kit.

*2.3. MagNA Pure LC DNA Isolation kit–Large Volume (cat. no. 3 310 515, Roche Diagnostics)**

1. MagNA Pure LC instrument and software.
2. MagNA Pure LC disposable plasticware.
   a. MagNA Pure LC sample cartridge (cat. no. 3 004 112).
   b. MagNA Pure LC reagent tub (small) (cat. no. 3 004 066).
   c. MagNA Pure LC medium reagent tub 20 (cat. no. 3 004 058).
   d. MagNA Pure LC medium reagent tub 30 (cat. no. 3 045 501).
   e. MagNA Pure LC reagent tub (large) (cat. no. 3 004 040).
   f. MagNA Pure LC tub lid (small, medium) (cat. no. 3 004 082).
   g. MagNA Pure LC tub lid (large) (cat. no. 3 004 074).
   h. MagNA Pure LC reaction tip (large) (cat. no. 3 004 171).
   i. MagNA Pure LC processing cartridge (cat. no. 3 004 147).
   j. MagNA Pure LC tip stand (cat. no. 3 004 155).
   k. MagNA Pure LC waste bag (cat. no. 3 004 201).
   l. MagNA Pure LC cartridge seal (optional) (cat. no. 3 118 827).
3. Wash buffer I.
4. Wash buffer II.
5. Wash buffer III.
6. Lysis/binding buffer.
7. Magnetic glass particles (MGP) suspension.
8. Elution buffer.
9. Proteinase K buffer II.
10. Proteinase K.

* All solutions listed are included in the MagNA Pure LC DNA Isolation kit.

3. Methods

3.1. Separation of Plasma from Whole Blood

1. Blood should be collected directly in potassium-EDTA–containing tubes.
2. Centrifuge 10–20 ml of blood at 1600*g* for 10 min.
3. Carefully aspirate the upper plasma layer and transfer to new Eppendorf tubes, avoiding the buffy coat and red cell layers.
4. Centrifuge the plasma at 16,000*g* for 10 min (*see* **Note 1**).
5. Carefully aspirate the plasma and transfer to new Eppendorf tubes, avoiding the pellets that may be present at the bottom of the tubes.
6. Store plasma samples at –70°C until further use (*see* **Note 2**).

3.2. Plasma DNA Isolation Using a Modified High Pure PCR Template Preparation kit

1. Dilute the needed amount of elution buffer (*see* **Note 3**) to a 20% solution by using ddH_2O, and prewarm at 70°C.
2. Centrifuge thawed plasma at 16,000*g* for 5 min.
3. Carefully aspirate 400 µl of plasma (*see* **Note 4**) and transfer to a new 2-ml Eppendorf tube, avoiding any cellular debris.
4. Add 400 µl of binding buffer (*see* **Note 5**) and vortex briefly to mix.
5. Add 40 µl of reconstituted proteinase K (*see* **Note 6**).
6. Vortex briefly to mix and incubate for 10 min at 70°C.
7. After incubation, add 200 µl of 100% isopropanol and vortex well.
8. Transfer 500 µl to the upper reservoir of a High Pure Filter collection tube provided in the kit.
9. Centrifuge at 8,000*g* for 1 min at room temperature.
10. Discard the flow-through and collection tube and combine the filter with a new collection tube.
11. Repeat **steps 8–10** until the entire sample has been loaded onto the filter.
12. Add 500 µl of inhibitor removal buffer (*see* **Note 5**) to the upper reservoir.
13. Centrifuge at 8,000*g* for 1 min at room temperature.
14. Discard the flow-through and collection tube and combine the filter with a new collection tube.
15. Add 500 µl of wash buffer to the upper reservoir.
16. Centrifuge at 8,000*g* for 1 min at room temperature.
17. Discard the flow-through and collection tube, and combine the filter with a new collection tube.
18. Repeat **steps 15–17** for a second wash.
19. Centrifuge at maximum speed (approximately 13,000*g*) for 10 s to remove any residual wash buffer.
20. Transfer the filter to a new 1.5-ml Eppendorf tube; place in the incubator at 70°C for 5 min to prewarm.
21. Add 100 µl (*see* **Note 7**) of prewarmed 20% (*see* **Note 8**) elution buffer carefully to the filter.
22. Place tube and filter in the incubator at 70°C and shake at low speed (approximately 400 rpm) for 5 min.
23. Centrifuge at 8,000*g* for 5 min to elute the DNA.
24. Store DNA preferably at 4°C before use or freeze at –70°C for longer periods.

3.3. Plasma DNA Isolation Using the Automated MagNA Pure LC Instrument

1. Aliquot 1,000 µl (*see* **Note 9**) of each plasma sample to a well of the sample cartridge (*see* **Note 10**).
2. Turn on the instrument and computer, click on "Start Program," and then on "Sample Ordering."

3. Under the "Protocol" menu item "DNA LV," select the protocol "DNA LV Blood_1000.blk." The program will calculate the amount of each solution needed for DNA isolation and determine the amount of plasticware required.
4. Enter the sample names according to their locations in the sample cartridge and then click on "Start Batch."
5. Set up all plasticware for the instrument as indicated by the program.
6. Fill the reagent tubs with the solutions (*see* **Note 11**) as indicated by the program, except for the MGP solution. Cover each tub with the appropriate lid after it has been filled.
7. Vortex the MGP solution thoroughly (*see* **Note 12**) and then add the indicated amount to the appropriate reagent tub and cap it with a tub lid.
8. Place the sample cartridge with the plasma samples in the instrument and close the door.
9. Confirm that all plasticware and solutions needed have been correctly loaded into the machine and that all the corresponding positions on the screen have been clicked to indicate that each component has been filled.
10. Click on the "OK" button that shows up after all the items have been confirmed to begin the processing.
11. When the run is completed, transfer the DNA to clean 1.5-ml tubes or alternatively, cover the sample cartridge with a cartridge seal.
12. Store DNA at 4°C and use within a week, or at –20°C for longer storage.

4. Notes

1. Plasma separation: Although the second, high-speed centrifugation should pellet out any cellular material from the plasma, try to avoid disturbing the buffy coat layer after the first, low-speed centrifugation. It is better to take less plasma than include buffy coat leukocytes in the sample.
2. Store plasma samples in smaller aliquots (500–1,000 μl) to avoid having to refreeze samples.
3. Manual DNA isolation: We try to avoid preparing more than 12 samples at one time, because processing >12 samples becomes tedious and leads to longer intervals between steps.
4. Manual DNA isolation: For an increased DNA yield, use 800 μl of plasma and double the volumes of binding buffer, proteinase K, and 100% isopropanol (*see* **Subheadings 3.2.4., 3.2.5–3.2.7.**). These solutions should be combined in a tube that can hold at least a 3-ml volume, because 2-ml Eppendorf tubes will not accommodate the final volume.
5. The binding buffer and inhibitor removal buffer from the manual kit, and the lysis/binding buffer and wash buffer I from the automated kit contain guanidine salts, and they should be always handled with gloves. Standard safety precautions should be followed.

6. Reconstituted proteinase K should be stored at –20°C in small aliquots to avoid refreezing or waste; 500-μl volumes are ideal (enough for 12 × 400-μl plasma samples or 6 × 800-μl samples).
7. Manual DNA isolation: We have found that eluting the DNA from the filters into 50-μl volumes, as is sometimes done, may result in less DNA isolated, because this volume is insufficient to saturate the filter. When identical plasma samples were prepared, the concentrations of DNA in the 100-μl elutions based on real-time PCR analysis were overall comparable with or greater than those in 50-μl volumes, and with overall less variability in concentrations across the samples (data not shown). Therefore, we recommend eluting into 100-μl volumes (*see* **Subheading 3.2.**, **step 21**).
8. Manual DNA isolation: We recommend eluting DNA from the filters using a diluted buffer solution (20%), which will permit higher volumes of DNA to be used for PCR in case of low concentrations.
9. Automated DNA isolation: Plasma volumes other than 1,000 μl may be used. Be certain to select the protocol that corresponds to the sample volume.
10. Automated DNA isolation: When adding plasma samples to the sample cartridges, take care to note that the wells are labeled horizontally from right to left (H ←A).
11. Automated DNA isolation: When adding solutions to the reagent tubs, always cover the tubs immediately after filling them to avoid contamination with other solutions.
12. Automated DNA isolation: Always vortex the MGP solution before adding it to the reagent tub to ensure complete suspension of the glass particles. In addition, always add it last to minimize alcohol evaporation.

References

1. Lo, Y. M., Corbetta, N., Chamberlain, P. F., et al. (1997) Presence of fetal DNA in maternal plasma and serum. *Lancet* **350,** 485–487.
2. Chiu, R. W. K., Poon, L. L. M., Lau, T. K., Leung, T. N., Wong, E. M. C., and Lo, Y. M. D. (2001) Effects of blood-processing protocols on fetal and total DNA quantification in maternal plasma. *Clin. Chem.* **47,** 1607–1613.
3. Huang, D. J., Zimmermann, B. G., Holzgreve, W., and Hahn, S. (2005) Use of an automated method improves the yield and quality of cell-free fetal DNA extracted from maternal plasma. *Clin. Chem.* **51,** 2419–2420.

16

Fetal DNA: *Strategies for Optimal Recovery*

Tobias J. Legler, Klaus-Hinrich Heermann, Zhong Liu, Aicha Ait Soussan, and C. Ellen van der Schoot

Summary

For fetal DNA extraction, in principle each DNA extraction method can be used; however, because most methods have been optimized for genomic DNA from leucocytes, we describe here the methods that have been optimized for the extraction of fetal DNA from maternal plasma and validated for this purpose in our laboratories. The use of the QIAamp DSP Virus kit (QIAGEN), the QIAamp DNA Blood Mini kit (QIAGEN), and the Magna Pure LC (Roche) is based on the kit components provided by the respective companies. However, we noticed that the yield of fetal DNA from maternal plasma can be increased when higher volumes are processed or some slight modifications of the protocols provided by the manufacturer are followed. Here, we also describe an in-house method that allows the specific capture of target molecules in an extremely low volume by using magnetic beads and magnetic tips. This method can be either performed by hand, or it can be adapted to a commercially available pipetting workstation.

Key Words: Fetal; DNA; recovery; maternal plasma.

1. Introduction

The nucleic acid extraction procedure is most important for the detection of fetal nucleic acids in maternal plasma. A European workshop in 2005 hosted by the Network of Excellence Special Noninvasive Advances in Fetal and Neonatal Evaluation Network revealed that the QIAamp DSP Virus kit (QIAGEN, Hilden, Germany), the High Pure PCR Template Preparation kit (Roche, Basel, Switzerland), and an in-house protocol using the QIAamp DNA blood Mini kit (QIAGEN) were useful for the extraction. The CST genomic DNA purification kit (Invitrogen, Paisley, U.K.), the Magna Pure LC (Roche), and the MDx and

From: *Methods in Molecular Biology, vol. 444: Prenatal Diagnosis*
Edited by: S. Hahn and L. G. Jackson © Humana Press, Totowa, NJ

the EZ1 (both instruments from QIAGEN) were the methods that could be used for the extraction of fetal DNA with maternal plasma. Furthermore, an in-house protocol using magnetic beads for manual and automated extraction was suitable for the extraction of fetal DNA from maternal plasma (*see* **Note 1**) *(1)*. Here, we describe these methods in more detail, which have been validated in our laboratories.

2. Materials

For the commercial kits, each component used is provided by the manufacturer. Storage conditions and expiratory date of the manufacturer should be followed.

2.1. QIAamp DNA Blood Mini Kit

1. QIAamp Mini Blood kit (QIAGEN, cat. no. 51104).
2. Lysis buffer (AL) (QIAGEN, cat. no. 19075).
3. Collection tubes (2 ml) (QIAGEN, cat. no. 19201).

2.2. QIAamp DSP Virus Kit

1. QIAamp DSP Virus kit (QIAGEN, cat. no. 60704).
2. Protease (QP; QIAGEN, cat. no. 19155).
3. Collection tubes (2 ml) (QIAGEN, cat. no. 19201).

2.3. MagNA Pure LC Kit

1. MagNA Pure LC Total Nucleic Acid kit-Large Volume (Roche, cat. no. 3 264 793). Protocol "Total NA Serum_Plasma."

2.4. Manual or Automated Tip Extraction

Well-designed oligonucleotides can hybridize effectively to both RNA and DNA at higher salt concentrations *(2)*. To eliminate contamination and interfering molecules, the template nucleic acid linked to the oligonucleotides is collected by a magnetic bead technique. These templates loaded onto the magnetic beads can directly be used in the polymerase chain reaction (PCR) (*see* **Note 2**).

1. Protease (QIAGEN, cat. no. 19157).
2. Dynabeads MyOne (Invitrogen, cat. no. 656.02).
3. 1 M Tris-HCl, pH 7.4.

4. Primers, high-performance liquid chromatography purified (Purimex, Grebenstein, Germany).
5. Reaction tubes (Sarstedt, Nümbrecht, Germany, cat. no. 60.9922.937).
6. Plastic sheets for tip-extraction (Heermann, Göttingen, Germany) (*see* **Note 3**).
7. Genesis Freedom 150/8 or 200/8 (Tecan, Männedorf, Switzerland) equipped with the software package Gemini (*see* **Note 4**).
8. Magnetic tips for manual tip extraction (Heermann) (*see* **Note 3**).
9. Magnetic tips for Automated tip extraction with Genesis Freedom.
10. Software for automated tip-extraction with Genesis Freedom and installed Gemini package.
11. Microplate rack for tip extraction.
12. Oligonucleotides.

The efficiency of hybrid capture depends predominantly on the length of the capturing primer, the hybridization temperature, and the stringency of the denaturation mixture. We obtained the best results with around 30-mer oligonucleotides. Oligomers >40 bases were avoided (*see* **Note 5**). We give an example how fetal *RHD* sequences for amplification of *RHD* exons 5 and 7 (for primers, *see* **Table 1**) can be captured from plasma of D-negative pregnant women.

a. Spin down tubes with lyophilized oligonucleotides.
b. Reconstitute in 10 mM Tris buffer, pH 7.4, to a concentration of 100 pmol/µl.
c. Incubate at room temperature (15–25°C) for 30 min.
d. Store aliquots at –20°C for up to 5 years.

13. Denaturation buffer: 6 M guanidinium hydrochloride, 1 mM dithiothreitol (DTT), and 20 mM Tris-HCl, pH 9.2 (*see* **Note 6**).
14. Protease: Reconstitute protease to a concentration of von 20 g/L in aq. dest. Store Aliquots at –20°C.
15. Binding buffer: 20 mM Tris-HCl, pH 7.4, and 1 mM DTT.

Table 1
Oligonucleotides for hybridization capture of *RHD* exons 5 and 7

RhD ex7 fish_a	5′-Biotin-GGAATATGGGTCTCACCTGCCAATCTGCTTATAA TAACACTTGTCCA
RhD ex7 fish_b	5′-Biotin-TGTTAAGGGGATGGGGGGTAAGCCCAGTGACCC ACATGCC
RhD ex5 fish_a	5′-Biotin-GCAGGAGTGTGATTCTGGCCAACCACCCTCTCT GGCC
RhD ex5 fish_b	5′-BiotinCCCTGAGATGGCTGTCACCACGCTGACTGCTACA GCATAGTAGG

3. Methods

3.1. QIAamp DNA Blood Mini Kit

This is a modification protocol which comes with the kit. We recommend comparing this protocol with the manufacturer's instructions, because changes in kit components might require changes of this protocol.

For isolation and purification of nucleic acids from 1,000 µl of EDTA plasma, follow the protocol below.

3.1.1. Things to Do before Starting

1. Equilibrate samples to room temperature (15–25°C) and ensure that they are well mixed.
2. Ensure that wash buffer 1 (AW1), wash buffer 2 (AW2), and QP have been prepared according to the manufacturer's instructions.
3. Set a heating block to 56°C for use in **step 3**.

3.1.2. Procedure

1. Pipet 20 µl of protease in a 50-ml lysis tube (LT) (Nunc-tube).
2. Add 1 ml of plasma and 1 ml of lysis buffer (AL) in the LT, close the lid, and vortex for ≥15 s.
3. Incubate LT for 10 ± 1 min at 56 ± 1°C in a thermo mixer or water bath.
4. Change gloves and open the LT carefully.
5. Add 1 ml of ethanol (96–100%) to the LT, close the lid, and mix thoroughly by pulse-vortexing for ≥15 s.
6. Apply 600 µl of the mixture from **step 5** to the QIAamp column (in a 2-ml collection tube) without wetting the rim; close the cap and centrifuge at 6000*g* for 1 min. Place the QIAamp column in a clean 2-ml collection tube and discard the tube containing the filtrate.
7. Repeat **step 6** until the mixture of **step 5** is finished.
8. Carefully open the QIAamp column and add 500 µl of AW1 (at room temperature, 15–25°C) without wetting the rim, close the cap, and centrifuge at 6000*g* for 1 min. Place the QIAamp column in a clean 2-ml collection tube and discard the tube containing the filtrate.
9. Carefully open the QIAamp column and add 500 µl of AW2 (at room temperature, 15–25°C), without wetting the rim, close the cap, and centrifuge at 20,000*g* for 3 min
10. Place the QIAamp column in a clean 2-ml collection tube and discard the tube containing the filtrate. Centrifuge the empty column again at 20,000*g* for 1 min and discard the tube, which can contain some filtrate (*see* **Note 7**).
11. Place the QIAamp column in a clean elution tube (ET) and add 60 µl of H_2O/AE buffer (at room temperature, 15–25°C) to the center of the membrane (*see* **Note 8**); avoid touching the membrane. Incubate at room temperature (15–25°C) for 5 min.

12. Centrifuge at 6,000*g* at 1 min to elute the DNA nucleic acids for the membrane.
13. Store DNA elute at 2–8°C and perform the PCR the same day as isolation (*see* **Note 9**).

3.2. QIAamp DSP Virus Kit

This is a modification of the vacuum protocol that comes with the kit *(3)*. We recommend comparing this protocol with the manufacturer's instructions, because changes in kit components might require changes of this protocol.

For isolation and purification of nucleic acids from 500 µl of EDTA- or citrate treated plasma and serum, follow the protocol below.

3.2.1. Things to Do before Starting

1. Separate plasma from cells by centrifugation at low speed, 2700*g* for 10 min without brake.
2. Separate plasma from cell fragments by centrifugation at 18,000*g* for 45 min (*see* **Note 10**).
3. Equilibrate samples to room temperature (15–25°C) and ensure that they are well mixed.
4. Ensure that AW1, AW2, and QP have been prepared according to the manufacturer's instructions.
5. Set a heating block to 56°C for use in **steps 4** and **14.**
6. Place ethanol on ice.

3.2.2. Procedure

1. Pipet 100 µl of QIAGEN reconstituted protease into a LT.
2. Add 500 µl of plasma or serum into a LT. Close the lid of the LT and mix by pulse-vortexing for 15 s to mix sample with protease.
3. Add 500 µl of AL to the LT, close the lid, and mix by pulse-vortexing for 15 s (*see* **Note 11**).
4. Incubate at 56 ± 1°C for 20 ± 1 min.
5. Centrifuge the LT for ≥5 s at full speed to remove drops from the inside of the lid.
6. Change gloves and open the LT carefully.
7. Add 600 µl of ethanol (96–100%) to the LT, close the lid, and mix thoroughly by pulse-vortexing for ≥15 s. Incubate for 10 ± 1 min at room temperature (15–25°C).
8. Centrifuge the LT for ≥5 s at full speed to remove drops from the inside of the lid.
9. Change gloves and carefully apply 570 µl of the mixture from **step 7** to the QIAamp column (in a 2-ml collection tube) without wetting the rim, close the cap, and centrifuge at 6,000*g* for 1 min. Place the QIAamp column in a clean

2-ml collection tube and discard the tube containing the filtrate. Repeat this step 2 times.

10. Carefully open the QIAamp column and add 600 µl of AW1 without wetting the rim, close the cap, and centrifuge at 6,000*g* for 1 min. Place the QIAamp column in a clean 2-ml collection tube and discard the tube containing the filtrate.
11. Carefully open the QIAamp column and add 750 µl of AW2 without wetting the rim, close the cap, and centrifuge at 18,000*g* for 3 min. Place the QIAamp column in a clean 2-ml collection tube and discard the tube containing the filtrate.
12. Carefully open the QIAamp column and add 750 µl of ethanol (96–100%) to the QIAamp column without wetting the rim. Avoid touching the QIAamp column membrane with the pipette tip, close the cap, and centrifuge at 18,000*g* for 3 min.
13. Place the QIAamp column in a clean 2.0-ml collection tube and discard the tube containing the filtrate. Centrifuge at 18,000*g* for 1 min (*see* **Note 7**).
14. Place the QIAamp column in a new wash tube (WT) and incubate with the lid open at 56°C for 3 min to evaporate any remaining liquid.
15. Place the QIAamp column in a clean ET, and discard the WT. Carefully open the lid of the QIAamp column and apply 40–60 µl of elution buffer (AVE) to the center of the membrane (*see* **Note 8**). Close the lid and incubate at room temperature (15–25°C) for 5 min.
16. Centrifuge at full speed (18,000*g*) for 1 min to elute the DNA (*see* **Note 9**).

3.3. MagNA Pure LC

For isolation of DNA proprietary magnetic particles and specialized reagent kits are used. DNA bind to magnetic glass particles (MGPs) in the presence of a chaotropic salt at a pH >7.0. MGPs have a glass (silica) surface and magnetic core. Nucleic acids are adsorbed to the silica surface of the MGPs in the presence of isopropanol and high concentrations of chaotropic salts, which remove water from hydrated molecules in solution. Polysaccharides and proteins do not adsorb to the beads, and they are removed by sequential washing steps. Pure nucleic acids are then eluted from beads by applying low-salt conditions and heat. Once bound on the surface of magnetic beads, the nucleic acids can be separated from the solution with a magnet. In addition to its use as a workstation for automated nucleic acid preparation, the MagNA Pure LC system can be used as a programmable pipetting robot, e.g., for set up of downstream PCR or reverse transcription-PCR reactions by using the previously isolated nucleic acids as template.

We recommend comparing this protocol with the manufacturer's instructions, because changes of kit components or instrument settings might require changes in this protocol.

1. During the prologue phase, reagents are transported from the reagent tubes into the wells of the processing cartridge.
2. Samples are lysed in the wells of the sample cartridge. The lysates are then transferred to the processing cartridge.
3. According to the reaction steps defined by the selected purification protocol, the reaction mix is transported to the next rows of wells containing the subsequently used reagents.
4. The complex of magnetic particles and bound nucleic acids is separated from the solution by applying the magnet. Proteins and other contaminating components are washed away by repeated separation and resuspension steps by using various wash buffers.
5. The complex of magnetic particles and bound nucleic acid is transported from the processing cartridge to the preset elution buffer in the elution cartridge. Nucleic acids are released from the magnetic particles by applying heat and low-salt conditions.
6. The eluted and purified nucleic acids are transferred from the Elution Cartridge into the Storage Cartridge.

3.4. Manual or Automated DNA Extraction with Magnetic Tips

1. In this protocol, *RHD* exon 5 and 7 are the targets for detection *(4)*. Dilute 0.5 μl of each fish-primer (*RhD* ex7 fish_a, *RhD* ex7 fish_b, *RhD* ex5 fish_a, and *RhD* ex5 fish_b, 100 pmol/μl) in 98 μl of aq dest. Mix 6 μl of diluted fish primers with 444 μl of denaturation buffer and add 50 μl of protease. Heat one incubator or water bath to 60°C and another incubator or water bath to 95°C. Automated: place samples with barcodes, denaturation buffer with primers and protease, pipetting tips, plastic sheets, and reaction tubes on the Tecan pipetting robot. The correct position is indicated on the screen when you start the software (further details come with the software). Manual: label a reaction tube.
2. Automated: start the Tecan software. Manual: pipet 500 μl of denaturation buffer with fish primers and protease into the reaction tube. Add 500 μl of plasma. Close the tubes, mix briefly by pulse-vortexing, and incubate at 60°C for 20 min.
3. Incubate the reaction tubes for 10 min at 95°C.
4. Incubate the reaction tubes for 30 min at 60°C.
5. Prepare a magnetic bead solution by mixing 12.5 μl of beads with 1,487.5 μl of binding buffer.
6. Automated: place the opened reaction tubes and the magnetic bead solution on the worktable and start automated pipetting (press "yes"). Manual: add 1,500 μl to each reaction tube, close the tube, and incubate for 45 min at room temperature (15–25°C) (*see* **Note 12**).
7. Centrifuge the tubes at 250*g* for 10 min without brake.
8. Place a PCR plate filled with all reagents for real-time PCR (we estimate that the beads come in a volume of 2 μl) in the microplate rack specially designed for this purpose.

9. Automated: place the tubes on the workstation and continue as described by the software (press "yes"). Manual: collect the beads with the magnet covered with a plastic sheet and transfer the beads first into a tube filled with 2 ml of binding buffer and subsequently to the PCR plate in a microplate rack. Remove the magnet and carefully place the plastic sheet in the PCR tube without touching the rim. After 5 min, move the PCR tubes down and place them into a thermocycler.

4. Notes

1. Based on U.S. patent 6,403,038 B1 and E.U. patent EP 0,995,096 B1 held by K.-H. Heermann, we have established a highly sensitive automated magnetic tip technique to isolate nucleic acids from plasma or serum. The technique is applied to large volumes with a minimum of manipulation steps. After immediate denaturation, biotinylated primers target nucleic acid sequences with high sensitivity and collect them onto magnetic beads. Subsequently, an immersed magnetic tip captures the bead-attached targets and allows their one-step transfer to standard amplification vials. Thereafter, the specific targets can be amplified by real-time PCR or used for other downstream applications.
2. Two different strategies can be used after magnetic bead capture of hybrids: the solid phase and the liquid phase technique. For the solid phase technique, oligonucleotides are covalently coupled to magnetic beads, which after template hybridization are removed from the extraction mixture. However, the liquid phase hybridization is much more sensitive. Biotinylated oligonucleotides are hybridized to the template, which could then bind to streptavidin-coated magnetic beads. A similar technology to separate nucleic acids was already described Ref. ***5*** and ***6***.
3. PD Dr. med. Dipl. Chem. Klaus-Hinrich Heermann, Dept. of Virology, University Clinics Göttingen, Kreuzbergring 57, 37075 Göttingen, Germany. Tel./Fax: +49-551-395800, E-mail: kheerma@gwdg.de.
4. So far, we have only experience with the eight-channel extraction techniques. Alternatively, a flexible multichannel robotic pipetting workstation of Hamilton (Bonaduz, Switzerland) can be used. The final robot configuration most likely will be as follows. The portal robot should have 16 channels, eight channels allowing high-throughput isolation of samples with magnetic tips and the other eight channels adding the reagents. The barcode reader provides increased automation, throughput, sample and reagent tracking, and information. A robot equipped as described can isolate nucleic acids from larger number of different samples or multiple templates from the same sample. The same software should control isolation and detection. The automated confirmation improves the process reliability. For a higher degree of automation a reaction block with incubation temperatures from 60°C to 95°C as well as a magnetic tool for the concentration of particles at the bottom of the tubes before the transfer with magnetic tips will be developed by Heermann in future.
5. With this protocol, we could detect about 15 genome equivalents with a 95% hit rate by real-time PCR. Moreover, the dynamic range was 10^{10}. At

concentrations >10^{11} genome equivalents, the concentration of capture primers limits the recovery of templates.

6. Initially, we chose guanidinium thiocyanate for fast biological inactivation and denaturation before nucleic acid extraction *(7)*. Because thiocyanate can be converted into the poisonous cyanogens in the presence of acids, we changed to guanidinium hydrochloride *(8)*. This treatment denatures, but it does not cleave proteins. However, certain concentrations can lead to precipitations in the sample which strongly influence the yield of the subsequent manipulation. Therefore, we have added a serine protease to digest proteins within higher concentration of guanidinium hydrochloride. In the final protocol, after the addition of proteases and guanidinium hydrochloride, the mixture is incubated at 95°C to inactivate the protease and thereby preserve the activity of enzymes (like polymerases) added later.
7. Omission of the dry centrifugation might lead to inhibition of the downstream assay.
8. The elution volume depends on the volume required for downstream applications.
9. The PCR should be performed at the same day as isolation. Otherwise, DNA will be degraded.
10. Plasma samples separated from cells and cell fragments can be stored at –70°C or less.
11. To ensure efficient lysis, it is essential that the sample and AL are mixed thoroughly to yield a homogeneous solution. Because AL has a high viscosity, be sure to add the correct volume of AL by pipetting carefully or by using a suitable pipette such as an Eppendorf multistep pipette or equivalent. Do not add QP directly to AL.
12. After sample treatment with 3 *M* guanidinium hydrochloride, this concentration is diluted to 1 M together with the addition of streptavidin-coated beads. This concentration does not affect the stability of the streptavidin *(9)*. Moreover, we use an excess of streptavidin over biotin to ensure the complete capture of all nucleic acid templates *(9,10)*. The concentration of magnetic beads does not interfere with real-time PCR when the light is passed through the vials vertically. In nonspecific extraction procedures for total nucleic acids per sample a large amount of adsorbent has to be used. This may well interfere with amplification and detection by real-time PCR.

References

1. Legler, T. J., Liu, Z., Mavrou, A., Finning, K., Hromadnikova, I., Galbiati, S., Meaney, C., Hultén, M. A., Crea, F., Olsson, M. L., Maddocks, D. G., Huang, D., Armstrong Fisher, S., Sprenger-Haussels, M., Ait Soussan, A. and van der Schoot, C. E. (2007) Workshop report on the extraction of fetal DNA from maternal plasma. *Prenat. Diagn.* **27,** 824–829.
2. Thompson, J. and Gillespie, D. (1987) Molecular hybridization with RNA probes in concentrated solutions of guanidine thiocyanate. *Anal. Biochem.* **163,** 281–291.

3. Liu, Z., Gutensohn, K., Hempel, M. and Legler, T. J. (2005) Optimisation of the QIA-amp DSP Virus Kit for the extraction of foetal DNA from maternal plasma. *Clin. Chem.* **51,** P73.
4. Liu, Z., Gutensohn, K., Hempel, M., Heermann K. H. and Legler, T. J. (2005) High-throughput fetal DNA separation from maternal plasma by hybridization capture and magnetic particles. *Clin Chem.* **51,** O28.
5. Uhlen, M. (1989) Magnetic separation of DNA. *Nature* **340,** 733–734.
6. Lambert, K. N. and Williamson, V. M. (1993) cDNA library construction from small amounts of RNA using paramagnetic beads and PCR. *Nucleic Acids Res.* **21,** 775–776.
7. Chomczynski, P. and Sacchi, N. (1987) Single-step method of RNA isolation by acid guanidinium thiocyanate-phenol-chloroform extraction. *Anal. Biochem.* **162,** 156–159.
8. Legler, T. J., Kohler, M. and Heermann, K. H. (1999) High-throughput extraction, amplification and detection (HEAD) of HCV-RNA in individual blood-donations. *J. Clin. Virol.* **13,** 95–103.
9. Heermann, K. H., Hagos, Y. and Thomssen, R. (1994) Liquid-phase hybridization and capture of hepatitis B virus DNA with magnetic beads and fluorescence detection of PCR product. *J. Virol Methods* **50,** 43–57.
10. Heermann, K. H., Seitz, H. and Thomssen, R. (1996) Capture and RT-PCR of hepatitis C virus RNA with safety primers. *J. Virol. Methods* **59,** 33–43.

17

Quantification of Circulatory Fetal DNA in the Plasma of Pregnant Women

Bernhard G. Zimmermann, Deborah G. Maddocks, and Neil D. Avent

Summary

The analysis of cell-free fetal DNA in the circulation of the pregnant woman plays the pivotal role in noninvasive prenatal research. Here, we describe an improved method for the quantification of male DNA, which is a valuable research tool for the quantification of fetal DNA. The quantification of fetal DNA serves two main purposes. First, the levels may indicate certain pregnancy-related disorders such as preeclampsia even before onset of the disease; thus, the quantification may serve as a marker for early detection. Second, extraction and enrichment strategies of the fetal DNA compartment are important factors in the development and implementation of clinical tests, such as detection of fetal sex, Rhesus D status, point mutations, and aneuploidies. In this context, the quantification of fetal DNA is an important tool for the evaluation of protocols.

Key Words: Real-time quantitative polymerase chain reaction (qPCR); absolute quantification; noninvasive prenatal diagnosis; circulatory free fetal DNA in plasma; DNA analysis; DYS14; limit of quantification (LOQ).

1. Introduction

Noninvasive prenatal medicine was pioneered by the discovery of "fetal DNA" (which is of placental origin) in the circulation of pregnant women *(1)*. The presence of fetal DNA was demonstrated by the polymerase chain reaction (PCR) amplification of Y-chromosome–specific sequences in the plasma of women pregnant with a male fetus. The fetal DNA comprises only a low percentage of the total DNA in the plasma *(2)*. However, due to the high maternal DNA background, only loci clearly distinct from maternal sequences

From: *Methods in Molecular Biology, vol. 444: Prenatal Diagnosis*
Edited by: S. Hahn and L. G. Jackson © Humana Press, Totowa, NJ

can be detected and quantified by real-time quantitative PCR *(3)*. These are the determination of fetal sex and Rhesus D status, which are already applied in clinical diagnosis. The genotyping of these fetal loci helps preventing invasive procedures in the case of X-linked disorders or the unnecessary administration of anti D-immunoglobulins in case of a Rhesus D-negative fetus *(4)*. In the past years, DNA of fetal origin was found in the maternal circulation as early as 5 weeks in pregnancy *(5)*. The finding that fetal DNA is predominantly short (<300 base pairs) as opposed to the longer maternal DNA opened a new approach for the diagnosis of more subtly different sequences such as point mutations *(6–8)*. The quantification by real-time PCR has shown that fetal DNA levels are elevated in certain pregnancy associated disorders, such as preeclampsia *(2,9–12)*, preterm labor *(13)*, and fetuses with trisomy 21 *(14,15)*. However, absolute quantification is difficult to achieve, and although these findings have great potential, proceedings need to be highly controlled. Here, we present, in our view at present, the optimal strategy of fetal DNA quantification for research application: the quantitative real-time PCR measurement of the multiple-copy sequence DYS14 on the Y-chromosome *(16)*. DYS14 is a sequence of approximately 50 copies, which are more or less identical to each other. The presented assay amplifies a subset of nine of these copies, with 100% sequence match and specificity, thus rendering the assay 10-fold more sensitive than generally used assays targeting the single-copy SRY gene *(17)*.

For research purposes, the quantification of male DNA from the male fetus is the most straightforward, because sequences clearly distinct from the maternal background are amplified. Even though it is only applicable to 50% of the pregnancies, the approach is much simpler and more quantitative than existing alternatives, allowing a higher throughput of samples. Currently, two approaches to measure fetal DNA independently from the Y-chromosome are being applied: polymorphisms *(18)* and placenta-specific methylation *(19)*; however, they are considerably more labor-intensive.

The quantification of fetal DNA is a relatively simple procedure, but protocols need to be followed meticulously and the exact setup validated before engaging in large projects. Otherwise, variability and imprecision will negatively affect the data.

After separation of plasma from the cellular fraction by centrifugation, the DNA from the plasma is extracted and analysed by real-time PCR. Although extracted DNA is generally stable for long periods, levels of fetal DNA have been shown to decrease over long storage periods; thus, the DNA should be analyzed speedily. The use of low-retention tubes for the storage may reduce such decrease, which is at least in part attributable to adsorption of DNA to the plastic tube *(20)*.

Optimal quantification begins with correct sample handling and extraction. These procedures are discussed in other chapters of this book. We emphasize that reproducibility and optimal recovery should always be assessed before engaging in a study, and if possible, during the study as well. Also, although the method performs very reproducibly, new users are encouraged to ensure that the data they generate is indeed quantitative. Such procedures were discussed extensively in a recent publication ***(16)***, and they are introduced at the end of the Methods (*see* **Subheading 3.**). They include the determination of coefficient of variation of replicate measurements, probit analysis to determine limit of detection (LOD), and limit of quantification (LOQ), or setting cut-offs for quantification below a certain copy number or C_T (threshold cycle) scatter of replicates.

One major hindrance for the broad assessment of fetal DNA quantification for clinical use is the discrepancies for the absolute levels between experiments and even more so between laboratories ***(21)***. One of the major biases is introduced by the quantification strategy: what calibrator is used and what points comprise the calibration curve. A simple way to reduce intrastudy differences is to assess samples in parallel. Theoretically, up to 90 samples can be measured simultaneously on a single 96-well plate, and the described method is intended for such sample numbers.

2. Materials

2.1. Performing Real-Time PCR

1. The described protocol was developed and evaluated with an ABI PRISM 7000 sequence detection system (Applied Biosystems, Foster City, CA, SDS7000).
2. For the PCR reactions, MicroAmp Optical 96-well reaction plates (Applied Biosystems, cat. no. 4306737) and the ABI PRISM Optical adhesive covers (cat. no. 4311971) were used.
3. The reactions are carried out with the TaqMan Universal PCR master mix (Applied Biosystems, cat. no. 4304437) containing AmpliTaq Gold DNA polymerase, Amperase uracil-*N*-glycosylase (UNG), dNTPs with dUTP, and passive Reference 1 (ROX fluorescent dye) (*see* **Note 1**).
4. Primers are high-performance liquid chromatography (HPLC) purified (*see* **Note 2**), sequences specific for DYS14 and Chr21 are listed in **Table 1**. They can be stored at –20°C as aliquots of 10 μ*M* concentration. Primers stored at 4°C are stable for several weeks.
5. We generally use TaqMan MGB probes (Applied Biosystems, cat. no. 43160324). These probes are 5′-labeled with the fluorescent dye (VIC or 5-carboxyfluorescein [FAM]) and 3'-labeled with a minor groove binder (MGB) and a nonfluorescent quencher (*see* **Note 3**). It is advisable to store aliquots of 5 μ*M* concentration at –20°C, where they are stable for >1 year. If used within 1 month, the probes can

Table 1
Primer and probe sequences for the measurement of fetal and total DNA. For completeness, primers and probes for the single copy Y-chromosome sequence SRY are also included

Target	Primer name	Sequence	Oligo length	Amplicon length
DYS14	DYS14_F	GGGCCAATGTTGTA TCCTTCTC	22	84
	DYS14_R	GCCCATCGGTCACTTA CACTTC	22	
SRY	SRT_F	TCCTCAAAAGAAAC CGTGCAT	21	78
	SRY_R	AGATTAATGGTTGCTA AGGACTGGAT	26	
Chromo 21	C21_F	CCCAGGAAGGAAGT CTGTACCC	21	80
	C21_R	CCCTTGCTCATTGCGCTG	26	
	Probe name			**MGB**
DYS14	DYS14_MGB	TCTAGTGGAGAGGTG CTC	18	Yes
DYS14	DYS14_Dual	CGAAGCCGAGCT GCCCATCA	20	No
SRY	SRY_MGB	TCCCCACAAC CTCTT	15	Yes
SRY	SRY_Dual	CACCAGCAGTAACTCC CCACAACCTCTTT	29	No
Chromo 21	Chr21_MGB	CTGGCTGAGCCATC	14	Yes
Chromo 21	Chr21_Dual	AGCCATCCTTCCC GGGCCTAGG	22	No

be stored at 4°C. Avoid exposure to light. Probe sequences of both formulations are listed in **Table 1**.

6. For calibration curves, we use reference DNA of known concentration in serial dilution (*see* **Note 4**). The DNA is extracted from blood or buffy coat and concentration determined with a spectrophotometer; alternatively it can be obtained from a commercial source. Dilutions are made with water or elution buffer (*see* **Note 5**), preferably in nonstick tubes (for example DNA loBind tube 1.5 ml, Eppendorf AG, Hamburg, Germany, cat. no. 2243 102-1). Store as aliquots at –20°C. Open aliquots can be used for 1 month if stored at 4°C. Once the slope of the standard curve is established, it suffices to amplify only one calibrator sample of adequate template copy number.

2.2. Analysis of Real-Time PCR Data

1. Data are analyzed with the Prism 7000 SDS software (Applied Biosystems).
2. Statistical software with probit regression analysis (predicted proportion positive; SPSS, SPSS Inc., Chicago, IL).

3. Methods

DNA extraction is discussed in other chapters of this book.

3.1. Quality Control of DNA Extraction

1. Manual extraction methods using filter tubes offer the possibility to extract the filter a second time (*see* **Note 6**) (with the same volume as in the first elution) and to quantify by quantitative (q)PCR the DNA content of the second eluate. This should not exceed 30% of the first eluate and should usually be <20%.
2. Replicate extractions of the same sample should be performed to evaluate extraction consistency. In general, it would be better to quantify multiple extractions of a sample each once by qPCR than the normal practice to qPCR one extraction in multiple replicates. This approach, however, increases demand on sample, which is usually scarce, and it is also more costly.

3.2. Determination of LOD and LOQ and Assessment of Protocol Validity

The limits of detection and quantification can be determined in a single experiment.

1. Amplify dilutions of standard DNA in a relatively large number of replicates. The copy number input should range from 100 to 0.01 genome equivalents (GE). An example of the number of reactions is given in **Table 2**.
2. The 95% LOD is determined by probit regression. The known template input of reference reactions is plotted against the fraction of positive PCRs per input amount, and the probit regression curve is fitted onto the data set. The amount of target necessary for 95% of the PCR reactions to be positive is defined as the 95% LOD. The LOQ is at 5 times the 95% LOD, and it should be $\approx$20 to 50 target copies per reaction.

3.3. Performing Real-Time PCR

1. Include at least one negative control DNA sample from plasma (*see* **Note 7**).
2. Include standard DNA for the generation of a calibration curve (*see* **Note 8**). The standard DNA for the calibration curve serves as positive control. It is preferable to work in units of GE (*see* **Note 9**).

Table 2
Proposed setup for the determination of LOD for DYS14 on one 96-well plate. If assessing a single copy sequence targeting assay, increase the input by fivefold (autosomal sequences = 2 copies per GE) or 10-fold (Y-chromosome sequences)

Input copy number in GE	Number of replicates
100.00	6
50.00	6
25.00	6
12.50	6
6.25	6
3.13	6
1.56	6
0.78	6
0.39	6
0.20	6
0.10	6
0.05	6
0.02	6
0.01	6
0	12

3. Measure all samples (patients and controls) on the same plate, if possible.
4. It is common practice to analyze each sample in duplicate or triplicate. Perform replicate reactions on separate plates, if necessary.
5. Prepare the real-time PCR reactions on ice or a cooler block (*see* **Note 2**).
6. The real-time PCR amplification is carried out in a single-plex reaction (i.e., only one primer/probe pair per reaction) (*see* **Note 10**) in a total volume of 20 μl (*see* **Note 11**), with concentrations of 300 nM of each primer and 200 nM of probe (*see* **Note 12**) at 1× concentration of the Universal PCR reaction mix.
7. First, pipette the PCR reaction mixture into the wells and then add 5 μl of the DNA sample. Carefully seal the reaction plate with the optical adhesive cover. Centrifuge at 2000*g* at 4°C for 1 min to spin down any droplets and remove air bubbles.
8. Immediately start the real-time PCR with the "emulation mode" off (*see* **Note 13**). After an initial incubation at 50°C for 2 min to permit AmpErase activity (*see* **Note 14**), 10 min at 95°C for activation of AmpliTaq Gold, and denaturation of the DNA, use the following cycle conditions: 50 cycles of 1 min at 60°C and 15 s at 95°C.

3.4. Analysis of Real-Time PCR Data

1. Perform the experimental analysis with the automatic baseline setting.
2. Set the threshold manually to a value where the signal increases exponentially for all amplifications (*see* **Note 15** and **Fig. 1**).
3. Check all amplification curves individually by sight to identify nonexponential signals that cross the threshold.
4. The copy numbers per reaction are determined automatically by the SDS software from the standard curve (*see* **Figs. 1 and 2**). However, negative wells are not included in the calculation of sample averages. Thus, quantities determined for replicate wells should be averaged using an Excel (Microsoft, Redmond, WA) spreadsheet.
5. Based on the copy number of a calibrator sample and the slope of the dilution curve (or the previously determined PCR efficiency) (*see* **Fig. 2**), the copy number in a sample also can be calculated with the following formula (*see* **Note 16**): Copies sample per PCR = copies (calibrator) × $E^{-\Delta CT}$, where E is PCR efficiency = $10^{-(1/\text{slope})}$ and $\Delta CT = C_T$ (sample) – C_T (calibrator).
6. The coefficient of variation is determined by dividing the standard error by the average quantity of a sample.
7. Determination of Validity of Data.

   a. The coefficient of variation should usually be in the range between 5 and 25%. It can be increased when replicates are run on different plates.

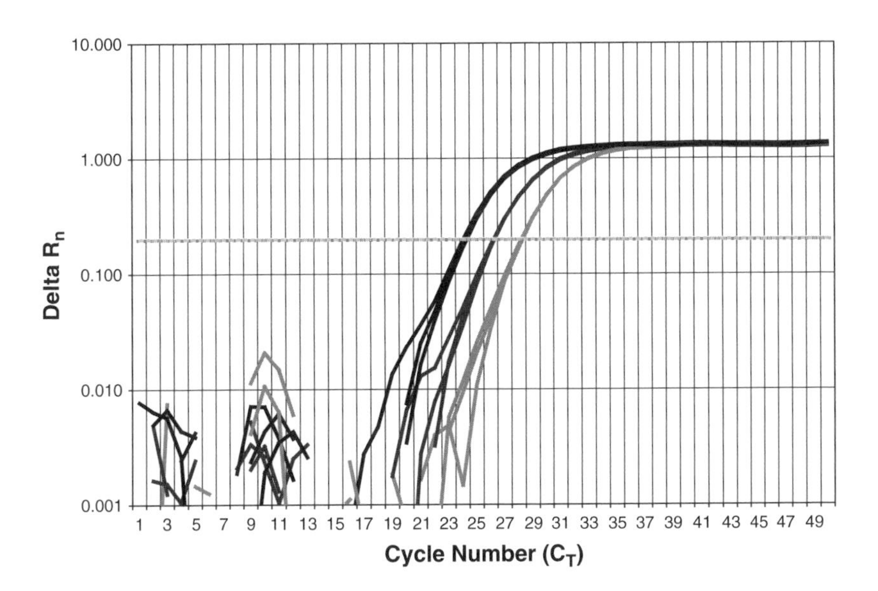

Fig. 1. Triplicate amplification plots of reference DNA with 80, 320, and 1280 GE input per reaction. The threshold for the determination of C_T values is set to 0.2 relative fluorescence units (ΔR_n).

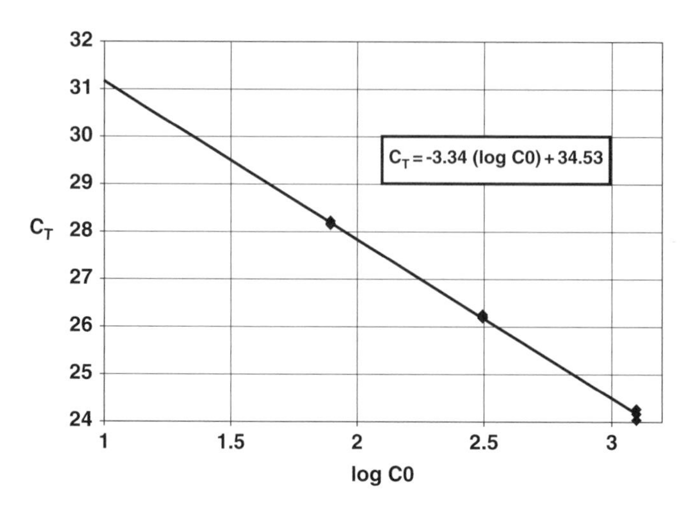

Fig. 2. Standard curve from amplifications in **Figure 1**. The slope of –3.34 indicates a PCR Efficiency of 1.99 (=$10^{-1/\text{slope}}$). The Intercept at 34.53 is the projected C_T value for a reaction with 1 GE input.

However, the proportionality between samples needs to be conserved if there are interplate differences.

b. Using the LOQ as a cut-off for accurate quantification helps to assess the quantitative validity of results. As alternative, a general copy number cut-off (>20 copies in reaction) or a maximal standard deviation between replicates (0.4 cycles = 30%).

4. Notes

1. ROX is a dye included in the 2× Universal master mix to provide an internal reference, to which the reporter-dye signal is normalized by the data analysis software. Normalization corrects for any signal fluctuations during the analysis.
2. It is important to avoid the amplification of any nonspecific products (for example primer dimers and misprimed sequences) to avoid both false-positive and false-negative results. Experimental preconditions to ensure optimal specificity are i) HPLC-purified primers, ii) reaction setup on ice, iii) use of a hot start polymerase, iv) use of AmpErase UNG, and v) single-plex reaction setup.
3. MGB-modified probes generate a strong signal, as the quenching of the fluorescence in the intact probe is very efficient due to their short length. This is of advantage to identify curves from unspecific probe cleavage with nonexponential fluorescence increase. Alternatively, dual-labeled probes can be used (5′-labeled with a fluorescent dye (VIC or FAM) and 3′-labeled with a suitable quencher (5-carboxytetramethylrhodamine or nonfluorescent quencher).

4. Standard DNA is usually amplified in fivefold dilutions, but 2- to 10-fold differences are also suitable. The number of reactions used for the standard curve depends on the objective of the experiment. If the purpose is only a qualitative assessment of sex or Rhesus D status, two single replicates of different standard copy numbers can serve as positive controls and simultaneously allow a crude copy number estimation. If the aim is accurate quantification of copy numbers, duplicate or triplicate amplifications of three to six known template numbers should be performed. The use of more standard samples generally results in a lower interexperiment variation. At template numbers of fewer than 20 copies, the variability of the measurements is increased. Consequently, to ensure accurate quantification, standards (and samples) should have higher template numbers than 20 copies.
5. DNA is more stable in the elution buffer provided by the DNA extraction kit (10 mM Tris buffer, pH 8.5) than in water. This effect is only observed if the DNA is stored for several months. Using water for the DNA elution allows higher input volumes into the PCR.
6. DNA recovery can be suboptimal when using reduced extraction volumes, especially 50 μl. Recovery can be examined by eluting the filter for a second time into a new Eppendorf tube (with 100-μl elution volume) and quantifying this second eluate by real-time PCR.
7. DNA extracted from female plasma is an optimal negative control. In this way, contamination of the reaction mix and also unspecific amplification from the background DNA can be excluded.
8. Any assay specific for an autosomal sequence can serve as control for successful DNA extraction and to measure total DNA in plasma. Unless accurate quantification of total DNA is required, one total DNA control per sample is sufficient. We propose an assay specific for the amyloid precursor protein locus on chromosome 21.
9. GE are the amount or quantity of DNA perceived to be the minimum amount to be representative of a genome that is 6.6 pg of DNA.
10. Reactions should be performed in single-plex. Because the levels of total DNA are ≈20-fold greater than the fetal fraction, the sensitivity and quantitative value of the fetal specific reactions would be markedly reduced by such multiplex analysis.
11. Reaction volumes between 20 and 50 μl can be used on the SDS7000. A volume of 50 μl is only necessary if a large sample input volume is needed.
12. Probe concentrations of 100 μM also work well; however, the fluorescence signal will be weakened.
13. Run the assay always at the same heating and cooling rates. We use the "emulation off" setting because it is faster. The emulation mode mimics the slower heating and cooling rate of the predecessor instrument, the SDS7700.
14. Amperase UNG is a component of the TaqMan Universal PCR master mix. Master mixes that contain the UNG component also contain dUTP as a substitute for dTTP in their dNTP mix. Therefore, dUTP is incorporated in place of dTTP in

the subsequent PCR products. UNG treatment before the PCR reaction prevents the amplification of PCR-carryover products by degrading any uracil containing DNA.

15. In **Fig. 1**, the threshold is set at 0.2 ΔR_n. At this value, the variation between replicates is minimal and the amplification curves are approximately a straight line.
16. The number of copies per milliliter of plasma can be calculated from the number of copies per PCR with the following formula: $C = copies_{PCR} \times DNA_{PCR} \times$ (plasma volume/elution volume), where C is copies per milliliter of plasma (in GE) and DNA_{PCR} is volume of DNA per reaction.

References

1. Lo, Y. M., Corbetta, N., Chamberlain, P. F., et al. (1997) Presence of fetal DNA in maternal plasma and serum. *Lancet* **350,** 485–487.
2. Zhong, X. Y., Laivuori, H., Livingston, J. C., et al. (2001) Elevation of both maternal and fetal extracellular circulating deoxyribonucleic acid concentrations in the plasma of pregnant women with preeclampsia. *Am. J. Obstet. Gynecol.* **184,** 414–419.
3. Hahn, S., Zhong, X. Y., Burk, M. R., Troeger, C., and Holzgreve, W. (2000) Multiplex and real-time quantitative PCR on fetal DNA in maternal plasma. A comparison with fetal cells isolated from maternal blood. *Ann. N Y Acad. Sci.* **906,** 148–152.
4. Rijnders, R. J., Christiaens, G. C., Bossers, B., van der Smagt, J. J., van der Schoot, C. E., and de Haas, M. (2004) Clinical applications of cell-free fetal DNA from maternal plasma. *Obstet. Gynecol.* **103,** 157–164.
5. Rijnders, R. J., Van Der Luijt, R. B., Peters, E. D., et al. (2003) Earliest gestational age for fetal sexing in cell-free maternal plasma. *Prenat. Diagn.* **23,** 1042–1044.
6. Li, Y., Zimmermann, B., Rusterholz, C., Kang, A., Holzgreve, W., and Hahn, S. (2004) Size separation of circulatory DNA in maternal plasma permits ready detection of fetal DNA polymorphisms. *Clin. Chem.* **50,** 1002–1011.
7. Li, Y., Di Naro, E., Vitucci, A., Zimmermann, B., Holzgreve, W., and Hahn, S. (2005) Detection of paternally inherited fetal point mutations for beta-thalassemia using size-fractionated cell-free DNA in maternal plasma. *J. Am. Med. Assoc.* **293,** 843–849.
8. Chan, K. C., Zhang, J., Hui, A. B., et al. (2004) Size distributions of maternal and fetal DNA in maternal plasma. *Clin. Chem.* **50,** 88–92.
9. Lo, Y. M., Leung, T. N., Tein, M. S., et al. (1999) Quantitative abnormalities of fetal DNA in maternal serum in preeclampsia. *Clin. Chem.* **45,** 184–188.
10. Zimmermann, B., Holzgreve, W., Zhong, X. Y., and Hahn, S. (2002) Inability to clonally expand fetal progenitors from maternal blood. *Fetal Diagn. Ther.* **17,** 97–100.
11. Swinkels, D. W., de Kok, J. B., Hendriks, J. C., Wiegerinck, E., Zusterzeel, P. L., and Steegers, E. A. (2002) Hemolysis, elevated liver enzymes, and low platelet

count (HELLP) syndrome as a complication of preeclampsia in pregnant women increases the amount of cell-free fetal and maternal DNA in maternal plasma and serum. *Clin. Chem.* **48,** 650–653.

12. Farina, A., Sekizawa, A., Iwasaki, M., Matsuoka, R., Ichizuka, K., and Okai, T. (2004) Total cell-free DNA (beta-globin gene) distribution in maternal plasma at the second trimester: a new prospective for preeclampsia screening. *Prenat. Diagn.* **24,** 722–726.
13. Farina, A., Le Shane, E. S., Romero, R., et al. (2005) High levels of fetal cell-free DNA in maternal serum: a risk factor for spontaneous preterm delivery. *Am. J. Obstet. Gynecol.* **193,** 421–425.
14. Zhong, X. Y., Burk, M. R., Troeger, C., Jackson, L. R., Holzgreve, W., and Hahn, S. (2000) Fetal DNA in maternal plasma is elevated in pregnancies with aneuploid fetuses. *Prenat. Diagn.* **20,** 795–798.
15. Lo, Y. M., Lau, T. K., Zhang, J., et al. (1999) Increased fetal DNA concentrations in the plasma of pregnant women carrying fetuses with trisomy 21. *Clin. Chem.* **45,** 1747–1751.
16. Zimmermann, B., El-Sheikhah, A., Nicolaides, K., Holzgreve, W., and Hahn, S. (2005) Optimized real-time quantitative PCR measurement of male fetal DNA in maternal plasma. *Clin. Chem.* **51,** 1598–1604.
17. Zimmermann, B. G., Holzgreve, W., Avent, N., and Hahn, S. (2006) Optimized real-time quantitative PCR measurement of male fetal DNA in maternal plasma. *Ann. N Y Acad. Sci.* **1075,** 347–349.
18. Daniels, G., Finning, K., Martin, P., and Summers, J. O. (2006) Fetal blood group genotyping. present and future. *Ann. N Y Acad. Sci.* **1075,** 88–95.
19. Poon, L. L., Leung, T. N., Lau, T. K., Chow, K. C., and Lo, Y. M. (2002) Differential DNA methylation between fetus and mother as a strategy for detecting fetal DNA in maternal plasma. *Clin. Chem.* **48,** 35–41.
20. Ellison, S., English, C., Burns, M., and Keer, J. (2006) Routes to improving the reliability of low level DNA analysis using real-time PCR. *BMC. Biotechnology* **6,** 33.
21. Lo, K.-W., Lo, Y M. D., Leung, S.-F., et al. (1999) Analysis of cell-free Epstein-Barr virus-associated RNA in the plasma of patients with nasopharyngeal carcinoma. *Clin. Chem.* **45,** 1292–1294.

18

Detection and Quantification of Fetal DNA in Maternal Plasma by Using LightCycler Technology

Yuditiya Purwosunu, Akihiko Sekizawa, and Takashi Okai

Summary

Since the demonstration of cell-free fetal DNA in maternal circulation, its aberrational quantification has been explored and demonstrated in numerous clinical situations. Moreover, several centers have begun to use detection and quantification of fetal DNA to diagnose fetal genetic status with high reliability, such as fetal *RhD*. One of the methods for analyzing cell-free fetal DNA is quantification by using LightCycler technology. With the advent of various fluorescent reporters and greater stability of kit's reagent, real-time quantitative polymerase chain reaction by using LightCycler technology is relatively simple and fast.

Key Words: Fetal free-cell DNA; DNA quantification; LightCycler.

1. Introduction

Fetal DNA in maternal plasma is a valuable means of noninvasive prenatal DNA diagnosis. Thus far, fetal DNA has been used to diagnose fetal gender *(1–3)*, *RhD* blood type *(4)*, and single-gene disorders, such as achondroplasia *(5)*. When attempting prenatal diagnosis by using maternal plasma, polymerase chain reaction (PCR) amplification with fluorescence-labeled probes has greater sensitivity than conventional PCR. Moreover, although the fetal DYS14 sequence has been detected using conventional PCR in 80% of plasma samples from male-bearing pregnant women during early gestation *(1)*, we reported that the detection rate approaches 98.2% of cases using LightCycler technology *(3)*. Thus, PCR amplification-based fluorescence-labeled probes have improved the sensitivity and accuracy of PCR diagnosis.

From: *Methods in Molecular Biology, vol. 444: Prenatal Diagnosis*
Edited by: S. Hahn and L. G. Jackson © Humana Press, Totowa, NJ

Furthermore, quantitative fetal DNA aberrations have been described in several pathological conditions of pregnancy, such as preeclampsia ***(6–8)***, fetal chromosomal aneuploidies ***(9)***, placenta acreta ***(10,11)***, and hyperemesis gravidarum ***(12,13)***. The availability of real-time quantitative PCR methods has made it easier to be performed.

All real-time PCR systems rely upon the detection and quantification of a fluorescent reporter, producing greater signal intensity in direct proportion to the amount of PCR product produced by a given reaction. There are several methods available. One method involves SYBR Green I, a dye that binds to double-stranded DNA (dsDNA) in a sequence-independent manner. Using PCR, we can prepare specific primer pairs for a gene to quantify the amount present. Using SYBR Green has several advantages: it is inexpensive, easy to use, and relatively sensitive. A disadvantage is that SYBR Green binds to all dsDNA in a reaction, including primer–dimers and other nonspecific reaction products, creating the potential for misdiagnosis and overestimation of the amount of targeted DNA. For single PCR product reactions with well-designed primers, SYBR Green can be useful, with nonspecific background fluorescence showing up only in very late cycles. However, late-cycle nonspecific background fluorescence can reduce the accuracy of quantification when used to detect genes normally present at low concentrations. Because fetal DNA is normally present at low concentrations in maternal plasma, SYBR Green quantification of fetal DNA has its limitations.

The two most popular alternative methods are TaqMan and molecular beacons, both of which use hybridization probes that rely on fluorescence resonance energy transfer (FRET) for DNA quantification. TaqMan probes are oligonucleotides with a fluorescent dye typically attached to their 5′ end and a quenching dye typically attached to their 3′ end. When irradiated, the excited fluorescent dye transfers energy to a nearby quenching dye molecule rather than fluorescing, producing a nonfluorescent substrate. TaqMan probes are designed to hybridize to an internal region of a given PCR product. During PCR, replication of a template to which a TaqMan probe is bound results in cleavage of the probe via 5′-exonuclease activity of the polymerase. This separates the fluorescent and quenching dyes, and FRET no longer occurs. Fluorescence increases with each cycle, proportional to the rate of probe cleavage. Thus, the TaqMan probe results in a highly specific assay by binding only to a specified cDNA sequence, while increasing the cost and complexity of the assay design (*see* **Note 1**).

Molecular beacons also contain fluorescent and quenching dyes, but FRET only occurs when the quenching dye is directly adjacent the fluorescent dye. Molecular beacons are designed to adopt a hairpin structure when free in solution, bringing the fluorescent dye and quencher into proximity. When a molecular beacon hybridizes to a target, the fluorescent and quencher dye molecules are separated, FRET does not occur, and the fluorescent dye emits light upon irradiation. Unlike TaqMan probes, molecular beacons are designed to remain intact during the amplification process, and they must rebind the target during every cycle for signal measurement. LightCycler technology usually uses this method to quantify the amount of target DNA.

TaqMan probes and molecular beacons allow multiple DNA species to be measured in the same sample (multiplex PCR), because fluorescent dyes with different emission spectra may be attached to different probes. Multiplex PCR allows internal controls to be coamplified and it permits allele discrimination in homogenous assays. These hybridization probes afford a level of discrimination beyond that obtained with SYBR Green, because they will only hybridize to true targets during PCR, and not to primer-dimers or other spurious products.

The basis of real-time quantitative PCR by using LightCycler technology is relatively simple. The LightCycler has two different components: a cycler and a fluorimeter. Unlike conventional PCR where cycling takes several hours, PCR analysis with the LightCycler takes only 20–30 min. The sample carousel holds 32 samples in glass capillaries (outer diameter, 1.55 mm; length, 35 mm) that also permit the reaction volume to be reduced to 10–20 μl. These glass capillaries also serve as cuvettes for flourimetric determination of the quantity of PCR products formed. The use of air as a heat-transfer medium contributes to the high-speed cycling capabilities of the LightCycler.

2. Materials

2.1. Sample Collection and DNA Recovery

1. Maternal peripheral blood (7 ml) is placed into EDTA-coated tubes.
2. QIAamp Blood Mini kit (QIAGEN, Hilden, Germany).

2.2. DNA Quantification (by Using LightCycler Technology)

1. LightCycler setup, LightCycler version 1.0 (Roche Diagnostics, Basel, Switzerland).
2. LightCycler-Fast Start DNA Master hybridization probes (Roche Diagnostics).
3. Male genomic DNA can be used as a control.

3. Methods

3.1. DNA Extraction from Maternal Plasma

1. Maternal blood samples (7 ml) should be collected into tubes containing EDTA. Within 3 h, the plasma can be separated by centrifugation at 3,000*g*. Plasma can

then be transferred into plain polypropylene tubes and stored at –20°C until it is used for DNA extraction.

2. DNA can be extracted from 1-ml samples of plasma supernatant using a QIAamp Blood Mini kit (QIAGEN) (*see* **Note 2**). Total DNA can be eluted from the columns upto 50 μl of water. This can be done in accordance with the manufacturer's "Blood and Body Fluid Protocol," with minor modifications as follows:

   a. Pipet 50 μl of QIAGEN protease (or protease K) into the bottom of a 15-ml tube.
   b. Add 1 ml of sample to the tube.
   c. Add 1 ml of lysis buffer (AL) to the sample. Mix by pulse-vortexing for 15 s.
   d. Incubate at 56°C for 10 min by using water bath.
   e. Briefly centrifuge the tube to remove drops from the inside of the lid.
   f. Add 1 ml of ethanol (96–100%) to the sample and mix again by pulse vortexing for 15 s. After mixing, briefly centrifuge the tube to remove drops from the inside of the lid.
   g. Carefully apply the mixture from **item f** to the QIAamp Spin column (in a 2-ml collection tube) without wetting the rim, close the cap, and centrifuge at 6.000*g* (8,000 rpm) for 1 min. Place the QIAamp Spin column sample into a clean 2-ml collection tube and discard the tube containing the filtrate.
   h. Carefully open the QIAamp Spin column and add 500 μl of buffer wash buffer 1 (AW1) without wetting the rim. Close the cap and centrifuge at 6,000*g* for 1 min. Place the QIAamp Spin column sample into a clean 2-ml collection tube (provided) and discard the collection tube containing the filtrate.
   i. Carefully open the QIAamp Spin column and add 500 μl of wash buffer 2 (AW2) without wetting the rim. Close the cap and centrifuge at full speed (20,000*g*; 14,000 rpm) for 3 min.
   j. Place the QIAamp Spin column sample into a new 2-ml collection tube and discard the collection tube with the filtrate. Centrifuge at full speed for 1 min to eliminate any chance of possible AW2 carryover.
   k. Place the QIAamp Spin column sample into a clean 1.5-ml microcentrifuge tube and discard the collection tube containing the filtrate. Carefully open the QIAamp Spin column and add 40 μl of distilled water. Incubate at room temperature for 5 min and then centrifuge at 6,000*g* (8,000 rpm) for 1 min.

3.2. PCR Quantification

1. LightCycler version 1.0 (Roche Diagnostics) can be used for real-time quantitative PCR assay of maternal plasma DNA.
2. Because women do not have Y chromosomes, the Y chromosome-specific DYS14 gene can be used as a molecular marker to quantify fetal cell-free DNA. The β-globin gene can be used for total DNA quantification. For detection of DYS14, the amplification primer DYS14-713F, and the dual-labeled fluorescent probe

DYS14-883T can be used. For β-globin, the primers GLB-F and GLB-R and the fluorescent probes GLB-LC and GLB-FT can be used (*see* **Table 1**).

3. All fluorogenic PCR reactions should be performed according to the manufacturer's instructions in a reaction volume of 20 µl. All components, except the fluorescent probe and amplification primers, can be obtained from a kit (LightCycler-Fast Start DNA Master hybridization probes, Roche Diagnostics). The 20-µl volume of reaction mixture in a glass capillary tube should contain 3.9 µl of H_2O, 1.6 µl of $MgCl_2$, 2 µl of Fast Start Reaction Mix hybridization probes, 1 µl of DYS14-713F, 1 µl of DYS14-880R, 0.5 µl of DYS14-883T, and 10 µl of DNA template (*see* **Note 3**). Primer–probe combinations can be designed using Primer Express software (PerkinElmer Life and Analytical Sciences, Boston, MA). Sequence data can be obtained from the GenBank sequence database. TaqMan probes can be custom-synthesized by Applied Biosystems (Foster City, CA).
4. Prepare a master mix (*see* **Note 2**) containing all the reaction reagents, except the DNA template. Pipette all reagents into 0.5- or 1.5-ml reaction tubes and mix by gentle vortexing. Place the capillary tubes (*see* **Note 4**) into adapters precooled in a cooling block. Pipette the reagent mix into a plastic container at the top of each capillary tube. Add the DNA template to each capillary tube and seal it with a plastic stopper.
5. To transfer the reaction mix from the plastic container at the top of each capillary tube into the capillaries, centrifuge the adapters containing the capillary tubes briefly in a standard benchtop centrifuge at 700*g*. Use only rotors designed to

Table 1
Primers

Primer	Sequences	Remarks
DYS14-713F	(5′-CAT CCA GAG CGT CCC TGG-3′)	Forward
DYS14-880R	(5′-TTC CCC TTT GTT CCC CAA A-3′)	Reverse
DYS14-883T	(5′-FAM-CGA AGC CGA GCT GCC CAT CA-TAMRA-3′)	FAM (6-carboxyfluorescein) and TAMRA (6-carboxymethylrhodamine)
GLB-F	(5′-ACA CAA CTG TGT TCA CTA GC-3′)	Forward
GLB-R	(5′-CAA CTT CAT CCA CGT TCA CC-3′)	Reverse
GLB-LC	(5′-LC Red-GAA GTC TGC CGT TAC TGC CCT G-Phosphate-3′)	
GLB-FT	(5′-CAT GGT GCA TCT GAC TCC TGA GG-FITC-3′)	

hold 2.0-ml reaction tubes. Do not exceed a centrifugation force of 700*g*. Remove the glass capillaries from the adapters and place them into the LightCycler sample carousel. Place the carousel containing the samples into the LightCycler apparatus and close the lid.

6. Initiate the thermal cycling for DYS14 PCR with denaturation at 95°C for 10 min, followed by 40 cycles of denaturation at 95°C for 10 s, annealing at 57°C for 20 s, and extension at 72°C for 20 s. Initiate the thermal cycling for β-globin PCR with denaturation at 95°C for 10 min, followed by 40 cycles at 95°C for 15 s, and then 57°C for 10 s, and 72°C for 10 s.
7. The number of genomic equivalents of male DNA present in each plasma sample can be determined by comparison with a standard dilution curve of male genomic DNA. For conversion to the amount of genome equivalents use 6.6 pg, as described previously *(2)*. Amplification data can be analyzed using LightCycler software (Roche Diagnostics) to calculate DYS14 and β-globin sequence concentrations (*see* **Fig. 1**). The mean quantity of each duplicate can be used for further calculation. The concentration, expressed in copies per milliliter, can be calculated using the following equation:

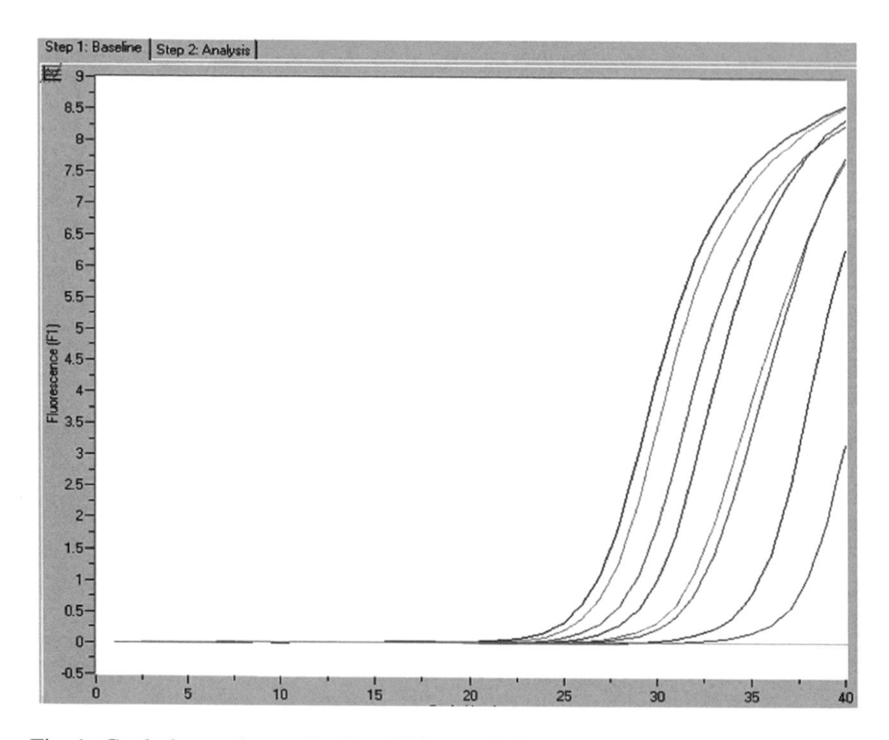

Fig. 1. Cycle-by-cycle monitoring. This graph shows increase in fluorescence during amplification of a target molecule.

$$C = Q \times \frac{V_{DNA}}{V_{PCR}} \times \frac{1}{V_{ext}},$$

where C is the target concentration in plasma or serum (copies per milliliter), Q is the target quantity (copies) determined by the PCR sequence detector, V_{DNA} is the total volume of DNA obtained after extraction, V_{PCR} is the volume of DNA solution used for PCR (typically 5–10 μl), and V_{ext} is the volume of plasma or serum extracted.

4. Notes

1. The TaqMan system can now be used to quantify fetal DNA concentrations in maternal plasma, because it provides greater stability than LightCycler technology.
2. Strict precautions need to be taken against contamination, and multiple negative-control water blanks should be included in every analysis. We recommend using a female staff member to perform all procedures, including sample preparation, DNA extraction, and PCR amplification.
3. Based on our experience, we recommend using no more than the following concentrations of template in each LightCycler assay: 50 ng of genomic DNA (for SYBR Green assay) or 500 ng of genomic DNA (for assay with hybridization probes).
4. If a capillary tube breaks or cracks while in the sample carousel, the broken glass must be removed and the capillary slot thoroughly cleaned. If a new capillary tube is re-inserted before the slot is cleaned, residual glass can damage the precision bore of the hole. Such damage may alter its optical alignment and affect the accuracy of future data. In place of the microbrushes provided by the LightCycler manufacturer, interdental brushes may be used.

References

1. Lo, Y. M., Corbetta, N., Chamberlain, P. F., et al. (1997). Presence of fetal DNA in maternal plasma and serum. *Lancet* **350,** 485–487.
2. Lo, Y. M., Tein, M. S., Lau, T. K., et al. (1998) Quantitative analysis of fetal DNA in maternal plasma and serum: implications for noninvasive prenatal diagnosis. *Am. J. Hum. Genet.* **62,** 768–775.
3. Sekizawa, A., Kondo, T., Iwasaki, M., et al. (2001) Accuracy of fetal gender determination by analysis of DNA in maternal plasma. *Clin. Chem.* **47,** 1856–1858.
4. Lo, Y. M., Hjelm, N. M., Fidler, C., et al. (1998) Prenatal diagnosis of fetal RhD status by molecular analysis of maternal plasma. *N. Engl. J. Med.* **339,** 1734–1738.
5. Saito, H., Sekizawa, A., Morimoto, T., Suzuki, M., and Yanaihara, T. (2000) Prenatal DNA diagnosis of a single-gene disorder from maternal plasma. *Lancet* **356,** 1170.

6. Sekizawa, A., Jimbo, M., Saito, H., et al. (2003) Cell-free fetal DNA in the plasma of pregnant women with severe fetal growth restriction. *Am. J. Obstet. Gynecol.* **188,** 480–484.
7. Sekizawa, A., Farina, A., Sugito, Y., et al. (2004) Proteinuria and hypertension are independent factors affecting fetal DNA values: a retrospective analysis of affected and unaffected patients. *Clin. Chem.* **50,** 221–224.
8. Lo, Y. M., Leung, T. N., Tein, M. S., et al. (1999) Quantitative abnormalities of fetal DNA in maternal serum in preeclampsia. *Clin. Chem.* **45,** 184–188.
9. Farina, A., LeShane, E. S., Lambert-Messerlian, G. M., et al. (2003) Evaluation of cell-free fetal DNA as a second-trimester maternal serum marker of Down syndrome pregnancy. *Clin. Chem.* **49,** 239–242.
10. Sekizawa, A., Jimbo, M., Saito, H., et al. (2002) Increased cell-free fetal DNA in plasma of two women with invasive placenta. *Clin. Chem.* **48,** 353–354.
11. Jimbo, M., Sekizawa, A., Sugito, Y., et al. (2003) Placenta increta: postpartum monitoring of plasma cell-free fetal DNA. *Clin. Chem.* **49,** 1540–1541.
12. Sekizawa, A., Sugito, Y., Iwasaki, M., Wet et al. (2001) Cell-free fetal DNA is increased in plasma of women with hyperemesis gravidarum. *Clin. Chem.* **47,** 2164–2165.
13. Sugito, Y., Sekizawa, A., Farina, A., et al. (2003) Relationship between severity of hyperemesis gravidarum and fetal DNA concentration in maternal plasma. *Clin. Chem.* **49,** 1667–1669.

19

Size Fractionation of Cell-Free DNA in Maternal Plasma and Its Application in Noninvasive Detection of Fetal Single Gene Point Mutations

Ying Li, Wolfgang Holzgreve, and Sinuhe Hahn

Summary

Recent studies have shown that cell-free fetal DNA in maternal plasma can be enriched by means of size fractionation. This technique makes use of the smaller size of fetal DNA fragments compared with maternal DNA fragments isolated simultaneously. On this basis, a highly improved detection of fetal single gene point mutations is permitted. Here, we introduce the use of agarose gel electrophoresis for the size fractionation of cell-free DNA from maternal plasma and the detection of fetal β-thalassemia mutations in the size-fractionated cell-free DNA by using peptide nucleic acid-clamping polymerase chain reaction (PCR) combined with an allele-specific real-time PCR assay. Matrix-assisted laser desorption ionization/time of flight mass spectrometry has been reliably used for detection of fetal single gene point mutations in maternal plasma. We also present its use for genotyping paternally inherited single-nucleotide polymorphism alleles in the size-fractionated cell-free DNA from maternal plasma.

Key Words: Cell-free DNA; size fractionation; single gene mutations; peptide nucleic acid (PNA) clamping; matrix-assisted laser desorption ionization/time of flight mass spectrometry (MALDI-TOF MS); prenatal diagnosis.

1. Introduction

Currently, detection of fetal single gene point mutations in maternal plasma is limited by the preponderance of maternal DNA in the background *(1)*. Recent studies have shown that the majority of cell-free fetal DNA exists in the maternal circulation as small fragments that are <300 bp, whereas maternal

From: *Methods in Molecular Biology, vol. 444: Prenatal Diagnosis*
Edited by: S. Hahn and L. G. Jackson © Humana Press, Totowa, NJ

origin DNA is mostly >300 bp *(2,3)*. This finding permits enrichment of cell-free fetal DNA by size selection. Further studies have indicated that such a selection can improve the detection of fetal single gene point mutations *(4,5)*. Here, we introduce size fractionation of cell-free fetal DNA from maternal plasma by using agarose gel-electrophoresis and detection of fetal single gene point mutations by using size-fractionated plasma DNA.

Agarose gel-electrophoresis provides the most commonly used means of separation of DNA fragments, and it is available in many clinical diagnostic laboratories. Cell-free DNA was extracted from 5 to 10 ml of maternal plasma, and it was separated by standard agarose gel-electrophoresis. The gel slice containing DNA of approximately 100–300 bp was carefully cut out according to DNA size markers. The DNA was eluted from the gel slice using a commercial gel extraction kit (*see* **Fig. 1A**). The eluted DNA containing highly concentrated fetal nucleic acid was used for the analyses of fetal single gene mutations.

Although many methods have been developed for the detection of single gene point mutations, genotyping of the trace amount of DNA in a relatively large background is very difficult *(6)*. Here, we first present a method based on peptide nucleic acid (PNA)-mediated polymerase chain reaction (PCR) clamping and allele-specific real-time PCR *(5)*. PNA is a DNA mimic in which the phosphoribose backbone is replaced by a peptide-like repeat of (2-aminoethyl)-glycine units *(7)*. Due to their different chemical nature, PNAs cannot serve as primers during PCR. However, PNAs can form PNA–DNA hybrids with much higher thermal stability than corresponding DNA–DNA hybrids. In addition, PNA–DNA hybrids are more susceptible to destablization when single base pair mismatches are present. These features permit the PNA sequence, which is complementary to wild-type DNA, to suppress amplification of the normal wild-type allele, and therefore to enrich the mutant allele.

Allele-specific PCR is a commonly applied method for detection of known single-base polymorphisms by using allele-specific primers to amplify and discriminate between two alleles of a gene simultaneously. Here, we use SYBR Green dye to monitor allele-specific PCR and to quantify wild-type maternal DNA and paternally inherited mutant sequences *(8)*.

The principle of our analysis is illustrated in **Figure 1**. We use three steps to enrich for fetal mutant alleles in the β-globin gene: (1) cell-free fetal DNA was selectively enriched from maternal plasma by size fractionation; (2) PNA probes, which are complementary to the maternal wild type sequence, inhibit the amplification of maternal DNA and further enrich the paternal mutant allele; and (3) allele-specific real-time PCR was used to amplify and quantify both mutant and wild-type alleles.

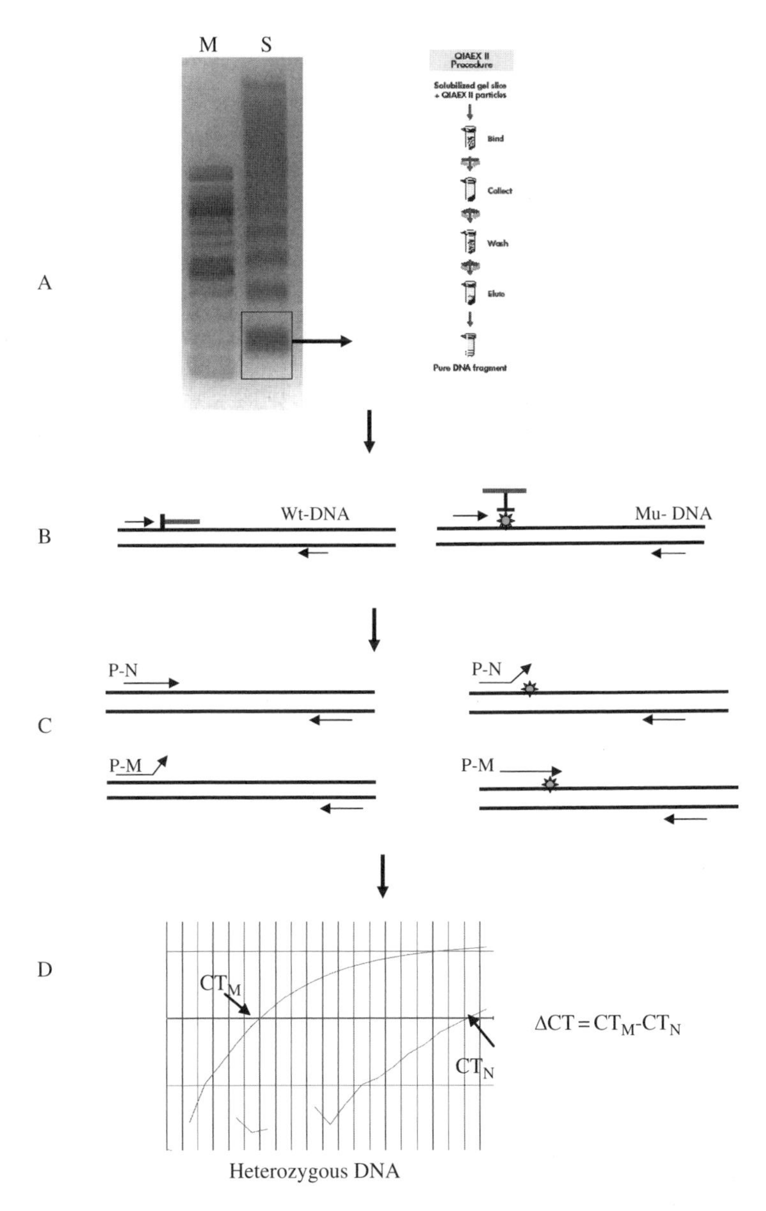

Fig. 1. Schematic illustration of the strategy of detection of β-thalassemia mutation in size fractionated maternal plasma DNA. **(A)** Size fractionation and purification of cell-free fetal DNA. **(B)** PNA-clamping PCR. **(C)** Allelic-specific real-time PCR. **(D)** Analysis of data. M, DNA size marker, S, plasma sample; Wt-DNA, wild-type DNA; Mu-DNA, mutant DNA; P-N, PCR primer 2 for normal allele; P-M, PCR primer 2 for mutant allele.

The homogenous MassEXTEND (hME) assay from Sequenom, Inc. (San Diego, CA) is a matrix-assisted laser desorption ionization/time of flight mass spectrometry ((MALDI-TOF MS)-based technology for single-nucleotide polymorphism (SNP) genotyping with high sensitivity and accuracy ***(9)***. This assay is based on an allele-specific primer extension reaction that allows for discrimination of single-nucleotide differences ***(10)***. Briefly, the target DNA sequence is amplified by PCR. The hME primer is then annealed to the target DNA adjacent to the SNP of interest, and it is extended through the polymorphic site. The extension reaction generates allele-specific extension products, each having a unique molecular mass. The resultant masses of the extension products are then analyzed by the MALDI-TOF MassARRAY software (for details, see Chap. 9). Here, we introduce the use of the hME assay for the detection of paternally inherited SNP alleles in size-fractionated maternal plasma DNA.

2. Materials

All solutions should be prepared with deionized or double-distilled water.

2.1. Size Fractionation of Cell-Free Maternal Plasma DNA

2.1.1. Maternal Plasma DNA Isolation

1. High Pure PCR Template Preparation kit (Roche, Basel, Switzerland, cat. no. 11796828001).
2. Isopropanol (molecular grade, Darmstadt, Germany).
3. Buffer AL (QIAGEN, Hilden, Germany, cat. no. 19075) (*see* **Note 1**).

2.1.2. Agarose Gel-Electrophoresis

1. AgaTabs (Eurgentec, Köln, Germany; cat. no. EP-0030-15) or agarose powder.
2. An electrophoresis apparatus and power supply (e.g., Bio-Rad, Hercules, CA, cat. no. 1704407).
3. 5× Tris borate-EDTA (TBE) buffer: 0.45 M Tris-borate, 0.01 M EDTA, pH 8.3.
4. 6× Blue/Orange Loading Dye (Promega. Madison, WI, cat. no. G1881).
5. 100-bp DNA size marker (New England Biolabs, Ipswich, MA, cat. no. N3231L).
6. Ethidium bromide stock solution (10 mg/ml) (Sigma, St. Louis, MO, cat. no. E1510).

2.1.3. Separation and Purification of Plasma DNA

1. QIAEX II gel extraction kit (QIAGEN, Hilden, Germany, cat. no. 20021).
2. Elution buffer: 10 mM Tris-HCl, pH 8.5.

2.2. Detection of Fetal β-Globin Gene Mutations in Size-Fractionated DNA

1. Primers used for PNA-clamping and allele-specific real-time-PCR (*see* **Tables 1** and **2**.
2. dNTPs mix (100 mM stock solution, each of dATP, dCTP, dDTP, and dTTP. Invitrogen, Frederick, MD, cat. no. 10297–018): Prepare mix of 4 dNTPs in double-distilled H_2O (ddH_2O) to a working concentration of 10 mM and store at –20°C.
3. AmpliTaq Gold DNA polymerase, supplied with 10× PCR buffer and 25 mM $MgCl_2$ solution (Applied Biosystems, Foster City, CA, cat. no. N8080247).
4. SYBR Green PCR core reagents (Applied Biosystems, cat. no. 4304886).

2.3. MALDI-TOF MS

2.3.1. Specialized Equipment and Hardware

For details, *see* **Subheading 2.1.** in Chap. 9.

2.3.2. PCR Reaction

1. Dissolve primers in ddH_2O to 10 μM each and store at –20°C.
2. 25 mM dNTPs mix.
3. HotStarTaq DNA polymerase, supplied with 10× PCR buffer containing 15 mM $MgCl_2$ and 25 mM $MgCl_2$ (QIAGEN, cat. no. 203205).

2.3.3. Shrimp alkaline phosphatase (SAP) Treatment

For details, *see* **Subheading 2.3.** in Chap. 9.

2.3.4. MassARRAY hME Primer Extension

For details, *see* **Subheading 2.4.** in Chap. 9.

2.3.5. Cleanup of Extension Products

For details, *see* **Subheading 2.5.** in Chap. 9.

2.3.6. Preparation of Chips for MS

For details, *see* **Subheading 2.6.** in Chap. 9.

Table 1
Primers and PCR conditions for PNA-clamping PCR

Mutation	Sequences of PCR primers and PNA probes (5-3)	Conditions of PNA clamping-PCR
IVSI-1	Primer-1: GTGAACGTGGATGAAGTTGGT Primer-2: TCTCCTTAAACCTGTCTTGTAACCTTCTAT PNA probe: GATACCAACCTGCCC	Incubate at 95°C for 10 min, followed by 25 cycles of 95°C for 15 s, 70°C for 1 min, 60°C for 15 s, and 72°C for 30 s; final extension at 72°C for 5 min
IVSI-6	Primer-1: GTGAACGTGGATGAAGTTGGT Primer-2: CTTAAACCTGTCTTGTAACCTTGA PNA probe: GATACCAACCTGCCC	Incubate at 95°C for 10 min, followed by 25 cycles of 95°C for 15 s, 70°C for 1 min, 60°C for 15 s, and 72°C for 30 s; final extension at 72°C for 5 min
IVSI-110	Primer-1: ACTCTTGGGTTTCTGATAGGCACT Primer-2: CAGCCTAAGGGTGGGAAAATAG PNA probe: TAGACCAATAGGC	Incubate at 95°C for 10 min, followed by 25 cycles of 95°C for 15 s, 71°C for 1 min, 62°C for 15 s, and 72°C for 30 s; final extension at 72°C for 5 min
Codon39	Primer-1: CTCTGCCTATTGGTCTATTTTCCC Primer-2: ATCCCCAAAGGACTCAAAGAACC PNA probe: ACCTCTGGGTCCA	Incubate at 95°C for 10 min, followed by 25 cycles of 95°C for 15 s, 72°C for 1 min, 63°C for 15 s, and 72°C for 30 s; final extension at 72°C for 5 min

Table 2
The primers and PCR conditions for allele-specific real-time PCR

Mutation	Sequences of PCR primers	Conditions of allele-specific real-time PCR
IVSI-1	Primer-1: GTGAACGTGGATGAAGTTGGT Primer-2N: TAAACCTGTCTTGTAACCTTGATACGAAC Primer-2M:TTAAACCTGTCTTGTAACCTTGATACGAAT	Incubate at 95°C for 10 min, followed by 40 cycles of 95°C for 15 s, 63.5°C for 15 s, and 72°C for 30 s
IVSI-6	Primer-1: GTGAACGTGGATGGAGTTGGT Primer-2N: CTTAAACCTGTCTTGTAACCTTCATA Primer-2M: CTTAAACCTGTCTTGTAACCTTCATG	Incubate at 95°C for 10 min, followed by 40 cycles of 95°C for 15 s, 62.5°C for 15 s, and 72°C for 30 s
IVSI-110	Primer-1: ACTCTTGGGTTTCTGATAGGCACT Primer-2N: CAGCCTAAGGGTGGGAAAATACACC Primer-2M: CAGCCTAAGGGTGGGAAAATACACT	Incubate at 95°C for 10 min, followed by 40 cycles of 95°C for 15 s, 59.5°C for 15 s, and 72°C for 30 s
Codon39	Primer-1:CTCTGCCTATTGGTCTATTTTCCC Primer-2N: ATCCCCAAAGGACTCAAAGAACCTGTG Primer-2M: ATCCCCAAAGGACTCAAAGAACCTGTA	Incubate at 95°C for 10 min, followed by 40cycles of 95°C for 15 s, 61.5°C for 15 s, and 72°C for 30 s

3. Methods

3.1. Size Fractionation of Cell-Free Maternal Plasma DNA

3.1.1. Maternal Plasma DNA Isolation

Blood samples were collected, and plasma was separated as described in Chap. 15.

Before starting DNA isolation, switch on a water bath and set temperature to 70°C.

1. Pipet 5–8 ml of maternal plasma into a 50-ml centrifuge tube.
2. Add equal sample volume of lysis buffer and mix briefly.
3. Add 500 μl of protease K (20 mg/ml) (*see* **Note 2**) and mix thoroughly by inverting the tube.
4. Incubate at 70°C for 10 min; mix the solution once per 2 min.
5. Add half of the sample volume of isopropanol to the solution and mix by vortexing.
6. Insert a column (Roche, High Pure PCR template preparation kit) into a rubber tube that is connected to a vacuum pump.
7. Pipet the solution onto the column and carefully pump the sample through the filter.
8. Place the filter column into a 2-ml collection tube (provided in the kit) and then close the cap and centrifuge at 8,000*g* (*see* **Note 3**) for 1 min. In the mean time, warm up the elution buffer to 70°C.
9. Without changing the collection tube, add 500 μl of inhibitor removal buffer onto the column and then centrifuge at 8,000*g* for 1 min.
10. Combine the column with a new collection tube, wash twice by adding 500 μl of wash buffer onto the column, and centrifuge for 1 min at 8,000*g*.
11. Combine the column with a new collection tube and centrifuge for 10 s at 13,000*g* to remove residual wash buffer.
12. Insert the column into a sterile 1.5-ml Eppendorf tube; add 50 μl of prewarmed elution buffer to the column and centrifuge at 8,000*g* for 1 min. The eluted DNA can be used directly for the next analysis or stored at –20°C.

3.1.2. Size Separation on Agarose Gel Electrophoresis

1. Prepare (*see* **Note 4**) a 1.0% agarose gel (*see* **Note 5**) containing 0.5% of ethidium bromide in 0.5× TBE buffer.
2. Mix 40 μl of plasma DNA with 6 μl of 6× loading buffer and carefully pipet all the solution into one gel well.
3. Load 100-bp ladder DNA marker (400 ng) into one well of the gel.
4. Carefully add 0.5× TBE buffer into the gel tank and close the gel tank (check whether the electrodes are in the correct direction!).
5. Running conditions: 100 V for 1 h.

6. Weigh a 1.5-ml sterile Eppendorf tube.
7. After electrophoresis, visualize the gel on an UV transilluminator (*see* **Note 6** and **7**). Carefully cut out the fragments between 100 and 300 bp according to the DNA size marker (*see* **Note 8**) and transfer them into the weighed Eppendorf tube. Weigh the tube again, now containing the gel slice.

3.1.3. Purification of DNA from the Agarose Gel Slice

For further details, refer to the handbook from the QIAGEN QIAEX II gel extraction kit.

1. Calculate the gel weight. Add 3 volumes of buffer QX1 to 1 volume of gel. For example, add 300 μl of buffer QXI to each 100 mg of gel (*see* **Note 9**).
2. Vortex the QIAEX II solution for 30 s to resuspend and then add 10 μl of QIAEX II to the sample.
3. Incubate at 50°C until the gel is solubilized. Mix by vortexing every 2 min to keep the QIAEX II in suspension.
4. Centrifuge the sample for 30 s (>10,000*g*) and wash the pellet with 500 μl of buffer QX1.
5. Wash the pellet twice with 500 μl of buffer PE.
6. Keep the tube open and air dry the pellet at room temperature (*see* **Note 10**).
7. Add 25 μl of 10 mM Tris-Cl, pH 8.5, and resuspend the pellet by vortexing. Incubate at room temperature for 10 min.
8. Centrifuge for 30 s at 13,000*g* and then carefully pipet the supernatant into a clean Eppendorf tube.
9. Repeat **steps 7** and **8** and combine the eluates; store the eluted DNA at –20°C

*3.2. Detection of Fetal β-Globin Gene Mutations in Size-Fractionated DNA: PNA-Clamping PCR (*see **Note 11***)*

*3.2.1. PNA-PCR Reaction Cocktail (*see **Note 12***)*

In a 0.5-ml sterile, thin-walled PCR tube, combine the following:

1. 3 μl of 10× PCR Gold buffer without $MgCl_2$.
2. 4.2 μl of 25 mM $MgCl_2$ (final concentration of magnesium in 30 μl of PCR reaction will be 3.5 mM).
3. 0.6 μl of 10 mM dNTPs mix (final concentration of each dNTPs will be 200 μM).
4. 0.6 μl each of 10 mM primers (final concentration will be 200 μM).
5. 0.2 μl of AmpliTaqGold DNA polymerase (5 U/μl).
6. 0.67 μM PNA probe (*see* **Notes 13–17**) for intervening sequence (IVS1)-I mutation (0.5 μM PNA for IVS1-6 mutation, 1 μM for IVS1-110 and codon39 mutation, respectively).

7. 8 μl of size-fractionated cell-free DNA, to which the appropriate volume of sterile distilled water has been added for a final volume of 30 μl. Mix gently and spin down briefly.

3.2.2. PNA-PCR Reactions

The PNA-clamping reactions are shown in **Table 1**. For each mutation, the additional step at 70, 70, 71, and 72°C, respectively, was chosen to allow the preferential annealing of PNA to DNA. The clamping reactions were performed in a thermal cycler.

3.2.3. Allele-Specific Real-Time PCR

1. Allele-specific real time PCR cocktail: prepare PCR reaction mixtures in 0.5-ml sterile, thin-walled PCR tubes, with one tube for amplifying the mutant allele and one tube for normal wild-type allele separately.
2. 25 μl of PCR reaction containing 160 nM of each primer (primer-N or primer-M and primer-F), 1× SYBR Green master mix (containing AmpliTaq Gold DNA polymerase) and 1 μl of the PCR-clamping product.
3. Allele-specific real-time PCR reactions: allele-specific real-time PCR reactions were performed with a PerkinElmer Applied Biosystems 7000 sequence detector (refer to **Table 2** for the reaction conditions).

3.2.4. Data Analysis

For each allele-specific real-time PCR, a threshold cycle value (CT) for the mutant allele (CT_M) and wild-type allele (CT_N) can be determined (*see* **Fig. 1D**). We used the difference in threshold values to distinguish mutant alleles from normal alleles: $\Delta CT = CT_M - CT_N$.

For each of the four mutations, a clear difference should be observed.

3.3. Detection of Fetal SNPs in Size-Fractionated Cell-Free DNA by MALDI-TOF MS

For details, see Chap. 9.

3.3.1. Amplification of Target Genes by PCR

1. Amplify (*see* **Note 18**) the target SNP (*see* **Note 19**) by PCR in 25-μl reaction volumes consisting of 8 μl of size-fractionated cell-free DNA, 200 nM of each primer, 50 μM dNTPs, 2.5 mM Mg^{2+}, and 0.5 U of HotStar Taq polymerase (*see* **Note 20**).
2. Initiate the PCR reaction by an initial incubation at 95°C for 15 min, followed by 50 cycles of 94°C for 20 s, 56°C for 30 s, and 72°C for 1 min, with a final incubation at 72°C for 5 min.

3.3.2. SAP Treatment

1. Transfer 10 μl of PCR product into a skirted 96-well plate.
2. Add 4 μl of SAP solution (0.34 μl of hME buffer, 0.6 μl of SAP enzyme, and 3.06 μl of water) to each 10 μl of PCR product.
3. Incubate at 37°C for 30 min, followed by 85°C for 5 min.

3.3.3. Primer Extension and Desalting

1. Extension of the SAP-treated PCR products: add 0.4 μl of appropriate terminator mix (containing 2′-3′-dideoxynucleoside triphosphates and dNTPs), 0.4 μl of 25 μM extension primer, and 0.04 μl of Thermo Sequenase enzyme (32 U/μl) to the PCR product.
2. Thermocycle using the following conditions in an Eppendorf Master PCR thermal cycler: an initial heating step of 94°C for 2 min, followed by 55 cycles of 95°C for 5 s, 52°C for 5 s, and 72°C for 5 s.
3. Add 32 μl of ddH_2O to each well containing extended products.
4. Add 6 mg of CLEAN resin to each well and then seal the plate. Slowly rotate it for 5–10 min.
5. Spin down the resin by centrifuging the plate at 1,600*g* for 5 min.

3.3.4. MALDI-TOF MS Data Acquisition and Analysis

1. Dispense approximately 15 nl of the desalted product solution onto a 384-format SpectroCHIP by using a MassARRAY nanodispenser.
2. Analyze the chip using the MassARRAY Analyzer Compact and MassARRAY software (*see* **Note 21**).

4. Notes

1. Buffer AL from QIAGEN can be replaced with lysis buffer from the Roche High Pure Template DNA purification kit.
2. Do not add proteinase K directly into lysis buffer (add only after plasma has been added first).
3. Centrifugation steps are carried out at room temperature.
4. To avoid contamination wash carefully the gel casting tray, comb and electrophoresis tank with ddH_2O. Before casting a gel, leave these tools under UV light for 2 min. Replace the buffer after each run. Use each gel for running only one sample to avoid sample cross-contamination.
5. When preparing the 1% agarose gel, make sure that the wells can each hold a 50-μl volume. We suggest making a 50-ml gel and using an eight-well comb when using a Mini-Sub Cell GT electrophoresis apparatus (Bio-Rad, cat. no. 1704467).
6. Ethidium bromide is a powerful mutagen, and it should be handled with care. Wear gloves while handling. All pipette tips, buffers, and so on must be disposed of appropriately.

7. Always wear protective eyewear when observing DNA on a transilluminator to avoid injury to the eyes and skin from UV light.
8. Usually, no bands can be seen or only the band ≈180 bp can be slightly observed on the agarose gel under UV light when the DNA was extracted from "fresh" plasma samples. If the whole blood samples undergo 24-h express courier delivery, the plasma DNA shows up as a smear and multiple bands due to apoptosis of maternal cells (*see* **Fig. 1A**).
9. Check the color of the mixture after the gel melts in buffer QXI. The color should be yellow. If it is orange or purple, add 3 μl of 3 M sodium acetate, pH 5.0, and mix; thereafter, the color should become yellow. Continue the incubation for a minimum of 5 min.
10. Do not over dry the pellet, because this might decrease the elution efficiency.
11. Maternal blood samples were collected from pregnant women carrying the β-thalassemia mutation, and the corresponding partner carrying a different mutation in the β-globin gene. This assay was used to exclude compound heterozygous β-thalassemia pregnancies. If the paternally inherited mutant allele is absent, invasive prenatal diagnosis could be avoided.
12. For each mutation, the concentration of primers and PNA probes and the reaction conditions, e.g., annealing temperature, should be optimized. Carefully adjust these factors such that the maximum distinction between mutant allele and wild-type allele can be achieved.
13. We recommend referring to the Guidelines for Sequence Design of PNA oligomers and online software for designing PNA probes. (www.appliedbiosystems.com/support/seqguide.cfm).
14. Depending on the yield of synthesis, add the appropriate volume of ddH_2O to prepare a 100 μM stock solution.
15. Leave the PNA solution at room temperature for at least 10 min with intermittent vortexing; then, warm it to 50°C for 10 min and vortex briefly.
16. Aliquot and dilute to the working concentrations, e.g., 10 μM, and store at 4°C.
17. Before preparing the PCR reaction, remove the PNA solution from the refrigerator and let it equilibrate to room temperature; then, warm it to 50°C for 10 min. Vortex and centrifuge briefly.
18. Avoidance of contamination should be strictly considered, especially with post-PCR procedures, e.g., prepare the PCR reaction and post-PCR steps in different places. Always set up reactions in a designated hood and turn on the UV light for a few minutes before and after use.
19. Instead of a multiplex reaction, we recommend to perform a single reaction for the analysis of maternal plasma DNA due to the low amount of input DNA.
20. HotStar Taq DNA Polymerase (QIAGEN) for the PCR reaction has been recommended by Sequenom, Inc. However, other Taq DNA polymerases, such as AmpliTaq Gold polymerase, also work well.
21. Interpret the results carefully when genotyping the fetal allele from maternal plasma. Always check the spectral views, because the findings there may not always correspond with the result report. We have sometimes found that a

reading of "no allele" or "low probability" allele appears in the report, although a peak, albeit of low intensity, is indeed present in the spectrum.

Acknowledgment

We thank Dr. D. J. Huang for helpful comments and proofreading.

References

1. Hahn, S. and Holzgreve, W. (2002) Prenatal diagnosis using fetal cells and cell-free fetal DNA in maternal blood: what is currently feasible? *Clin. Obstet. Gynecol.* **45,** 649–656.
2. Chan, K. C., Zhang, J., Hui, A. B., et al. (2004) Size distributions of maternal and fetal DNA in maternal plasma. *Clin. Chem.* **50,** 88–92.
3. Li, Y., Zimmermann, B., Rusterholz, C., Kang, A., Holzgreve, W., and Hahn, S. (2004) Size separation of circulatory DNA in maternal plasma permits ready detection of fetal DNA polymorphisms. *Clin. Chem.* **50,** 1002–1011.
4. Li,Y., Holzgreve, W., Page-Christiaens, G. C., Gille, J. J., and Hahn, S. (2004) Improved prenatal detection of a fetal point mutation for achondroplasia by the use of size-fractionated circulatory DNA in maternal plasma—case report. *Prenat. Diagn.* **24,** 896–898.
5. Li, Y., Di Naro, E., Vitucci, A., Zimmermann, B., Holzgreve, W., and Hahn, S. (2005) Detection of paternally inherited fetal point mutations for beta-thalassemia using size-fractionated cell-free DNA in maternal plasma. *J. Am. Med. Assoc.* **293,** 843–849.
6. Mike, M. G. (2004) PCR-based detection of minority point mutations. *Hum. Mutat.* **23,** 406–412.
7. Murdock, D. G., Christacos, N. C., and Wallace, D. C. (2000) The age-related accumulation of a mitochondrial DNA control region mutation in muscle, but not brain, detected by a sensitive PNA-directed PCR clamping based method. *Nucleic Acids Res.* **28,** 4350–4355.
8. Shively, L., Chang, L., LeBon, J. M., Liu, Q., Riggs, A. D., and Singer-Sam, J. (2003) Real-time PCR assay for quantitative mismatch detection. *Biotechniques* **34,** 498–502.
9. Tost, J. and Gut, I.G. (2005) Genotyping single nucleotide polymorphisms by MALDI mass spectrometry in clinical applications. *Clin. Biochem.* **38,** 335–350.
10. Jurinke, C., Oeth, P., and van den Boom, D. (2004) MALDI-TOF mass spectrometry: a versatile tool for high-performance DNA analysis. *Mol. Biotechnol.* **26,** 147–164.

20

MALDI-TOF Mass Spectrometry for Analyzing Cell-Free Fetal DNA in Maternal Plasma

Chunming Ding

Summary

The discovery of cell-free fetal DNA in the plasma and serum of pregnant women has opened a new window for noninvasive prenatal diagnosis. Robust detection and quantification have been achieved when the fetal DNA sequence of interest does not have a maternal counterpart (e.g., Y chromosomal DNA, *RhD* gene when the mother is *RhD* negative) by techniques such as real-time polymerase chain reaction (PCR). However, detection of subtle fetal mutations is difficult due to the overwhelming maternal DNA background. A method combining PCR, base extension reaction, and matrix-assisted laser desorption ionization/time of flight mass spectrometry (MALDI-TOF MS) allowing DNA detection with single base specificity and single DNA molecule sensitivity is described. DNA sequence is amplified by PCR first. Then, a third primer (extension primer) is designed to anneal to the region immediately upstream of the mutation site. Depending on the specific mutation and the ddNTP/dNTP mixtures used, either one or two bases are added to the extension primer to produce two extension products from the wild-type DNA and the mutant DNA. Last, the two extension products are detected by high-throughput MALDI-TOF MS. In addition, with an improved base extension method called single allele base extension reaction, fetal DNA can be robustly detected even when overwhelming maternal background DNA is present.

Key Words: Cell-free fetal DNA; plasma, serum; DNA mutation; single-nucleotide polymorphism (SNP); DNA quantification; matrix-assisted laser desorption ionization/time of flight mass spectrometry (MALDI-TOF MS); base extension; single allele base extension reaction (SABER); prenatal diagnosis.

From: *Methods in Molecular Biology, vol. 444: Prenatal Diagnosis*
Edited by: S. Hahn and L. G. Jackson © Humana Press, Totowa, NJ

1. Introduction

The discovery of cell-free fetal DNA in the plasma and serum of pregnant women has opened a new window for noninvasive prenatal diagnosis ***(1)***. Cell-free fetal DNA constitutes approximately 3.4% of all DNA in the maternal plasma, whereas intact fetal cells constitute only approximately 0.001% of all cells in the maternal blood ***(2)***. This enrichment of approximately 3400-fold may allow us to directly analyze the total cell-free DNA from the maternal plasma for prenatal diagnosis purposes. Robust detection and quantification have been achieved when the fetal DNA sequence of interest does not have a maternal counterpart (e.g. Y chromosomal DNA, *RhD* gene when the mother is *RhD* negative) by techniques such as real-time PCR ***(2,3)***. However, detection of subtle fetal mutations is difficult due to the overwhelming maternal DNA background ***(4)***. Enrichment of fetal DNA based on the size difference between the maternal and fetal DNA in the maternal plasma ***(5)*** and single allele base extension reaction (SABER) have been shown to improve the detection of fetal-derived subtle mutations ***(6,7)***.

Matrix-assisted laser desorption ionization/time of flight mass spectrometry (MALDI-TOF MS) is extremely precise in determining the mass-to-charge ratios with a resolution approximately 100 times higher than capillary sequencing for DNA analysis ***(8)***. MALDI-TOF MS is extremely fast (a few seconds for each sample), and it has been highly automated ***(9–11)***. As a result, MALDI-TOF MS is most widely used in high-throughput qualitative analyses, such as protein sequencing ***(12)*** and single-nucleotide polymorphism (SNP) genotyping ***(10,13)***.

The sensitivity and specificity of DNA analysis by MALDI-TOF MS are further enhanced by two amplifications steps. DNA is first amplified by polymerase chain reaction (PCR) (*see* **Fig. 1**). Then, an additional linear amplification step called base extension reaction is performed with a third primer (extension primer) designed to anneal to the region immediately upstream of the mutation site. Depending on the specific mutation introduced and the ddNTP/dNTP mixtures used, either one or two bases are added to the extension primer to produce two extension products from the wild-type DNA and the mutant DNA. The single base specificity is achieved in this base extension step by the polymerase (Thermo Sequenase), which typically has a misincorporation rate around 10^{-7} ***(14)***. As a result, it is likely that the base extension reaction can distinguish single base mutations better than hybridization-based techniques such as TaqMan probes.

Here, examples and detailed protocols are provided to demonstrate the utility of MALDI-TOF MS for cell-free fetal DNA detection in the maternal plasma. First, detection of fetal Y chromosomal *SRY* gene is used to demonstrate the concept and basic procedure for the combined method of PCR, base extension

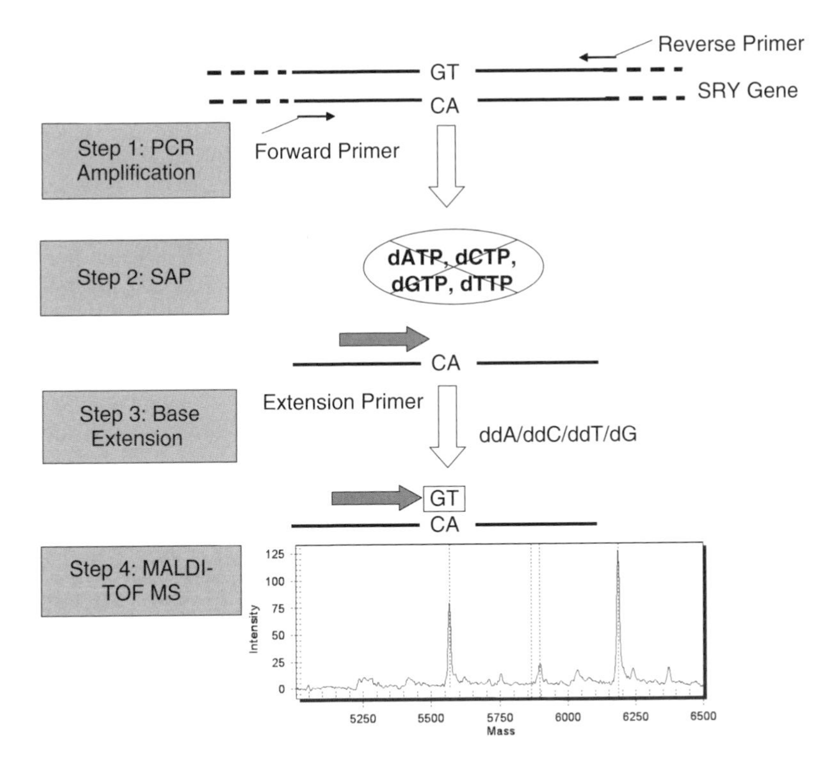

Fig. 1. Flowchart for *SRY* gene detection with PCR, base extension reaction, and MALDI-TOF MS. Cell-free DNA from the maternal plasma is first amplified by PCR. Then, SAP treatment is used to remove the remaining dNTPs. Afterward, base extension reaction is carried out. In base extension reaction, an extension primer is annealed to the amplicon and is extended by two bases. This extension primer increases the specificity as unspecifically amplified products cannot serve as the template. The extension product (and sometimes, unused extension primer) is detected by MALDI-TOF MS in an automated mode.

reaction, and MALDI-TOF MS. Second, SABER method is described for the robust detection of paternally inherited single nucleotide polymorphism in the maternal plasma.

2. Materials

1. QIAamp DNA Blood Mini kit (QIAGEN, Valencia, CA).
2. Water (DNase- and RNase-free, 0.1 μM filtered, Sigma-Aldrich, St. Louis, MO, W4502).
3. HotStar *Taq* Polymerase (QIAGEN).

4. Shrimp alkaline phosphatase (SAP) (Sequenom, Inc., San Diego, CA).
5. Thermo Sequenase (Sequenom, Inc.).
6. dNTP mixture (Invitrogen, Carlsbad, CA) 25 mM each.
7. ddNTP/dNTP mixtures (Sequenom, Inc.). Typically, three different ddNTPs and one dNTP are mixed together at equal molar concentrations.
8. PCR and extension primers (Integrated DNA Technologies, Coralville, IA).
9. Homogenous MassEXTEND (hME) buffer (Sequenom, Inc.).
10. 384- or 96-well microplate (Marsh Biomedical Products, Rochester, NY) and plate seal (Applied Biosystems, Foster City, CA, cat. no. 4306311).
11. 12-channel electronic pipette (0.5–10 μl, cat. no. E12-10, Rainin).
12. 96-well SpectroCLEAN plate, SpectroCLEAN resin, and scraper (Sequenom, Inc.).
13. 384-format SpectroCHIP prespotted with 3-hydroxypicolinic acid (Sequenom, Inc.).
14. SpectroPOINT nanodispenser (Sequenom, Inc.).
15. Modified MALDI-TOF MS compact with Quantitative Gene Analysis (formerly called Allelotyping) software (Sequenom, Inc.).

3. Methods

The methods described here include plasma DNA extraction, PCR, SAP treatment, base extension reaction and single allele base extension reaction, post-PCR sample processing, nanodispensing, and MALDI-TOF MS analysis. It is recommended to use the commercially available MassARRAY (Sequenom, Inc.) system bundled with all necessary software for most efficient and accurate analyses. Expert users also may be able to replicate the applications in any MALDI-TOF MS-based platform.

3.1. DNA Extraction from Maternal Plasma

1. Venous blood (2.5 ml) is drawn into EDTA tubes.

3.1.1. Plasma Collection for DNA Extraction

1. Centrifuge the blood sample at 1,600*g* for 10 min at room temperature.
2. Use a sterile pipette to transfer the plasma to a sterile 1.5-ml microcentrifuge tube, avoid taking out any cells.
3. Centrifuge the plasma at 16,000*g* for 10 min at room temperature.
4. Carefully pipette out the plasma without taking any of the cells.
5. Store the plasma as 400-μl aliquots in 1.5-ml microcentrifuge tubes at –20°C until DNA extraction.

3.1.2. Plasma DNA Extraction Using Blood and Body Fluid Spin Protocol of the QIAamp DNA Blood Mini Kit

1. Thaw 2× 400 μl of plasma aliquots. Pipette 40 μl of QIAGEN Protease into each tube. Vortex for 10 s.
2. Add 400 μl of buffer AL to each of the sample. Vortex for 30 s.
3. Incubate at 56°C for 10 min. (We use a dry bath for incubation.)
4. Briefly centrifuge the 1.5-ml microcentrifuge tubes to spin down the drops from the inside of the lid due to evaporation in the last step.
5. Add 400 μl of cold absolute ethanol to the sample and mix again for 30 s. After mixing, briefly centrifuge the 1.5-ml microcentrifuge tube at room temperature to spin down the drops from the inside of the lid.
6. Apply the mixture from **step 5** (approx. 630 μL at one time) to the QIAamp spin column (in a 2-ml collection tube) without wetting the rim, close the cap, and centrifuge at 9,000*g* for 1 min at room temperature.
7. Discard the liquid waste in the 2-ml collection tube, transfer the spin column to a clean 2-ml collection tube and repeat **step 6** until all mixture is loaded.
8. Open the QIAamp spin column and add 630 μl of buffer AW1 without wetting the rim. Close the cap and centrifuge at 9,000*g* for 1 min. Place the QIAamp spin column in a clean 2-ml collection tube and discard the collection tube containing the filtrate.
9. Open the QIAamp spin column and add 630 μl of buffer AW2 without wetting the rim. Close the cap and centrifuge at full speed (16,000*g*) for 3 min.
10. (Optional) Transfer the column to a new collection tube and centrifuge at 16,000*g* for 1 min. This optional step is to eliminate any chance of possible buffer AW2 carryover.
11. Place the QIAamp spin column in a sterile 1.5-ml microcentrifuge tube. Open the QIAamp spin column and add 50 μl of sterile water (be sure to add the water to the middle of the column, avoid leaving it at the wall of the column). Incubate at room temperature for 5 min and then centrifuge at 16,000*g* for 1 min.
12. Label the sample and store the DNA solution at –20°C until further analysis.

3.2. PCR and Base Extension Reaction Assay Design

3.2.1. SRY Gene Detection

This section provides a generic design for detecting any genomic DNA sequence (e.g., to detect the presence of a Y-chromosome sequence). The main task of the assay design is to identify a short sequence of approximately 60–80 bp within the DNA sequence of interest that satisfies the following criteria (*see* **Note 1**):

1. The target sequence should preferably not have a high GC content so that it is easily amplifiable by PCR.

2. Two PCR primers (forward and reverse primers) that anneal to the two ends of the sequence. The PCR primers are typically 20 base pairs and have a melting temperature (Tm) above 58°C. A 10-base tag (5′-ACGTTGGATG-3′) is added to the 5′ end of the PCR primers so that these primers will not interfere in mass spectra.
3. An artificial mutation can be introduced roughly in the middle of the sequence. This artificial mutation is created to satisfy the AssayDesigner software (Sequenom, Inc.) file format requirement (*see* **Note 1**).
4. Directly adjacent to the mutation site (either on the forward or reverse strand) an extension primer (*see* **Note 2**) of 16–25 bases (approx 4800 to 7500 Da), with a Tm 55°C or above can be identified.

The selected *SRY* sequence for the analysis (with an artificial SNP) is 5′-CGCATTTTCAGGACAGCAGTAGAGCA[G/A]TCAGGGAGGCAGAT CAGCAGGGCAAGTAGTCAACGTTAC-3′. The PCR primers and the extension primer sequences are as follows: 5′-ACGTTGGATGCGCATTTTTC AGGACAGCAG-3′, 5′-ACGTTGGATGGTAACGTTGACTACTTGCCC-3′, and 5′-CAGGACAGCAGTAGAGCA-3′ (molecular mass 5566.6 Da). The extension mixture is ACT (ddATP/ddCTP/ddTTP/dGTP). The extension product for the *SRY* gene is 5′-CAGGACAGCAGTAGAGCA**Gt**-3′ (molecular mass 6184.1 Da), where G is dG and t is ddT.

3.2.2. SABER Design for Paternal SNP Detection

This design applies to situations where a researcher wants to detect the mutant sequence in the presence of an overwhelming (>95%, or even >99%) wild-type sequence, such as in the case of detecting paternally inherited mutations in the maternal plasma for prenatal exclusion of β-thalassemia major *(7)*.

In SABER, only one ddNTP is supplied such that only the mutant sequence can serve as the template for base extension (*see* **Fig. 2**). The specificity of the assay relies solely on the Thermo Sequenase (*see* **Note 3**). It is likely that SAP may not remove all remaining dNTPs from the PCR step. Thus, the leftover dNTP may still be used in the base extension by the wild-type DNA template. For this reason, the assay design needs to be carefully carried out such that all these issues can be dealt with. The following example with SNP rs2187610 (dbSNP ID) will be used to illustrate the assay design principles.

The genomic sequence surrounding SNP rs2187610 is the following: 5′-GGAAAAATAAAGAAGTGAGGCTACATCAAACTAAAAAATTTCCA CACAAAAAA[C/G]AAAACAATGAACAAATGAAAGGTGAACCATGAA ATGGCATATTTGCAAACCAAATATTTCTTAAATATTTTGGTTAATAT CCAAAATA-3′. The mutation is in the format [C/G]. AssayDesigner software is used to design the PCR primers and the extension primer. The output PCR primers are as follows: 5′-ACGTTGGATGATGCCATTTCATGGTTCACC-3′

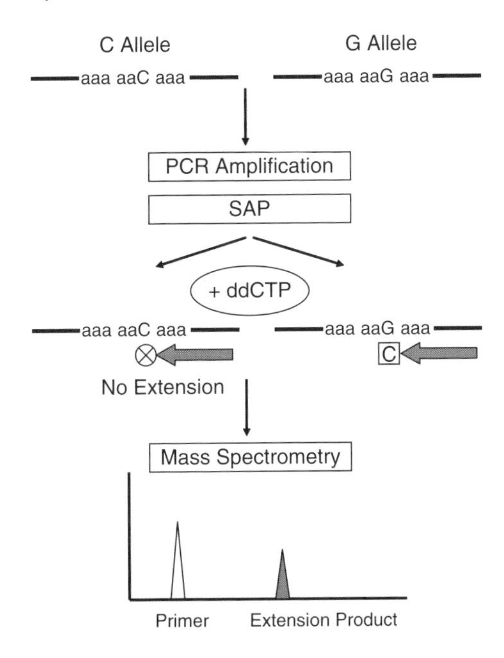

Fig. 2. Schematic illustration of the SABER assay. Maternal plasma detection of the paternally inherited SNP allele by using SNP rs2187610 is presented as an illustrative example. Maternal plasma is first amplified by PCR. The PCR products are subjected to the SABER protocol. The SABER method only extends the fetal specific mutant allele since only ddCTP is supplied. Some of the extension primer may not be used (white triangle).

and 5′-ACGTTGGATGGAAGTGAGGCTACATCAAAC-3′. The output extension primer is 5′-CCTTTCATTTGTTCATTGTTTT-3′ in the reverse direction. To detect the mutation, only ddCTP is added so that the extension product is 5′-CCTTTCATTTGTTCATTGTTTTc-3′ where the small letter c represents the addition of a ddC. If there is leftover dNTP from incomplete SAP treatment, the extension primer will be extended by a dG by using the wild-type sequence as the template and the leftover dGTP. But the molecular masses of dG and ddC extended products differ by 56 Da, so they can be readily separated in MALDI-TOF MS (*see* **Note 4**).

3.3. SRY Detection

3.3.1. PCR

A protocol for carrying out PCR reactions in a 384-well microplate is provided in this section. Generally, you can scale up or down. In this protocol, I have provided two reaction conditions. In condition 1, 1 μl of DNA is used in a

Table 1
PCR cocktail setup

Reagent	Final conc.	5-μl Reaction (μl)	10-μl Reaction (μl)
Water (*see* **Subheading 2.**)	N/A	2.24	1.48
10× HotStar *Taq* PCR buffer, containing 15 mM $MgCl_2$	1×buffer, containing 1.5 mM $MgCl_2$	0.5	1
25 mM $MgCl_2$	1 mM $MgCl_2$	0.2	0.4
dNTP mix, 25 mM each	200 μM each	0.04	0.08
HotStar *Taq* polymerase, 5 U/μl	0.1 U/reaction	0.02	0.04
Forward and reverse PCR primers (1 μM each)	200 nM	1	2
DNA		1	5
Total volume		5.00	10

N/A, not applicable.

5-μl PCR reaction. In condition 2, 5 μl of DNA is used in a 10-μl PCR reaction. At the end, 5 μl of DNA in a 10-μl PCR reaction gives robust detection of *SRY* gene in the maternal plasma.

1. Prepare a PCR cocktail as specified in **Table 1** (*see* **Note 5**).
2. Add 5 or 10 μl of PCR reaction into each well of a 384-well microplate. Seal the plate and centrifuge it for 3 min at 2,000 rpm.
3. Perform PCR as follows:

   a. 95°C, 15 min.
   b. 95°C, 20 s.
   c. 56°C, 30 s.
   d. 72°C, 1 min.
   e. go to **step b**, 44 times.
   f. 72°C, 3 min.
   g. 4°C, forever.

3.3.2. SAP Treatment

This step is used to neutralize the remaining dNTPs from the PCR reactions so that they will not interfere with the subsequent base extension reaction.

1. Prepare the SAP reaction solution by mixing 1.53 μl of water, 0.17 μl of hME buffer, and 0.30 μl of SAP. Make sure the solution is mixed well, because this

solution can be viscous due to the high concentration of glycerol. For the 10 μl PCR reactions, take 5 μl for this step.

2. Add 2 μl of the above-mentioned solution into each well of the 384-plate from the PCR reaction. Each well now contains 7 μl of solution.
3. Seal the plate and centrifuge it at 2,000 rpm for 3 min.
4. Thermocycle the sample plate as follows:

   a. 37°C, 20 min.
   b. 85°C, 5 min.
   c. 4°C, forever.

3.3.3. Base Extension Reaction

In this step, an extension primer is annealed to the region immediately upstream of the mutation site. The Thermo Sequenase, three different ddNTPs and one dNTP are used to extend the primer for one or two bases (terminated by ddNTP). This step produces extension products (typically 18–25 bases).

Prepare base extension cocktail by mixing 1.674 μl of water, 0.2 μl of 10× hME Buffer with 2.25 mM d/ddNTPs each, 0.108 μl of extension primer (100 μM stock), and 0.018 μl of Thermo Sequenase (32 U/μl stock). Make sure the solution is mixed well before aliquoting.

1. Add 2 μl of the above-mentioned solution into each well of the 384-plate from the SAP reaction. Each well now contains 9 μl of solution.
2. Seal the plate and centrifuge at 2000 rpm for 3 min.
3. Thermocycle the sample plate as follows:

   a. 94°C, 2 min.
   b. 94°C, 5 s.
   c. 52°C, 5 s.
   d. 72°C, 5 s.
   e. go to **step b**, 39 times.
   f. 4°C, forever.

3.3.4. Cation Removal

This step is used to remove cations in the solution because cations might interfere in mass spectra. For high-throughput analyses, please refer to "Cleaning Up the hME Reaction Products" in the MassARRAY Liquid Handler User's Guide (Sequenom, Inc.).

1. Place sufficient SpectroCLEAN resin onto a 96-well SpectroCLEAN plate and then use a scraper to spread the resin into the wells of the SpectroCLEAN plate. Next, scrape off the excess resin.

2. Place a clean, 96-well microplate upside-down over the SpectroCLEAN plate and align all the wells. Flip the two plates over so that the resin is transferred from the SpectroCLEAN plate to the 96-well plate. Tap on the SpectroCLEAN plate, if necessary, to fully release the resin.
3. Add 70 µl of water into each well of the 96-plate with resin and thoroughly mix with the resin.
4. Add 16 µl of resin/water solution into each well of the 384-plate from the base extension reaction. Seal the plate and place on a rotator for 15 min at room temperature to ensure that the resin and the solution are mixed thoroughly.
5. Centrifuge the plate for 3 min at 2,000 rpm.

3.3.5. SpectroCHIP Spotting

This step transfers approximately 10 nl of the final reaction solution after cation removal onto a 384-format SpectroCHIP. For the most accurate results, it is recommended to spot a 384-plate onto 2–4 384-format SpectroCHIPs. Please refer to "Dispensing MassEXTEND Reaction Products onto SpectroCHIPs" in the MassARRAY Nanodispenser User's Guide (Sequenom, Inc.) for instructions (all Sequenom, Inc., manuals mentioned come with the Sequenom, Inc. MALDI-TOF MS system).

3.3.6. MALDI-TOF MS

The Quantitative Gene Analysis software is bundled with the MALDI-TOF MS as a complete package. The software performs data acquisitions, mass spectrometric peak identifications, noise normalizations and peak area analyses. Please refer to the "SpectroACQUIRE" chapter in the MassARRAY Typer User's Guide for MALDI-TOF MS operations.

3.3.7. Results

The level of the *SRY* gene and the β-globin gene were quantified by real-time PCR, as described previously *(2)*. Because 800 µl of plasma was used for DNA extraction and a final elution volume of 50 µl were used, each microliter of extracted DNA corresponded to 16 µl of plasma. The amount of the *SRY* DNA input for the 5-µl PCR system (with 1 µl of DNA) and the 10 µl of PCR system (with 5 µl of DNA) were calculated and shown in **Table 2**. As shown in the table, for some samples, 1 µl of DNA contains, on average, less than 1 copy of the *SRY* DNA. Thus, for sample 13 and 16, false-negative results were obtained. However, when using 5 µl of DNA for the 10-µl PCR system, the input DNA for all samples was more than 1 copy. As a result, 100% sensitivity was achieved. Additionally, all pregnant women carrying a female fetus were negative for the *SRY* test by using the MS assay (as well as the real-time PCR

Table 2
SRY detection by MALDI-TOF MS

Sample	1 μl of DNA	SRY copy no. with 1 μl of DNA	5 μl of DNA	SRY copy no. with 5 μl of DNA	SRY (copies/ml)[a]	β-Globin (copies/ml)
1	2/4[b]	2.2	2/2	11.0	137.5	526.8
2	4/4	0.7	2/2	3.5	44	164.5
3	4/4	7.8	2/2	38.9	486	6674.5
4	4/4	1.0	2/2	5.2	65.5	555.0
5	2/4	0.4	2/2	2.2	27	662.5
6	3/4	0.5	2/2	2.7	34	521.0
7	0/4	0.0	0/2	0.0	0	237.9
8	2/4	0.3	2/2	1.6	19.4	300.1
9	1/4	0.2	2/2	1.2	15.1	390.1
10	0/4	0.0	0/2	0.0	0	760.9
11	2/4	0.6	2/2	2.8	34.6	187.4
12	4/4	0.6	2/2	3.0	37.5	505.0
13	0/4	0.3	2/2	1.7	21.8	257.8
14	4/4	0.3	2/2	1.7	21.1	408.4
15	0/4	0.0	0/2	0.0	0	561.6
16	0/4	0.3	2/2	1.4	17	164.3
17	0/4	0.0	0/2	0.0	0	612.8
18	0/4	0.0	0/2	0.0	0	362.5

[a]DNA copies per milliliter of plasma.
[b]2/4: two assays give positive results out of four independent assays.

assay). It is notable that for quite a few samples, even with 5 μl of DNA input, the *SRY* copy number was only one to five copies (samples 2, 5, 6, 8, 9, 11, 12, 13, 14, and 16). Yet, the MS detection was 100% sensitive. Robust sensitivity of single copy DNA has been reported before with the MS technique *(**15**)*. Two factors contribute to this extremely high sensitivity. First, the amplicon is designed to be very short (67 bp excluding the two 10-bp tags), which is likely to be amplified with high efficiency. Short amplicon is of extreme importance given the observation that most fetal DNA in the maternal plasma is shorter than 200 bp *(**5**)*. Additionally, the base extension reaction (40 cycles) provides a linear amplification step, which further improves the sensitivity as well. The sensitivity of this assay may be further improved in a 25- or 50-μl PCR system with 12.5 or 25 μl of DNA input. This improvement may be useful for very early gestational cases.

3.4. Paternal SNP Detection by Using SABER

3.4.1. Experimental Protocol

All steps are the same as in **Subheading 3.3.**, with the following notables:

1. The 10-μl PCR is used with 5 μl of DNA input.
2. For base extension, only one ddNTP is added for the extension using only the mutant DNA.
3. For SAP treatment, the incubation time at 37°C is extended to 45 min to ensure more complete removal of dNTPs.

3.4.2. Results

Sixteen plasma DNA samples were selected for the analysis (**Table 3**). For samples with maternal/fetal genotype pair CC/GC, SABER was carried out to detect the G allele (presumably derived from the paternally inherited SNP of the fetus). In all six cases, the G allele was detected, demonstrating a good sensitivity of the system (no false negatives). In all 10 cases where the maternal/fetal genotype pair is CC/CC, the results were all negative for G allele detection, demonstrating a good specificity of the system (no false positives).

Table 3
Detection of paternal-inherited SNP allele by SABER

Sample	Genotype (maternal/fetal)	Detection of G allele in maternal plasma
1	CC/GC	+
2	CC/CC	–
3	CC/CC	–
4	CC/GC	+
5	CC/CC	–
6	CC/CC	–
7	CC/CC	–
8	CC/CC	—
9	CC/CC	–
10	CC/GC	+
11	CC/CC	–
12	CC/GC	+
13	CC/CC	–
14	CC/GC	+
15	CC/CC	—
16	CC/GC	+

4. Notes

1. The AssayDesigner software (Sequenom, Inc.) is tailored for designing such assays. In practice, a user only needs to create any artificial mutation (such as C/G) in the middle of the DNA sequence of interest and input this information into the software. The AssayDesigner also can handle multiplex assays, ensuring that the molecular mass values of all primers (and the extension products from them) involved do not interfere with each other in a mass spectrum. A sample input file format for the software is in text file format with the content such as the following:

   ID<TAB>Sequence
   SRY<TAB>TGGCGATTAAGTCAAATTCGCATTTTTCAGGACAGCAGT
   AGAGCA[G/A]TCAGGGAGGCAGATCAGCAGGGCAAGTAGTCAAC
   GTTACTGAATTACCATGTTTTGCTTGAGAATGAATACATTGTCAG
   GGTACTAGGGGG

   The first line contains "ID," followed by a TAB, then "Sequence." The second line contains *SRY* (the ID of the sequence), followed by a TAB, then the actual sequence of interest, followed by ENTER (to create a new line). The actual sequence of interest normally is about 200 bp so that the program can select most appropriate primers with some freedom.
2. The quality of extension primers is crucial for accurate quantifications. Specifically, the (n-1), (n-2), and other by-products resulting from incomplete oligonucleotide synthesis should only be present at a small proportion (preferably <10% of the desired oligonucleotide). Integrated DNA Technologies (http://www.idtdna.com) is recommended since they routinely quality control their oligonucleotide products with MALDI-TOF MS.
3. Misincorporation of nucleotides by DNA polymerase does occur, although typically at a very low rate of about 10^{-7} *(**14**)*. But, I recommend some control experiment with only the wild-type DNA sequence to make sure the assay is specific.
4. Salt adduct peaks (with sodium ion) can often occur in mass spectra with a molecular mass of N+23, where N is the expected molecular mass of the extension product. Thus, even when the expected molecular mass of the unintended extension product from the wild-type DNA is approximately 23 Da smaller than the extension product from the mutant DNA, the salt adduct peak of the first extension product may be mistakenly identified as the second extension product, potentially causing a false-positive result. For this reason, I recommend that the unintended extension product from the wild-type DNA having a molecular mass larger than the extension product from the mutant DNA and a good separation of their molecular mass (preferably >40 Da difference, at least 25-Da difference).
5. When analyzing multiple samples, it is highly recommended that a supermix cocktail is prepared first. For example, in a supermix for 10 reactions, all reagents (except the DNA samples) were mixed first, with each individual reagent at

11 times of the volume of the single reaction. The extra amount was used to compensate potential pipetting loss in aliquoting the mixtures.

Acknowledgments

I thank Dennis Lo and Rossa Chiu for generously agreeing to let me include the unpublished protocol and data on SRY detection here. I was previously supported by the Tung Wah Group of Hospitals and am currently supported by the Research Fund for the Control of Infectious Disease from the Health, Welfare and Food Bureau of the Hong Kong SAR Government.

References

1. Lo, Y. M., Corbetta, N., Chamberlain, P. F., et al. (1997) Presence of fetal DNA in maternal plasma and serum. *Lancet* **350,** 485–487.
2. Lo, Y. M., Tein, M. S., Lau, T. K., et al. (1998) Quantitative analysis of fetal DNA in maternal plasma and serum: implications for noninvasive prenatal diagnosis. *Am. J. Hum. Genet.* **62,** 768–775.
3. Lo, Y. M., Hjelm, N. M., Fidler, C., et al. (1998) Prenatal diagnosis of fetal RhD status by molecular analysis of maternal plasma. *N. Engl. J. Med.* **339,** 1734–1738.
4. Chiu, R. W., Lau, T. K., Leung, T. N., Chow, K. C., Chui, D. H., and Lo, Y. M. (2002) Prenatal exclusion of beta thalassaemia major by examination of maternal plasma. *Lancet* **360,** 998–1000.
5. Chan, K. C., Zhang, J., Hui, A. B., et al. (2004) Size distributions of maternal and fetal DNA in maternal plasma. *Clin Chem.* **50,** 88–92.
6. Li, Y., Di Naro, E., Vitucci, A., Zimmermann, B., Holzgreve, W., and Hahn, S. (2005) Detection of paternally inherited fetal point mutations for beta-thalassemia using size-fractionated cell-free DNA in maternal plasma. *J. Am. Med. Assoc.* **293,** 843–849.
7. Ding, C., Chiu, R. W., Lau, T. K., et al. (2004) MS analysis of single-nucleotide differences in circulating nucleic acids: application to noninvasive prenatal diagnosis. *Proc. Natl. Acad. Sci. USA* **101,** 10762–10767.
8. Tang, K., Shahgholi, M., Garcia, B. A., et al. (2002) Improvement in the apparent mass resolution of oligonucleotides by using 12C/14N-enriched samples. *Anal. Chem.* **74,** 226–231.
9. Koster, H., Tang, K., Fu, D. J., et al. (1996) A strategy for rapid and efficient DNA sequencing by mass spectrometry [see comments]. *Nat. Biotechnol.* **14,** 1123–1128.
10. Tang, K., Fu, D. J., Julien, D., Braun, A., Cantor, C. R., and Koster, H. (1999) Chip-based genotyping by mass spectrometry. *Proc. Natl. Acad. Sci. USA* **96,** 10016–10020.
11. Van Ausdall, D. A. and Marshall, W. S. (1998) Automated high-throughput mass spectrometric analysis of synthetic oligonucleotides. *Anal Biochem.* **256,** 220–228.
12. Mann, M., Hendrickson, R. C., and Pandey, A. (2001) Analysis of proteins and proteomes by mass spectrometry. *Annu. Rev. Biochem.* **70,** 437–473.

13. Haff, L. A. and Smirnov, I. P. (1997) Single-nucleotide polymorphism identification assays using a thermostable DNA polymerase and delayed extraction MALDI-TOF mass spectrometry. *Genome Res.* **7,** 378–388.
14. Diehl, F., Li, M., Dressman, D., et al. (2005) Detection and quantification of mutations in the plasma of patients with colorectal tumors. *Proc. Natl. Acad. Sci. USA* **102,** 16368–16373.
15. Ding, C. and Cantor, C. R. (2003) Direct molecular haplotyping of long-range genomic DNA with M1-PCR. *Proc. Natl. Acad. Sci. USA* **100,** 7449–7453.

21

Isolation of Cell-Free RNA from Maternal Plasma

Xiao Yan Zhong, Wolfgang Holzgreve, and Dorothy J. Huang

Summary

The discovery of cell-free RNA in plasma and serum samples provides possibilities for noninvasive prenatal diagnosis. Quantitative alterations of cell-free placental-derived mRNA in the maternal circulation are associated with many pregnancy-related disorders, such as preeclampsia and preterm labor. Obtaining circulating cell-free RNA is the first and often the most critical step for analyzing the placental-derived mRNA in maternal blood. We have compared different protocols for the extraction of cell-free RNA from plasma samples, and we have found the protocol using TRIzol LS reagent (Invitrogen) as lysis buffer combined with RNeasy Mini kit (QIAGEN) to be the optimal method for extracting high-quality cell-free RNA in the highest quantities possible. This method is also amenable to the simultaneous extraction of cell-free RNA from many different samples, by use of the QIAGEN Vacuum Manifold QIAvac 24 Plus.

Key Words: Cell-free RNA; maternal plasma; RNA extraction; noninvasive prenatal diagnosis; TRIzol LS reagent; manifold.

1. Introduction

Recovery of cell-free DNA in maternal blood offers an opportunity for noninvasive prenatal diagnosis of fetal Rhesus D status, sex status, and fetal inherited mutant alleles from an affected father *(1,2)*. The level of cell-free fetal DNA in the maternal circulation is elevated in cases of pathological pregnancies; therefore, analysis of cell-free DNA might also be used in detecting and monitoring complications of pregnancy *(3–5)*. However, fetal DNA-based noninvasive prenatal diagnosis is limited by the lack of fetal-specific genetic markers that would allow to monitor all pregnancies. Specific tissue-derived mRNA in plasma or serum samples has been introduced as a noninvasive

From: *Methods in Molecular Biology, vol. 444: Prenatal Diagnosis*
Edited by: S. Hahn and L. G. Jackson © Humana Press, Totowa, NJ

tool for detection and surveillance of several conditions. Placenta-specific expressed cell-free human placental lactogen, beta-subunit human chorionic gonadotropin, and corticotropin-releasing hormone mRNA have been found in maternal circulation *(6)*. Quantitative assays showed that high levels of cell-free placenta expressed mRNA transcripts in maternal circulation are associated with pathological pregnancies *(7)*. Because instability of mRNA and contamination of DNA affect detection of rare placenta origin mRNA in maternal circulation, choosing optimal protocols for extracting mRNA from maternal plasma becomes very important. In our lab, we compared several protocols and commercial mRNA isolation kits, including RNeasy Mini kit (QIAGEN, Hilden, Germany), High Pure RNA Isolation kit (Roche, Basel, Switzerland), and PAX gene blood collection system, in an effort to select an mRNA isolation procedure suitable for use on large-volume plasma and serum samples. Methods were evaluated in terms of fast and simple procedures and high yield of mRNA, which was quantified by performing Taqman real-time polymerase chain reaction for 18S-DNA. We used QIAGEN Vacuum Manifold QIAvac 24 Plus to remove the flow-through buffer during the binding and washing procedures instead of centrifugation, thereby eliminating the tedious loading and unloading of spin columns into a centrifuge. Our experience and data showed that the combination of using TRIzol LS reagent (Invitrogen, Carlsbad, CA) as lysis buffer, the RNeasy Mini kit (except lysis buffer), and the Vacuum Manifold QIAvac 24 Plus is a rapid and reliable procedure for simultaneous mRNA extraction from many samples and from large volumes of samples with high quantity. In our lab, this system has been recommended as the optimal protocol for detecting placenta mRNA in maternal plasma fraction.

2. Materials

2.1. Separation of Plasma from Whole Blood

1. Potassium-EDTA–containing tubes.

2.2. Extraction of RNA from Plasma Samples

1. TRIzol LS Reagent (Invitrogen, cat. no. 15596-018).
2. RNeasy Mini kit (QIAGEN, cat. no. 74904) (includes RNeasy Mini Spin columns, wash buffer RW1, and wash buffer RPE). Prepare the buffers before use according to the recommendation of manufacturer).
3. RNase-Free DNase set (QIAGEN, cat. no. 79254).
4. RNase-free water.
5. QIAvac 24 Plus vacuum manifold (QIAGEN, cat. no. 19413).

3. Methods

3.1. Separation of Plasma from Whole Blood

For details refer to Chapter 15.

3.2. Plasma RNA Extraction (see Notes 1 and 2)

1. Add 2 ml of TRIzol LS reagent (*see* **Note 3**) to 1.6 ml of plasma sample and vortex vigorously. The mixture can be stored at –70°C until further use (*see* **Notes 4** and **5**).
2. Add 0.4 ml of chloroform to the mixture and vortex vigorously.
3. Centrifuge the mixture at 12,000*g* for 15 min at 4°C.
4. Insert 1-24 RNeasy Mini Spin columns into the vaccum manifold (*see* **Note 6**).
5. Prepare DNase for on-column DNase digestion (*see* **Note 7**) with the RNase-Free DNase set according to manufacturer's instructions (*see* **step 9**). For each sample, mix 5 μl of DNase I stock solution with 35 μl of buffer RDD (*see* **Notes 8** and **9**).
6. Transfer the upper aqueous layer from centrifugation from **step 3** into new tubes (*see* **Note 10**). Add 1 volume of 70% ethanol to 1 volume of the aqueous layer. The final mixture should contain 35% ethanol and briefly mixed by vortexing.
7. Turn on the vacuum and keep adding the mixture from **step 6** into the spin columns continually until the whole liquids are drawn efficiently through the spin columns. This processing lasts several seconds.
8. Turn off the vacuum; add 350 μl of wash buffer RW1 (*see* **Note 3**) and turn on the vacuum until the wash buffer is drawn efficiently through the spin columns.
9. Turn off the vacuum, transfer 40 μl of RNase-free DNase mixed in **step 5** onto the column exactly on the silica-gel membrane, and incubate for 15 min at room temperature.
10. Add 350 μl of wash buffer RW1 and turn on the vacuum until the wash buffer is drawn efficiently through the spin columns (*see* **Notes 4** and **5**).
11. Turn off the vacuum, add 500 μl of RPE buffer (*see* **Note 11**) to the column, and turn on the vacuum until the wash buffer is drawn efficiently through the spin columns.
12. Repeat **step 11**.
13. Transfer the column to a new collection tube. Centrifuge the column at maximum speed for 2 min.
14. Transfer the column to a new RNase-free tube.
15. To elute RNA, add 30 μl of RNase-free water onto the column exactly on the filter (*see* **Note 12**).
16. Incubate for 5 min at room temperature and centrifuge the column at maximum speed for 1 min.
17. Re-elute the RNA by repeating the elution steps (**steps 15** and **16**) by using the first elute (*see* **Notes 1** and **10**). After extraction of RNA, put the tubes directly on ice.
18. Store the extracted RNA at –70°C until further use.

4. Notes

1. Prevent any contamination of DNA, RNA, and RNase.
2. Due to the instability of RNA, it is very important to reduce the processing time. Therefore, it is necessary to prepare the buffers, label the tubes, and insert the RNeasy Mini Spin columns into the vacuum manifold shortly before starting the RNA extraction. It is recommended to perform **steps 6–18** (*see* **Subheading 3.2.**) without any break and to finish each step as soon as possible. We process maximally six samples at a time to avoid longer intervals between the processing steps.
3. TRIzol and buffer RW1 are irritants. Appropriate safety measures and wearing gloves are required when handling these materials.
4. Preparing samples: After separating the plasma from whole blood, directly add 2 ml of TRIzol LS reagent to 1.6 ml of plasma sample and process the sample immediately. For future use, the mixture can be stored at –70°C.
5. Frozen plasma samples also can be used for RNA extraction. We usually pipette 2 ml of TRIzol LS reagent to 5-ml Falcon tubes and keep the tubes on ice before thawing the samples. The plasma samples stored in –70°C should be thawed on ice. Then, 1.6 ml of this thawed plasma should be immediately transferred and added to the prepared TRIzol LS reagent-containing tubes.
6. The vacuum manifold should be cleaned immediately after use and should only be used for RNA samples.
7. DNase digestion: Performing the on column DNase digestion for DNase digestion is required.
8. Do not shake the mixture of DNase I stock solution too strong. Mixing should only be carried out by gently inverting the tube.
9. The DNase I stock solution remaining should be divided into single-use aliquots and stored at –20°C. Avoid refreezing.
10. **Step 6** should be conducted on ice.
11. Wash buffers: Buffer RPE is supplied as a concentrate. Before the first use, add 4 volumes of ethanol (96–100%) to obtain the final solution. The volume of wash buffer and wash steps cannot be reduced.
12. When performing **steps 17** and **18** (*see* **Subheading 3.2.**), 30 μl of the elution buffer should be pipetted precisely in the middle of the filter to cover the silica-gel membrane area completely.

References

1. Zhong, X. Y., Holzgreve, W., and Hahn, S. (2001) Risk free simultaneous prenatal identification of fetal Rhesus D status and sex by multiplex real-time PCR using cell free fetal DNA in maternal plasma. *Swiss Med. Wkly.* **131,** 70–74.
2. Li, Y., Di Naro, E., Vitucci, A., Zimmermann, B., Holzgreve, W., and Hahn, S. (2005) Detection of paternally inherited fetal point mutations for beta-thalassemia using size-fractionated cell-free DNA in maternal plasma. *J. Am. Med. Assoc.* **293,** 843–849.

3. Zhong, X. Y., Laivuori, H., Livingston, J. C., et al. (2001) Elevation of both maternal and fetal extracellular circulating deoxyribonucleic acid concentrations in the plasma of pregnant women with preeclampsia. *Am. J. Obstet. Gynecol.* **184,** 414–419.
4. Zhong, X. Y., Burk, M. R., Troeger, C., Jackson, L. R., Holzgreve, W. and Hahn S. (2000) Fetal DNA in maternal plasma is elevated in pregnancies with aneuploid fetuses. *Prenat. Diagn.* **20,** 795–798.
5. Zhong, X. Y., Holzgreve, W., Li, J. C., Aydinli, K., and Hahn, S. (2000) High levels of fetal erythroblasts and fetal extracellular DNA in the peripheral blood of a pregnant woman with idiopathic polyhydramnios: case report. *Prenat. Diagn.* **20,** 838–841.
6. Tsui, N. B., Ng, E. K., and Lo, Y. M. (2006) Molecular analysis of circulating RNA in plasma. *Methods Mol. Biol.* **336,** 123–134.
7. Zhong, X. Y., Gebhardt, S., Hillermann, R., Tofa, K. C., Holzgreve, W. and Hahn, S. (2005) Parallel assessment of circulatory fetal DNA and corticotropin-releasing hormone mRNA in early- and late-onset preeclampsia. *Clin. Chem.* **51,** 1730–1733.

22

A Microarray Approach for Systematic Identification of Placental-Derived RNA Markers in Maternal Plasma

Nancy B. Y. Tsui and Y. M. Dennis Lo

Summary

Circulating fetal RNA in maternal plasma has offered a new approach for noninvasive prenatal diagnosis and monitoring. Circulating fetal RNA markers could potentially be used for all pregnant women without being limited by fetal–maternal genetic polymorphisms and fetal gender. Over the past few years, encouraging findings have been reported on the detection and possible clinical applications of circulating fetal RNA. Placental-derived RNA has been shown to be easily detectable in maternal plasma during pregnancy and rapidly cleared after delivery. Such observations suggest that the placenta is an important organ for releasing fetal RNA into maternal plasma. Noninvasive prenatal gene expression profiling of the placenta also has been demonstrated to be feasible by analyzing the circulating placental RNA in maternal plasma. Thus, circulating placental RNA is a potentially useful tool for noninvasive investigation of the placenta. Here, we describe a systematic method for efficient development of new placental-specific RNA markers that could be detected in maternal plasma. The method is based on the use of oligonucleotide microarray (Affymetrix, Santa Clara, CA) technology to simultaneously analyze >39,000 RNA transcripts in the placenta. This development has implication for the development of new markers for studying disease conditions associated with placental pathology, such as preeclampsia.

Key Words: Circulating fetal RNA, circulating placental RNA, microarray, fetal marker identification, noninvasive gene expression profiling, noninvasive prenatal diagnosis.

1. Introduction

The detection of fetal RNA in maternal plasma offers new possibilities for noninvasive prenatal investigation. Several placental-derived RNA species, such as the transcripts for human placental lactogen, β-subunit of human

From: *Methods in Molecular Biology, vol. 444: Prenatal Diagnosis*
Edited by: S. Hahn and L. G. Jackson © Humana Press, Totowa, NJ

chorionic gonadotropin (βhCG) *(1)*, and corticotropin-releasing hormone (*CRH*) *(2)*, have been shown to be readily detectable in maternal plasma. The rapid clearance of these placental RNAs from maternal plasma after delivery revealed their pregnancy specificity. The clinical value of circulating fetal RNA detection also has been demonstrated. For example, a significant elevation of circulating *CRH* mRNA has been detected in the plasma of preeclamptic pregnant women *(2)*. In fetal trisomy 18, the βhCG mRNA concentration in maternal serum has been shown to be significantly reduced *(3)*.

The placenta has been suggested to be an important organ for releasing RNA transcripts into maternal plasma *(1)*. In fact, the relative abundance of the placental transcripts in maternal plasma has been shown to be correlated well with their gene expression levels in the placenta *(4)*. This observation demonstrates that placental gene expression plays an important role in determining the placental RNA levels in maternal plasma; therefore, the detection and measurement of fetal RNA in maternal plasma is a potentially useful tool for noninvasive prenatal placental gene expression profiling.

High-density oligonucleotide microarray analysis (e.g., using the Affymetrix platform *(5)*) is a powerful technology for global gene expression profiling. The GeneChip probe array (Affymetrix, Santa Clara, CA) contains hundreds of thousands of oligonucleotide probes packed at extremely high densities that permit the detection of >39,000 human RNA species simultaneously. With regard to the placental RNA marker identification, the use of microarray allows the genome-wide survey of a large number of transcripts expressed in the placenta. Without being restricted by the knowledge of the genes, novel RNA markers that are specifically expressed in the placenta could be identified simply and rapidly. This new strategy would greatly increase the number of RNA markers that can be evaluated for their potential detection in maternal plasma and for studying their potential applications in the diagnosis and monitoring of pregnancy-associated disorders.

Here, we describe the use of microarray technology for identification of new placental RNA markers that are detectable in maternal plasma. The identification strategy is summarized in **Figure 1**. The strategy is based on two hypotheses: First, the placenta is a major organ for releasing the placental-expressed RNA into the maternal plasma *(1)*. Second, hematopoietic cells have previously been shown to be a predominant source of circulating DNA in the plasma of normal individuals *(6)*. Based on this observation, we hypothesize that much of the background maternal nucleic acids in maternal plasma also may originate from the hematopoietic compartment. The ultimate aim of our identification strategy is to identify placental expressed transcripts that are fetal specific, amongst the circulating RNA molecules in maternal plasma. To achieve this, the gene expression profiles of both the placentas and their

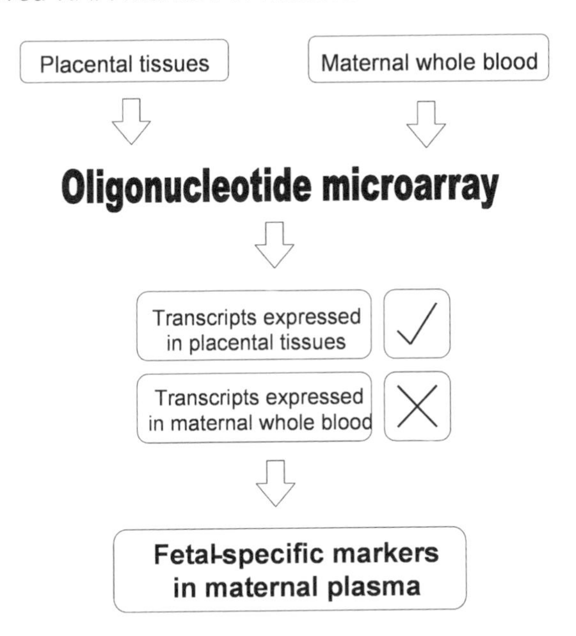

Fig. 1. Outline of the strategy used for the systematic identification of pregnancy-specific placental-expressed RNA markers in maternal plasma. Paired placenta and maternal whole blood samples are collected and subjected to microarray analysis. Candidate markers are selected from transcripts with increased expression in the placenta relative to the maternal whole blood.

corresponding maternal whole blood samples are compared. Circulating fetal RNA markers are identified by screening the transcripts that are expressed specifically in the placenta but not in the maternal blood. These candidate fetal RNA markers could then be further evaluated in maternal plasma for their detectability and possible association with disease conditions.

2. Materials

2.1. Sample Collection

1. RNAlater stabilization reagent (Ambion, Austin, TX) for placental tissue collection.
2. PAXgene Blood RNA tubes (PreAnalytiX, Hilden, Germany) for maternal whole blood collection.

2.2. RNA Extraction

1. Homogenizer.
2. TRIzol reagent (Invitrogen).

3. RNeasy Mini kit (QIAGEN, Hilden, Germany).
4. PAXgene Blood RNA kit (PreAnalytiX).
5. RNase-Free DNase set (QIAGEN).
6. Spectrophotometer.
7. Agilent 2100 Bioanalyzer (Agilent, Amstelveen, The Netherlands).
8. RNA 6000 Nano LabChip kit (Agilent).

2.3. Gene Expression Profiling by Microarray

2.3.1. cDNA Synthesis

1. T7-$(dT)_{24}$ primer (Affymetrix).
2. SuperScript II reverse transcriptase (Invitrogen, Carlsbad, CA).
3. *Escherichia coli* DNA polymerase I (Invitrogen).
4. *E. coli* DNA ligase (Invitrogen).
5. *E. coli* RNase H (Invitrogen).
6. T4 DNA polymerase (Invitrogen).
7. GeneChip Sample Cleanup Module (Affymetrix).

All of the cDNA synthesis reagents can alternatively be purchased as a kit in One-Cycle cDNA Synthesis kit (Affymetrix).

2.3.2. cRNA Labeling

1. GeneChip IVT Labeling kit (Affymetrix).

2.3.3. Hybridization and Scanning

1. Bovine serum albumin (BSA) (Invitrogen).
2. Herring sperm DNA (Promega. Madison, WI).
3. 5 M sodium chloride (NaCl) (Ambion).
4. Morpholinoethane sulfonic acid (MES) hydrate (Sigma-Aldrich, St. Louis, MO).
5. MES sodium salt (Sigma-Aldrich).
6. EDTA disodium salt (Sigma-Aldrich).
7. Dimethyl sulfoxide (DMSO) (Sigma-Aldrich).
8. Tween 20 (Pierce Chemical, Rockford, IL).
9. R-Phycoerythrin streptavidin (Molecular Probes, Invitrogen).
10. 20× sodium chloride + sodium phosphate + ethylenediaminetetraacetic acid (Ambion).
11. Goat IgG, reagent grade (Sigma-Aldrich).
12. Anti-streptavidin antibody (goat), biotinylated (Vector Laboratories, Burlingame, CA).
13. GeneChip Eukaryotic Hybridization Control kit (Affymetrix).
14. GeneChip Human Genome U133A and U133B Arrays (Affymetrix) (*see* **Note 1**).
15. Hybridization oven (Affymetrix).
16. GeneChip Fluidics Station (Affymetrix).
17. GeneArray scanner (Affymetrix).

2.3.4. Data Analysis

1. GeneChip Microarray Suite 5.0 (MAS 5.0) (Affymetrix) (*see* **Note 2**).

3. Methods

3.1. Sample Collection

For each pregnant woman, paired placenta and maternal whole blood samples are collected. Placental sample is collected immediately after delivery. Alternatively, the placental sample could be collected by chorionic villus sampling (CVS) at early pregnancy. Maternal whole blood sample is collected several minutes before collection of the placenta. To minimize expression variation due to individual biological samples, we usually included samples from five pregnant women for each microarray analysis (*see* **Note 3**).

3.1.1. Collection of Placenta

1. Process the placenta within 10 min after placental delivery.
2. Excise a rectangular segment from the placenta. The excision site for every placental sample should be consistent (*see* **Note 4**).
3. Rinse the dissected section thoroughly in RNase-free water and eliminate any blood vessels and blood clots.
4. Cut out five portions (approx 27 mm^3 each) from the placental section. Store each portion in 1 ml of RNAlater stabilization reagent.
5. Store the sample at 4°C overnight.
6. Remove the RNAlater reagent from the placenta.
7. Store the RNAlater-treated placenta at –80°C until RNA extraction.

For collection of CVS, store three to four strands of CVS tissue in 1 ml of RNAlater reagent at 4°C overnight. Afterward, remove the RNAlater reagent and stored the CVS at –80°C until RNA extraction.

3.1.2. Collection of Maternal Whole Blood

1. Collect 5 ml of whole blood in two PAXgene Blood RNA tubes. Mix the blood with stabilizing reagent in the tubes immediately.
2. Stored the blood tubes at room temperature overnight.
3. Proceed directly to RNA extraction.

3.2. RNA Extraction

3.2.1. RNA Extraction from Placenta

1. Homogenize the RNAlater-treated placenta in 3 ml of TRIzol reagent (or CVS in 0.8 ml of TRIzol reagent) by using a homogenizer.

2. Centrifuge the homogenate at 12,000*g* for 10 min at 4°C. Transfer the supernatant into a new tube.
3. Add 0.2 ml of chloroform per 1 ml of TRIzol added in step 1 and mix vigorously. Incubate the mixture for 2–3 min at room temperature. Centrifuge the mixture at 12,000*g* for 15 min at 4°C.
4. Transfer the upper aqueous layer into a new tube. To precipitate RNA, add 0.5 ml of isopropyl alcohol per 1 ml of TRIzol used in **step 1** (*see* **Note 5**). Incubate the mixture for 10 min at room temperature. Centrifuge the mixture at 12,000*g* for 10 min at 4°C.
5. Discard the supernatant. The RNA precipitate appears as a gel-like pellet at the bottom of the tube.
6. Wash the RNA pellet by adding 1 ml of 75% ethanol. Mix by vortexing. Centrifuge the mixture at 7,500*g* for 5 min at 4°C.
7. Discard the supernatant. Dissolve the RNA pellet in 100 μl of RNase-free water.
8. Clean up the RNA with RNeasy Mini column. Add 350 μl of RLT buffer into the dissolved RNA and mix.
9. Add 250 μl of absolute ethanol to the mixture and mix. Apply the mixture to an RNeasy Mini column.
10. Wash the column twice with RPE buffer according to the manufacturer's recommendations.
11. Elute RNA with 30 μl of RNase-free water.
12. Store the RNA at –80°C.

3.2.2. RNA Extraction from Maternal Whole Blood

Total RNA from the maternal whole blood is extracted a using PAXgene column following manufacturer's instruction. The procedures are briefly summarized as follows:

1. Centrifuge the PAXgene Blood RNA tube at 4,000*g* for 10 min at room temperature to pellet nucleic acids.
2. Discard the supernatant. Resuspend the pellet in 5 ml of RNase-free water and centrifuge at 4,000*g* for 10 min at room temperature.
3. Discard the supernatant. Resuspend the pellet in 360 μl of buffer BR1. Add 300 μl of buffer BR2 and 40 μl of proteinase K. Incubate the mixture for 10 min at 55°C in a shaker-incubator.
4. Centrifuge the mixture at 16,000*g* for 3 min.
5. Collect the supernatant. Mix the supernatant with 350 μl of absolute ethanol. Apply the mixture to a PAXgene RNA spin column and wash according to manufacturer's instruction.
6. Perform the "On-Column DNase Digestion" with the RNase-Free DNase Set according to the manufacturer's recommendations.
7. Elute RNA with 30 μl of RNase-free water.
8. Store the RNA at –80°C.

3.2.3. Determination of RNA Yield and Integrity

Total RNA extracted from the placenta and the whole blood sample is quantified by spectrophotometer at 260 nm. It is established that RNA solution with an optical density of 1.0 has a concentration of 40 mg/ml. The purity of the RNA is estimated by the ratio of the readings at 260 and 280 nm, which is 1.9–2.1 optimally. The integrity of the total RNA can be assessed on an Aglient 2100 Bioanalyzer. Examples of a good quality of total RNA sample can be found in GeneChip Expression Analysis Technical Manual (Affymetrix). Only RNA samples with good qualities are subsequently analyzed in microarray experiment.

3.3. Gene Expression Profiling by Microarray

The procedures for the microarray experiment are detailed in GeneChip Expression Analysis Technical Manual–Eukaryotic Sample and Array Processing from Affymetrix. Biotinylated cRNA is prepared from total RNA according to 'One-Cycle Target Labeling Assay' procedures as described in the manual. In brief, single-strand cDNA is synthesized using T7-oligo(dT) promoter primer follows by second-strand cDNA synthesis. The second-strand cDNA serves as a template in the subsequent in vitro transcription (IVT) reaction. The IVT reaction generates biotinylated cRNA in the presence of T7 RNA polymerase and a biotinylated nucleotide analog/ribonucleotide mix. The labeled cRNA is finally fragmented and hybridized to GeneChip expression array.

3.3.1. cDNA Synthesis

1. Prepare RNA/primer mix according to **Table 1**. Incubate the mix at 70°C for 10 min and then chill on ice.
2. Prepare first-strand reaction master mix according to **Table 1**. Add 7 μl of the reaction master mix to the RNA/primer mix prepared in step 1. Incubate the mixture at 42°C for 2 min.
3. Add 2 μl of SuperScript II reverse transcriptase to the mixture and incubated the reaction at 42°C for 2 h.
4. Prepare second-strand reaction master mix according to **Table 2**. Add 130 μl of the master mix to the first-strand cDNA synthesized in **step 3**. Incubate the reaction for at 16°C for 2 h.
5. Add 2 μl of T4 DNA Polymerase and incubate at 16°C for 5 min.
6. After incubation, add 10 μl of 0.5 M EDTA.
7. Proceed to "Cleanup of Double-Stranded cDNA" by the GeneChip Sample Cleanup Module according to the manufacturer's instructions. In brief, follow these steps:

    a. Add 600 μl of cDNA binding buffer to the double-stranded cDNA.
    b. Apply the mixture to the cDNA Cleanup spin column.

Table 1
First-strand cDNA reaction master mix

Reagent	Amount
RNA/primer mix	
Sample RNA	10 µg
T7-oligo(dT) primer, 50 uM	2 µl
RNase-free water	to 11 µl final volume
Total volume	11 µl
First-strand reaction master mix	
5× first strand buffer	4 µl
Dithiotreitol, 0.1 M	2 µl
dNTP, 10 mM	1 µl
Total volume	7 µl

c. Wash the column with 750 µl of cDNA wash buffer.
d. Centrifuge the column for 5 min at maximum speed.
e. Elute the cDNA with 14 µl of cDNA elution buffer.

3.3.2. cRNA Labeling

1. Prepare IVT reaction according to **Table 3** using reagents provided in the GeneChip IVT Labeling kit. Incubate the reaction at 37°C for 16 h in a thermal cycler.
2. After the reaction, proceed to "Cleanup of Biotin-Labeled cRNA" by the GeneChip Sample Cleanup Module following manufacturer's instruction. In brief, follows these steps:

Table 2
Second-strand cDNA reaction master mix

Reagent	Amount
RNase-free water	91 µl
5× second strand buffer	30 µl
dNTP, 10 mM	3 µl
E. coli DNA Ligase	1 µl
E. coli DNA Polymerase I	4 µl
RNase H	1 µl
Total volume	130 µl

Table 3
IVT reaction master mix

Reagent	Amount (μl)
Double-strand cDNA	6
RNase-free water	14
10× IVT labeling buffer	4
IVT labeling NTP mix	12
IVT labeling enzyme mix	4
Total volume	40

a. Add 60 μl of RNase-free water, 350 μl of IVT cRNA binding buffer, and 250 μl of absolute ethanol to the IVT reaction in sequence.
b. Apply the mixture to the IVT cRNA Cleanup spin column.
c. Wash the column with 500 μl of IVT cRNA wash buffer and then 500 μl of 80% ethanol.
d. Centrifuge the column for 5 min at maximum speed.
e. Elute the cRNA with 11 μl of RNase-free water.
f. Perform the second elution with 10 μl of RNase-free water.

3. Quantify the purified cRNA by using a spectrophotometer.
4. Calculate the adjusted cRNA yield with the following equation (*see* **Note 6**):

Adjusted cRNA yield
= (measured cRNA amount)
  – (starting total RNA amount) (fraction of cDNA used for IVT)
= (measured cRNA amount) – (10 μg) (0.5)

5. Save a cRNA aliquot for determining the size distribution of the IVT product by the Bioanalyzer. Refer to GeneChip Expression Analysis Technical Manual for an example spectrum.
6. Fragment all of the labeled cRNA with 5× fragmentation buffer provided in the GeneChip Sample Cleanup Module. Prepare fragmentation master mix according to **Table 4**.
7. Incubate the reaction at 94°C for 35 min in a thermal cycler.
8. Aliquot 30 μl of the fragmented cRNA in each of the new tubes. Store the fragmented cRNA at –80°C until hybridization.
9. Save an aliquot of fragmented cRNA for analysis on the Bioanalyzer. Refer to GeneChip Expression Analysis Technical Manual for an example spectrum.

3.3.3. Hybridization and Scanning

Hybridization and scanning are performed according to the procedures described in "Eukaryotic Arrays: Washing, Staining and Scanning" of the

Table 4
cRNA fragmentation master mix

Reagent	Amount
cRNA[a]	x μg
5× fragmentation buffer	2x/5 μl
RNase-free water	to 2x μl final volume
Total volume	2x μl

[a]The cRNA concentration should be 0.5 μg/μl in the final fragmentation reaction. The values in the table are assuming the adjust cRNA yield is x μg.

GeneChip Expression Analysis Technical Manual (Affymetrix). Refer to the manual for the handling and operation of the array chips, hybridization oven, fluidic station, and scanner.

1. Before starting the experiment, prepare buffers and reagents following the instructions described in the GeneChip Expression Analysis Technical Manual (Affymetrix).
2. Prepare hybridization cocktail according to **Table 5**. Incubate the hybridization cocktail at 99°C for 5 min, follows by 45°C for 10 min.
3. Wet the arrays with 1× hybridization buffer according to manufacturer's instruction.

Table 5
Hybridization cocktail

Reagent	Amount (μl)
Fragmented cRNA	30
Control oligonucleotide B2[a]	5
20× Eukaryotic hybridization controls	15
Herring sperm DNA	3
BSA	3
2× hybridization buffer	150
DMSO	30
RNase-free water	64
Total volume	300

[a]The hybridization control transcripts (control oligonucleotide B2 and 20× eukaryotic hybridization controls) are provided in GeneChip Eukaryotic Hybridization Control kit (Affymetrix).

4. Hybridize the fragmented cRNA sample to the GeneChip U133A or U133B Arrays at 45°C for 16 h according to the instructions described in the GeneChip Expression Analysis Technical Manual (Affymetrix).
5. After hybridization, wash and stain each array in the GeneChip Fluidics Station by using the "Antibody Amplification for Eukaryotic Targets" fluidics protocol.
6. Scan each array with the GeneArray scanner. The gene expression data are automatically collected and analyzed by the MAS 5.0 software.
7. Inspect the data quality of each array (*see* **Note 7**). Only samples with good data qualities are included for further analysis.

3.4. Identification of Circulating Placental RNA Markers

We aim to identify placental-expressed RNA markers that are 1) potentially detectable in maternal plasma and 2) fetal-specific amongst the large amount of background maternal RNA in maternal plasma. The identification is achieved by analyzing and comparing the gene expression data of both the placenta and the maternal whole blood by using the MAS 5.0 software (*see* **Note 2**).

3.4.1. Identification of Placental-expressed Transcripts That Are Potentially Detectable in Maternal Plasma

We have previously demonstrated that transcripts with relatively higher microarray signal intensities were more readily detectable in maternal plasma *(4)*. Moreover, these placental-expressed transcripts could be robustly detected in maternal plasma provided that their expression levels exceed a threshold microarray signal *(4)*. Thus, in the first part of the identification strategy, we aim to identify a panel of transcripts that are highly expressed in the placenta. These transcripts represent the candidate RNA that are potentially detectable in maternal plasma.

1. Perform global normalization for all of the placenta and whole blood expression arrays. The overall intensity of each array is globally scaled to a target intensity value of 500.
2. For each transcript, calculate the median of the scaled expression intensities in all of the five placental samples.
3. Select transcripts that are called "present" by the MAS 5.0 software in all of the five placental samples.
4. Sort the selected transcripts in descending order according to the medians of their expression signals.
5. Transcripts with expression signal intensities that exceed a defined threshold value are identified. This threshold intensity value varies depending on the sensitivity of follow-up measures used for plasma RNA detection. In our experience, transcripts with gene expression intensities >10,000 in the placentas could

generally be detected in maternal plasma by one-step real-time quantitative reverse transcriptase-polymerase chain reaction (RT-PCR) *(4)*.

3.4.2. Identification of Circulating Placental-Expressed Markers That Are Fetal-Specific

We have hypothesized in the previous section that a significant proportion of circulating RNA in maternal plasma is originated from the maternal hematopoietic compartment. Thus, among the placental-expressed RNA identified in **Subheading 3.4.1.**, transcripts that are expressed in the maternal blood cells to a higher or similar degree as that of the placenta should be eliminated. The transcripts remain represent the placental-expressed RNA markers that are specifically deviated from the fetus.

1. Compare the expression profile of each placenta with its corresponding maternal whole blood by the MAS 5.0 software, by using the maternal blood as baseline. Five pairs of comparisons are thus performed.
2. For each transcript, calculate the median of -fold changes in all of the five comparisons.
3. Within the RNA panel identified in **Subheading 3.4.1.**, select transcripts whose expression levels are ''increased" (up-regulated) in the placentas compared with the corresponding whole blood in all of the five comparisons.
4. Sort the selected transcripts in descending order according to the medians of their -fold changes.
5. Transcripts at the top of this panel represent potential fetal-specific plasma RNA markers. These fetal RNA markers are potentially less susceptible to interference by the maternal hematopoietic background RNA in maternal plasma and thus they would more truly reflect the expression pattern in the placenta noninvasively.
6. These candidate RNA markers could next be validated in maternal plasma by more precise and accurate measures such as real-time quantitative RT-PCR *(7,8)*.

The microarray identification strategy described here represents a basic approach for identifying new circulating placental RNA markers. This development has particular implication for the generation of new markers for the studying of conditions in which placental abnormalities have been reported, such as trisomy 21 *(9)* and preeclampsia *(10)*.

4. Notes

1. The gene expression profiling can also be performed by a more recently developed GeneChip Human Genome U133 Plus 2.0 Array (Affymetrix), which has the advantage of combining transcripts represented in the U133A and U133B Arrays on one single array.

2. GeneChip Microarray Suite 5.0 (MAS 5.0) is the software provided by Affymetrix for processing and analyzing scanned chip images. The processed data could optionally be further analyzed by various microarray data mining software packages. These software tools are either commercially available, such as GeneSpring (SilconGenetics, Redwood City, CA) or freely accessible online such as the Bioconductor packages (www.bioconductor.org) ***(11,12)***.
3. Biological variation is one of the major sources of data variability in microarray experiments ***(13,14)***. It is caused by the influence of genetic factors, environmental factors, or a combination on the expression patterns of different individuals. To account for the variation, several biological replicates from different individuals are used. We usually include five biological replicates in one microarray analysis. The use of more biological replicates would allow us to better deal with such variability. A minimum of three biological replicates has been suggested by Affymetrix and others ***(14,15)***.
4. The placental gene expression pattern has been shown to be nonuniform across different regions within the placenta ***(16)***. This may due to different cell type composition, villous maturation, fibrin deposition, oxygenation among different placental regions, or a combination. To obtain a consistent gene expression profiles, the sampling site should be the similar for all placental samples. Our usual practice is to excise a longitudinal placental section within 5 cm from the umbilical cord insertion site. Both of the layers along the basal plate and the chorionic surface are removed, remaining the middle layer for further processing.
5. For RNA extraction from CVS or small amount of placental tissue, add 10 μg of glycogen to the aqueous phase before the addition of isopropyl alcohol. Addition of glycogen aids in visualization of the pellet and may increase recovery.
6. The minimal adjusted cRNA yield required for each sample is 30 μg, which is enough for hybridizing on the U133A and U133B Arrays. If the adjusted cRNA yield is insufficient, perform a second IVT reaction with the remaining double-strand cDNA sample.
7. Quality control data of each array could be inspected in the report generated by the MAS 5.0 software. All quality control parameters are detailed in the GeneChip Expression Analysis Technical Manual (Affymetrix).

   a. Background: the background intensity of the array, which is typically 20–100.
   b. Noise (RawQ): A measure of electrical noise of the scanner. The values should be comparable among all the arrays.
   c. Scale Factor: The factor for data of an array to globally scale to a target intensity of 500. The scale factors should be similar among replicated samples.
   d. Percent Present: Percentage of the number of probe sets call "Present." Similar values should be obtained for replicated samples.
   e. (3′/5′) of housekeeping controls: Ratio of 3′ probe set intensity to the corresponding 5′ probe set intensity of a particular housekeeping gene. High (3′/5′) value indicates RNA degradation or inefficient cDNA and cRNA synthesis. The (3′/5′) values of *GAPDH* and *β-actin* genes should optimally be less than 3.

f. Hybridization controls (*BIOB*, *BIOC* , *BIODN* , and *CREX*): They are external controls used for evaluating sample hybridization efficiency. *BIOB* should be "Present" 50% of the time, whereas *BIOC*, *BIODN* , and *CREX* should always be "Present."

Acknowledgments

This project is supported by the Innovation and Technology Fund (ITS/195/01) and an Earmarked Research Grant from the Hong Kong Research Grants Council (CUHK 4474/03M).

References

1. Ng, E. K., Tsui, N. B., Lau, T. K., et al. (2003) mRNA of placental origin is readily detectable in maternal plasma. *Proc. Natl. Acad. Sci. USA* **100,** 4748–4753.
2. Ng, E. K., Leung, L. N., Tsui, N. B., et al. (2003) The concentration of circulating corticotropin-releasing hormone mRNA in maternal plasma is increased in preeclampsia. *Clin. Chem.* **49,** 727–731.
3. Ng, E. K., El-Sheikhah, A., Chiu, R. W., et al. (2004) Evaluation of human chorionic gonadotropin beta-subunit mRNA concentrations in maternal serum in aneuploid pregnancies: a feasibility study. *Clin Chem.* **50,** 1055–1057.
4. Tsui, N. B.., Chim, S. S., Chiu, R. W., et al. (2004) Systematic micro-array based identification of placental mRNA in maternal plasma: towards non-invasive prenatal gene expression profiling. *J. Med. Genet.* **41,** 461–467.
5. Lipshutz, R. J., Fodor, S. P., Gingeras, T. R., and Lockhart, D. J. (1999) High density synthetic oligonucleotide arrays. *Nat. Genet.* **21,** 20–24.
6. Lui, Y. Y., Chik, K. W., Chiu, R. W., Ho, C. Y., Lam, C. W., and Lo, Y. M. (2002) Predominant hematopoietic origin of cell-free DNA in plasma and serum after sex-mismatched bone marrow transplantation. *Clin Chem.* **48,** 421–427.
7. Ng, E. K., Tsui, N. B., Lam, N. Y., et al. (2002) Presence of filterable and nonfilterable mRNA in the plasma of cancer patients and healthy individuals. *Clin. Chem.* **48,** 1212–1217.
8. Bustin, S. A. (2000) Absolute quantification of mRNA using real-time reverse transcription polymerase chain reaction assays. *J. Mol. Endocrinol.* **25,** 169–193.
9. Gross, S. J., Ferreira, J. C., Morrow, B., et al. (2002) Gene expression profile of trisomy 21 placentas: a potential approach for designing noninvasive techniques of prenatal diagnosis. *Am. J. Obstet. Gynecol.* **187,** 457–462.
10. Reimer, T., Koczan, D., Gerber, B., Richter, D., Thiesen, H. J., and Friese, K. (2002) Microarray analysis of differentially expressed genes in placental tissue of pre-eclampsia: up-regulation of obesity-related genes. *Mol. Hum. Reprod.* **8,** 674–680.
11. Xia, X., McClelland, M., and Wang, Y. (2005) WebArray: an online platform for microarray data analysis. *BMC Bioinformatics* **6,** 306.

12. Irizarry, R. A., Wu, Z., and Jaffee, H. A. (2006) Comparison of Affymetrix GeneChip expression measures. *Bioinformatics* **22**, 789–794.
13. Churchill, G. A. (2002) Fundamentals of experimental design for cDNA microarrays. *Nat. Genet.* **32 (Suppl.),** 490–495.
14. Yang, Y. H. and Speed, T. (2002) Design issues for cDNA microarray experiments. *Nat. Rev. Genet.* **3,** 579–588.
15. Lee, M. L., Kuo, F. C., Whitmore, G. A., and Sklar, J. (2000) Importance of replication in microarray gene expression studies: statistical methods and evidence from repetitive cDNA hybridizations. *Proc. Natl. Acad. Sci. USA* **97,** 9834–9839.
16. Wyatt, S. M., Kraus, F. T., Roh, C. R., Elchalal, U., Nelson, D. M., and Sadovsky, Y. (2005) The correlation between sampling site and gene expression in the term human placenta. *Placenta* **26,** 372–379.

23

A Novel Method to Identify Syncytiotrophoblast-Derived RNA Products Representative of Trisomy 21 Placental RNA in Maternal Plasma

Attie T. J. J. Go, Allerdien Visser, Marie van Dijk, Monique A. M. Mulders, Paul Eijk, Bauke Ylstra, Marinus A. Blankenstein, John M. G. van Vugt, and Cees B. M. Oudejans

Summary

A novel in vitro method is described wherein gene expression profiling is reflective and informative for the way how syncytiotrophoblast cells shed RNA products in vivo in maternal plasma. After controlled denudation, RNA is obtained selectively from the syncytiotrophoblast cells of trisomy 21 placentae. cDNA copies are subsequently analyzed by microarray profiling and cDNA cloning with sequencing. Given the preponderance of 5′ mRNA fragments lacking a poly-A tail, the placental RNA products are amplified after polymerase A-mediated tailing by using a method originally designed for small-sized microRNAs. This approach, when combined with cDNA library or microarray expression screening, is a novel in vitro method to screen for syncytiotrophoblast-derived RNA products representative of trisomy 21 placental RNA as present in vivo in maternal plasma.

Key Words: Placenta; maternal plasma; RNA, syncytiotrophoblast; trisomy 21; denudation.

1. Introduction

Noninvasive prenatal diagnosis of Down syndrome by quantitative analysis of placental RNA or DNA isolated from maternal plasma should be feasible when three conditions are met: (1) detectable in maternal plasma, (2) informative for chromosome 21, and (3) changed in copy number when a third copy

From: *Methods in Molecular Biology, vol. 444: Prenatal Diagnosis*
Edited by: S. Hahn and L. G. Jackson © Humana Press, Totowa, NJ

of chromosome 21 is present. RNA expressed from genes within the Down syndrome critical region on chromosome 21 and overexpressed in the syncytiotrophoblast of trisomy 21 placentae are candidate markers for screening by using RNA *(1)*. DNA sequences (i.e., CpG islands) from genes on chromosome 21 with differential modification (such as methylation) being restricted to the placenta are candidate markers for screening by using DNA *(2)*. Both approaches have intrinsic merits, but each critically depends on the identification of markers on chromosome 21, that fulfill all criteria of detectability, informativity, and predictive power.

We showed chromosome 21-encoded mRNA (LOC90625, now called C21orf105) to be detectable in maternal plasma during normal pregnancies *(1)*. By gene expression profiling using microarray, this gene was found to be up-regulated in placentae with trisomy 21 compared with placentae with a normal phenotype *(3)*. However, marker profiling of affected tissues using whole tissue fragments does not correct for the fact that the predominant, if not exclusive, source of placental RNA shed into the maternal circulation is derived from the syncytiotrophoblast *(4)*. Genes with additional or restricted placental expression in cells other than syncytiotrophoblast cells, such as villous stromal cells or cytotrophoblast stem cells, will generate false-positive signals in expression profiling experiments when whole tissue fragments are used. In vitro methods should be designed and tested, where gene expression profiling is reflective and informative for the syncytiotrophoblastic origin of the placental RNA products found in maternal plasma. Baczyk et al. *(5)* recently developed a floating villous explant culture model in which denuded first trimester villi spontaneously regenerate syncytiotrophoblast after 48 h of culture. Besides the biological importance of their data, showing the existence of bipotential trophoblast progenitor cells in the first trimester placenta, their method of controlled selective removal of syncytiotrophoblast can be of great use for plasma marker profiling. The in vitro manipulation method allows comparison to the way how the syncytiotrophoblasts shed DNA/RNA microparticles in vivo. After controlled denudation, we selectively obtained RNA from the syncytiotrophoblast of trisomy 21 placentae that was subsequently analyzed by microarray expression profiling and sequencing after cDNA cloning (*see* **Fig. 1**). Given that placental RNA in maternal plasma is associated with a preponderance of 5′ mRNA fragments lacking a poly-A tail *(6)*, the placental RNA products were subjected to polymerase A-mediated tailing, 5′-adapter ligation and reverse transcription-polymerase chain reaction (RT-PCR) with extended oligo(dT) primers by using a method originally designed for microRNAs *(7)*.

This novel approach, for several reasons, is clinically useful and biologically relevant: (1) The RNA recovered after denudation is representative of

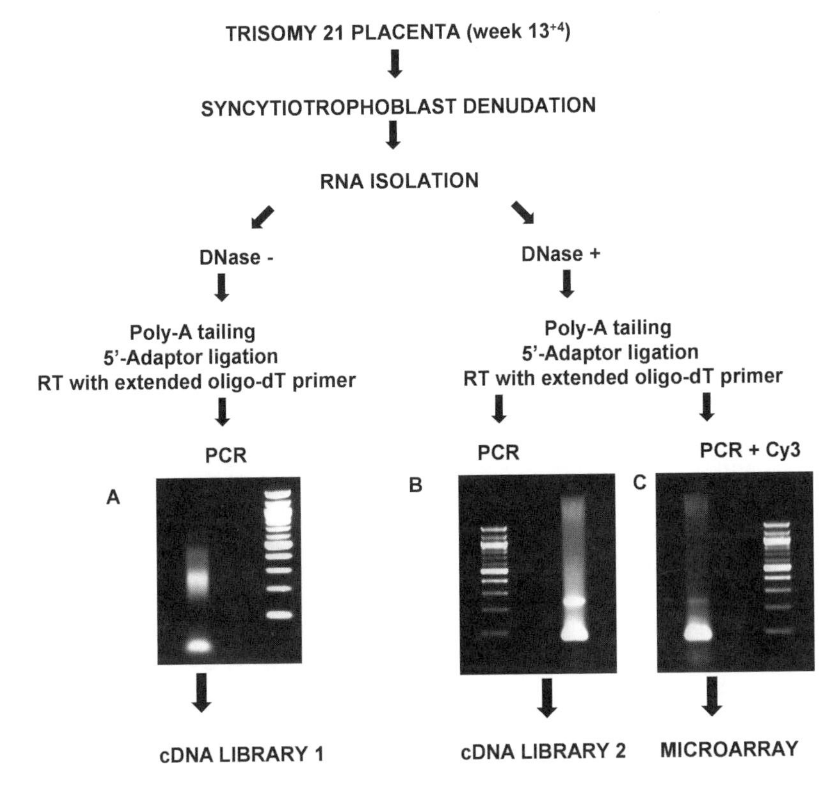

Fig. 1. Schematic representation of the strategy used to identify syncytiotrophoblast-derived RNA products representative of trisomy 21 placental RNA in maternal plasma.

the RNA expressed by and released from the placental syncytiotropoblast as indicated by the presence of both high- and low-abundance targets, i.e., hPL and C21orf105, recovered reliably before and after amplification. (2) The RNA isolated can be used for cDNA synthesis, including adequate recovery of small-sized RNAs. The 95-bp microRNA precursor of hsa-miR-141 *(8)* was correctly and consistently identified after cDNA synthesis and cloning. (3) The amplicons generated after cyanine (Cy)3 labeling can be used for genome-wide expression screening using microarray approaches. To ensure complete genome-wide coverage, future screening strategies should correct for the fact that placental RNA is enriched for 5′ mRNA fragments *(6)*, whereas most expression libraries, including the library used in the present study, are enriched for 3′ mRNA fragments. (4) Application of this method to placental RNA isolated from maternal plasma during first trimester could permit a genome-wide expression strategy for fetal RNA marker screening in clinical samples.

2. Materials

2.1. Cell Culture and Lysis

1. Hanks' balanced salt solution (HBSS) (Invitrogen, Carlsbad, CA).
2. F-12 nutrient mixture (Ham's) with GlutaMAX I (Invitrogen) supplemented with 10% fetal bovine serum and penicillin/streptomycin (100 μg/ml).
3. Trypsin-EDTA (10×).

   a. 0.5% (w/v) trypsin in EDTA (2 g/l) (Invitrogen).

2.2. RNA Isolation (Virus Min-Elute System)

1. Carrier RNA: 1 μg/l in AVE buffer (QIAGEN, Valenica, CA). Store diluted carrier RNA as aliquots at –20°C.
2. Lysis buffer: 9.04 mg/l carrier RNA in AL buffer (QIAGEN). Prepare fresh and mix gently before use. Store AL buffer at room temperature and do not shake before use.
3. Protease solution: Dilute dry protease (QIAGEN) in 4.4 ml of protease resuspension buffer (QIAGEN). Store protease solution as aliquots at –20°C.
4. QIAVac 24 (QIAGEN): Connect QIAVac to a vacuum pump producing –800 to –900 mBar. Connect the MinElute columns (stored at 4°C) (QIAGEN) to the QIAVac by using the VacConnectors (QIAGEN). Apply the extension tube on top of the column.
5. AW1 washing buffer (QIAGEN). Store at room temperature and shake before use.
6. AW2 washing buffer (QIAGEN). Store at room temperature and shake before use.
7. AVE elution buffer (QIAGEN). Store at room temperature.
8. DNase: Dissolve dry DNase (1500 Kunitz units) (QIAGEN) in 550 μl of RNAse-free water and mix gently. Do not vortex. Divide in aliquots of 10 μl and store at –20°C.
9. RDD buffer (QIAGEN). Store at 4°C.

2.3. Q-RT-PCR (EZ RTth System)

1. TaqMan EZ buffer (5×): 250 mM bicine, 575 mM potassium acetate, 0.05 mM EDTA, 300 nM passive reference, 1.40% (w/v) glycerol, pH 8.2 (Applied Biosystems, Foster City, CA).
2. 25 mM manganese acetate (Applied Biosystems).
3. Deoxynucleotides: 10 mM dATP, 10 mM dCTP, 10 mM dGTP, and 20 mM dUTP.
4. 2.5 U/μl RTth DNA polymerase (Applied Biosystems).
5. 1 U/μL AmpErase uracil-*N*-glycosylase (UNG) (Applied Biosystems).
6. TaqMan probes (50 μM; all from Eurogentec).

   a. hPL 5′-FAM-TTC TGT TGC GTT TCC TCC ATG TTG G-TAMRA-3′.
   b. C21orf105 5′-FAM-CGC CTA CTG GCA CAG ACG TG-TAMRA-3′.

7. Primers (50 μM; all from Eurogentec).
   a. hPL-F 5′-CAT GAC TCC CAG ACC TCC TTC-3′.
   b. hPL-R 5′-TGC GGA GCA GCT CTA GAT TG-3′.
   c. C21orf105-F 5′-TGC ACA TCG GTC ACT GAT C-3′.
   d. C21orf105-R 5′-GGG TCA GTT TGG CCG ATA-3′.
8. MicroAmp Optical 96-well reaction plates (Applied Biosystems).
9. ABI Prism Optical adhesive covers (Applied Biosystems).
10. ABI 7300 Real-Time PCR system (Applied Biosystems).

2.4. cDNA Synthesis

1. 1.5 U/μl Poly(A) polymerase (Takara, Gennevilliers, France).
2. 10 mM ATP (Roche, Mannheim Germany).
3. 40 U/μl RNasin (Promega, Madison, WI).
4. 10× reaction buffer: 0.5 M Tris-HCl, pH 8.0, 2.5 M NaCl, and 100 mM $MgCl_2$.
5. RNABee (Tel-Test Inc., Friendswood, TX).
6. Precipitation mix.
   a. Prepare a mix of ethanol/NaOAc by mixing 95 ml of 100% ethanol with 4 ml of 3 M NaOAc, pH 4.8, and 1 ml of sterile water.
   b. Store at –20°C.
7. 20 mg/mL glycogen (Roche).
8. 10× T4 RNA ligase buffer (Promega).
9. 10 U/μl T4 RNA ligase (Promega).
10. 40 U/μl RNasin (Promega).
11. 40% (w/v) polyethylene glycol (Sigma-Aldrich, St. Louis, MO).
12. 5′-Adapter RNA (5′-CGA CUG GAG CAC GAG GAC ACU GAC AUG GAC UGA AGG AGU AGA AA'-3, 100 ng/μl; Eurogentec).

2.5. Reverse Transcription

1. RT-primer: 5′-ATT CTA GAG GCC GAG GCG GCC GAC ATG-$d(T)_{30}$ (A, G, or C) (A, G, C, or T)-3′ (2 μM; Eurogentec).
2. Deoxynucleotides (all 10 mM; dATP, dCTP, dGTP, dTTP).
3. 10× reverse transcriptase buffer (200 mM Tris-HCl, pH 8.4, and 500 mM KCl.
4. 25 mM $MgCl_2$.
5. 0.1 M Dithiothreitol (DTT).
6. 40 U/μl RNaseOUT.
7. 200 U/μL SuperScript III RT (Invitrogen).

2.6. Polymerase Chain Reaction

1. Taq DNA polymerase.
2. 10× PCR buffer II: 100 mM Tris-HCl, pH 8.3, 500 mM KCl (Applied Biosystems).

3. Deoxynucleotides (all 2 mM; dATP, dCTP, dGTP, dTTP).
4. 25 mM $MgCl_2$.
5. 5 M (5×) betaine (Fluka, Buchs, Switzerland): Dilute 33.79 g of betaine in 50 ml of Milli-Q (Millipore Corporation, Billerica, MA) water. Store as aliquots at –20°C.
6. Forward PCR primer (5′-ATT CTA GAG GCC GAG GCG GCC GAC ATG T-3’ (50 pmol/µl; Eurogentec).
7. Reverse PCR primer (5′-GGA CAC TGA CAT GGA CTG AAG GAG TA-3′ (50 pmol/µl; Eurogentec).

2.7. Cloning

1. QIAquick gel extraction kit (QIAGEN).
2. TOPO TA cloning kit (Invitrogen).
3. TOPO vector (Invitrogen).
4. Transformation SOC medium: 2% (w/v) tryptone, 0.5% (w/v) yeast extract, 10 mM NaCl, 2.5 mM KCl, 10 mM $MgCl_2$, 10 mM $MgSO_4$, 20 mM glucose.
5. Competent cells F10’ (Invitrogen).
6. Luria Bertani (LB) plates containing 1.0% (w/v) tryptone, (w/v) 0.5% yeast extract and 1.0% NaCl, 50 µg/ml ampicillin, 40 mg/ml 5-bromo-4-chloro-3-indolyl-β-D-galactoside (X-Gal), and 100 mM isopropyl β-D-thiogalactoside (IPTG).

2.8. Sequencing

1. Big Dye Terminator v3.1 cycle sequencing kit (Applied Biosystems).
2. Primers (Eurogentec).
   a. M13-F: 5′-GTAAAACGACGGCCAGT-3′.
   b. M13-R: 5′- CAGGAAACAGCTATGAC-3′.
3. Primers (Eurogentec).
   a. PCR-F: 5′-ATT CTA GAG GCC GAG GCG GCC GAC ATG T-3′.
   b. PCR-R: 5′-GGA CAC TGA CAT GGA CTG AAG GAG TA-3′.

2.9. Microarray

All technical details can also be found at http://www.vumc.nl/microarrays.

2.9.1. Preparation of Labeled cDNA

1. Fluorolink Cy3 Monofunctional Dye 5-pack (GE Healthcare, Chalfont, St. Giles, UK). Dissolve dry pellet in 20 µL of DMSO (Sigma). Aliquot 2 µl into 10 single-use tubes that are then dried in a speed vac. Store desiccated at 4°C. NHS-ester-conjugated Cy dye is rapidly hydrolyzed in water; therefore, do not store in dimethyl sulfoxide (DMSO) or water.

2. 1 M $NaHCO_3/Na_2CO_3$, pH 9.0.
3. Deoxynucleotides (10 mM dATP, 8 mM dCTP, 10 mM dGTP, 10 mM dTTP, and 2 mM Cy3-dCTP).
4. QIAquick PCR purification kit (QIAGEN).

2.9.2. Hybridization of Labeled cDNA to Oligonucleotide-DNA Slides

1. 10 μg/μl $pd(A)_{40-60}$ (GE Healthcare).
2. 9.2 μg/μl tRNA (Sigma-Aldrich).
3. 1.0 μg/μl human Cot-1 DNA (Invitrogen).
4. 3 M NaAc, pH 5.2.
5. 10% sodium dodecyl sulfate (SDS).
6. 20× standard saline citrate (SSC) (Sigma-Aldrich).
7. Dextran sulfate.
8. 100% formamide.
9. Compugen Hum Libnr 2 slides containing 29,134 oligos of 28,830 genes.

3. Methods

3.1. Syncytial Denudation of Trisomy 21 Placenta

Placental tissue fragments (week 13^{+4}) are obtained by currettement of trisomy 21 pregnancies. The diagnosis is confirmed by karyotyping. Informed consent should be obtained before use.

1. Within 10 min after removal, wash placental tissue fragments extensively in cold HBSS.
2. Perform syncytial denudation (*see* **Note 1**) by digestion for 5 min at 37°C with 0.125% trypsin in phosphate-buffered saline. Use 2 ml per well. Immediately after this denudation, the medium containing DNA and RNA released selectively from the syncytiotrophoblast cells is subjected to RNA isolation.

3.2. RNA Isolation

1. Add to 75 μl of protease, 400 μl of the medium containing shedded DNA/RNA, and 400 μl of lysis buffer. Mix by pulse-vortexing.
2. Incubate for 15 min at 56°C and centrifuge briefly.
3. Add 500 μl of 100% ethanol and mix by pulse-vortexing.
4. Incubation for 5 min at room temperature.
5. Apply the lysate to the extension tube of the QIAamp MinElute column and apply vacuum of –800 mbar.
6. Switch off the pump when all lysate has passed the column. In the next washing steps, the pump is also used at –800 mbar to get the fluid through the column.
7. Wash column with 600 μl of AW1 buffer.

8. Apply a mix of 10 μl of DNAse and 70 μl of RDD buffer (do no vortex) on the column and incubate for 30 min.
9. Wash column with 750 μl of AW2 buffer.
10. Wash with 750 μl of 100% ethanol.
11. Remove the column from the QIAVac system and centrifuge for 3 min at 20,000*g*.
12. Incubate the column for 5 min with 50 μl of AVE buffer and centrifuge for 1 min at 20,000*g*.
13. Store the eluate in aliquots of 10 μl at –80°C.

3.3. Quantitative RT-PCR

1. Dilute a standard curve from 0 to 10^7 copies /10 μl in Milli-Q water for hPL or C21orf105.
2. Prepare the master mix for the number of samples needed by adding per sample: 1× TaqMan EZ buffer; 3 mM manganese acetate; 300 μM each of dATP, dCTP, dGTP, 600 μM UTP; 0.1 U/μl RTth enzyme; 0.01 U/μl AmpErase UNG; 300 nM each of forward and reverse primer (hPL or C21orf105); and 100 nM TaqMan probe for hPL or 400 nM TaqMan probe for C21orf105 in a volume of 40 μl.
3. Pipette the mix in the reaction plate and add 10 μl of sample. The total volume is 50 μl. Cover the plate.
4. Run the RT-PCR under the following conditions: 2 min at 50°C, 30 min at 60°C, 5 min at 50°C followed by 50 cycles for 20 s at 94°C, and 1 min at 56°C (hPL) or 59°C (C21orf105).

3.4. cDNA Synthesis

1. Prepare a mix of 10 μl of RNA, 5 U of poly(A) polymerase, 100 nM ATP, 40 U of RNasin, and 1× reaction buffer.
2. Add Milli-Q water to a total volume of 50 μl.
3. Incubate at 37°C for 30 min.
4. The polyadenylation reaction is stopped by phenol/chloroform extraction.
5. Add 1 volume of RNABee and 0.1 volume of chloroform.
6. Shake vigorously and incubate for 5 min on ice.
7. Spin down for 15 min at maximum speed at 4°C.
8. Transfer the aqueous phase to a clean tube and add 2.5 volumes of cold ethanol/NaOac mix and 2 μl of glycogen.
9. Mix and incubate at –80°C for 30 min.
10. Centrifuge for 25 min at maximum speed at 4°C.
11. Wash twice with 80% cold ethanol.
12. Air dry the pellet and resuspend in 34.5 μl of Milli-Q water.
13. The 5′ adapter RNA is ligated to poly(A)-tailed RNA in the reaction mix containing: 34.5 μl of RNA, 1× ligase buffer, 120 U of RNasin, 20% polyethylene glycol (w/v), 30 U of T4 RNA ligase, and 750 ng of 5′ adapter RNA.
14. Incubate at 37°C for 30 min.

15. The reaction is stopped by phenol/chloroform extraction and ethanol precipitation.
16. The pellet (*see* **Note 2**) is diluted in 8 μl of Milli-Q water.

3.5. Reverse Transcription

1. To the RNA pellet in 8 μl of Milli-Q water, add 2 nmol of RT-primer and 1 nmol of dNTP mix.
2. Incubate the sample mix for 5 min at 65°C and chill subsequently on ice.
3. Mix the reverse transcription mix to contain 1× RT-buffer, 100 nM $MgCl_2$, 2 μM DTT, 40 U of RNaseOUT, and 200 U of SuperScript III RT enzyme.
4. Add this mix to the sample mix and incubate for 1 h at 50°C.
5. Inactivate this reaction by 15 min at 70°C and store the cDNA at –20°C.

3.6. PCR

1. Amplify the cDNA with a PCR mix containing 10 μl of cDNA, 1× buffer II, 0.2 mM dNTP, 10 μM forward PCR primer, 10 μM reverse PCR primer, 1.5 μM $MgCl_2$, 0.2 U/μl Taq polymerase, and 1 M betaine in a reaction volume of 50 μl.
2. The PCR conditions are 4 min at 95°C, 25 cycles of 1 min at 95°C, 1 min at 50°C, and 2 min at 72°C followed by 10 min at 72°C.

3.7. Cloning

1. Separate the PCR-amplicon by running a 2% gel electrophoresis and extract the different fragments of the amplicon with the QIAquik gel extraction kit.
2. The eluate volume is 30 μl.
3. Clone the isolated amplicons into the Topo-TA vector by performing the ligation and transformation reaction.
4. For this, mix of 4 μl of isolated amplicon with 1 μl of salt solution and 1 μl of TOPO vector.
5. Mix gently and incubate for 30 min at room temperature. Store at –20°C.
6. The ligation mix is transformed into TOP 10 F' competent cells by adding 2 μl of ligation mix to 50 μl of F10' competent cells.
7. Incubate for 30 min on ice.
8. Heat shock the mixture for 30 s at 42°C and chill on ice.
9. Add 250 μl of SOC medium (room temperature) and incubate gently by shaking for 1 h at 37°C.
10. Spread 100 μl of the mix on prewarmed (37°C) LB plates with ampicillin, X-Gal, and IPTG.
11. Incubate overnight at 37°C.
12. The white colonies are screened by colony PCR. The colony is picked with a sterile toothpick and incubated for 5 h at 37°C in 100 μl of LB media with ampicillin. PCR is done as described above with 10 μl of inoculated LB mix by using PCR or M13 primers.

3.8. Sequencing

1. The sequence reaction is performed on the amplicons by adding 1 μl of terminator ready reaction mix, 1 μl of amplicon, 3.2 pmol of primer (M13 forward or M13 reverse), and 1× sequence buffer in a reaction volume of 20 μl of Milli-Q water.
2. The temperature profile is 25 cycles of 10 s at 96°C, 5 s at 50°C, and 4 min at 60°C.
3. Precipitation of the product is done by isopropanol.
4. Sequence analysis is performed on the ABI3100 (Applied Biosystems).

3.9. Microarray

3.9.1. RT-PCR

For microarray screening, RT-PCR is done in the presence of allylamine-dCTP.

1. Amplify the cDNA with a PCR mix containing 10 μl of cDNA, 1× buffer II, 0.2 mM dATP, 0.2 mM dGTP, 0.2 mM dTTP, 0.16 mM dCTP, 0.04 mM aminoallyl dCTP, 10 μM forward PCR primer, 10 μM reverse PCR primer, 1.5 μM $MgCl_2$, 0.2 U/μl Taq polymerase, and 1 M betaine in a reaction volume of 50 μl.
2. The PCR conditions are 4 min at 95°C, 25 cycles of 1 min at 95°C, 1 min at 50°C, and 2 min at 72°C followed by 10 min at 72°C.0

3.9.2. Purification of RT-PCR Product

The PCR product is purified with the QIAquick PCR purification kit protocol.

1. Add 500 μl of buffer PB to the PCR product, mix, and transfer it to a QIAquick column.
2. Spin at maximum speed for 1 min and discard flow through.
3. Wash the column with 750 μl of buffer PE and spin at maximum speed for 1 min and discard flowthrough.
4. Spin again to completely dry the column.
5. Elute in a fresh 1.5-ml tube by adding 50 μl of EB buffer directly onto the filter and incubate for 2 min.
6. Spin at maximum speed for 1 min.
7. Dry the eluate in the speed vac for 1 h.

3.9.3. Labeling and Purification of the RT-PCR Product

1. Dissolve the pellet in 9 μl of 50 mM $NaHCO_3/Na_2CO_3$, pH 9.0 (prepared fresh from 1 M stock).
2. Incubate for 15 min at room temperature to dissolve.
3. Transfer the DNA to a tube containing 2-μl aliquots of Cy3 as dry pellet.
4. Mix well and transfer to original tube and incubate for 1 h at room temperature in the dark.
5. Block the Cy3 dyes by adding 4.5 μl of 4 M hydroxylamine and incubate for 15 min at room temperature in the dark.

6. Add 70 μl of Milli-Q water to the labeled DNA and remove uncoupled dye material by performing the QIAquick PCR purification kit protocol (*see* **Subheading 3.5.2.**).
7. Elute in 30 μl of buffer EB.

3.9.4. Hybridization of Probe to Microarray

1. Prepare 126.7 μl of hybridization mix by adding 90 μl of Milli-Q water, 1.2 μl of $pd(A)_{40-60}$, 6.5 μl of tRNA, and 24 μl of Cot-1 DNA to 30 μl of labeled DNA.
2. Precipitate this by using 1.2 volumes of 3 M NaAc, pH 5.2, and 2.5 volumes of 100% ethanol, mix by inversion, and incubate on ice for 5 min.
3. Spin down for 10 min at 12,000*g* at 4°C.
4. Remove the supernatant with a pipette and air dry the pellet for 5–10 minutes.
5. Carefully dissolve the pellet in 40.7 μl of Milli-Q water and 2.5 μl of 10% SDS (avoid foam).
6. Incubate at room temperature for at least 15 min.
7. Prepare a master mix with 1 g of dextran sulphate, 5.3 ml of formamide 100%, 0.7 ml of Milli-Q water, and 1.0 ml of 20× SSC, pH 7.0.
8. Add 83.5 μl of mastermix to the dissolved pellet and mix gently.
9. Denature the hybridization solution at 70°C for 10 min and incubate on ice/water for 1min.
10. Subsequently, incubate for 1 h at 37°C.
11. The mastermix with labeled DNA is ready to be hybridized on the hybridization station.
12. Scanning is done using the Agilent DNA microarray scanner.

4. Notes

1. The completeness of the syncytiotrophoblast cell removal can be checked by using microscopic analysis with phase contrast. If necessary, optimize using immunohistochemistry as described in **ref. 5**.
2. The completeness and specificity of the cDNA synthesis reaction can be checked by the presence of sequences representative for hsa-miR-141 ***(8)***. In addition, the presence of intronic sequences indicates incomplete DNA removal.

Acknowledgment

M.v.D. is supported by the SAFE Network Project LSHB-CT-2004-503243).

References

1. Oudejans, C. B. M., Go, A. T. J. J., Visser, et al. (2003) Detection of chromosome 21- encoded mRNA of placental origin in maternal plasma. *Clin. Chem.* **49,** 1445–1449.

2. Chim, S. S., Tong, Y. K., Chin, R. W. K., et al. (2005) Detection of placental epigenetic signature of the maspin gene in maternal plasma. *Proc. Natl. Acad. Sci. USA* **102,** 14753–14758.
3. Gross, S. J., Ferreira, J. C., Morrow, B., et al. (2002) Gene expression profile of trisomy 21 placentas: a potential approach for designing noninvasive techniques of prenatal diagnosis. *Am. J. Obstet. Gynecol.* **187,** 457–462.
4. Ng, E. K. O., Tsui, N. B. Y., Lau, T. K., et al. (2003) mRNA of placental origin is readily detectable in maternal plasma. *Proc. Natl. Acad. Sci. USA* **100,** 4748–4753.
5. Baczyk, D., Dunk, C., Huppertz, B., et al. (2006) Bi-potential behaviour of cytotrophoblast in first trimester chorionic villi. *Placenta* **27,** 367–374.
6. Wong, B. C. K., Chiu, R. W. K., Tsui, N. B. Y., et al. (2005). Circulating placental RNA in maternal plasma is associated with a preponderance of 5′ mRNA fragments: implications for noninvasive prenatal diagnosis and monitoring. *Clin. Chem.* **51,** 1786–1795.
7. Fu, H., Tie, Y., Xu, C., et al. (2005). Identification of human fetal liver miRNAs by a novel method. *FEBS Lett.* **579,** 3849–3854.
8. Kiriakidou, M., Nelson, P. T., Kouranov, A., et al. (2004) A combined computational-experimental approach predicts human microRNA targets. *Genes Dev.* **18,** 1165–1178.

24

Method for Extraction of High-Quantity and -Quality Cell-Free DNA from Amniotic Fluid

Olav Lapaire, Kirby L. Johnson, and Diana W. Bianchi

Summary

Circulating cell-free fetal deoxyribonucleic acids (cffDNAs) are promising biomarkers with various potential clinical applications. Second and third trimester amniotic fluid (AF) is a rich source of cffDNAs. Further improvements to the original protocol for the extraction of cffDNAs from AF supernatant resulted in statistically significant higher yields of high-quality cffDNAs, allowing for a substantial majority of samples to be analyzed with subsequent molecular methods (e.g., comparative genomic hybridization microarrays) to further assess for genetic abnormalities. Several advantages have been realized with the optimized protocol. In addition to an improved yield from a greater proportion of samples compared with the original protocol, the current method, using large silico-membranes, allows for the extraction of cffDNAs from up to 10 samples in <3 h. The replacement of the original lysis buffer eliminates the need for a heating bath during the lysis step, and fewer overall steps are involved in the protocol (e.g., to reduce potential contamination). The improvements in the yield with the current protocol make it possible to augment current standard of care through the analysis of this previously unappreciated source of genetic material. Furthermore, the improvements allow for exploration of widely unknown genetic, pathophysiological, and kinetic issues of cell-free fetal DNA in AF.

Key Words: Prenatal diagnosis; cell-free fetal DNA; prenatal screening; pregnancy.

1. Introduction

Since the detection of cell-free fetal nucleic acids (cffDNAs) in maternal plasma and serum, many investigators have focused on its biology and potential role as a genetic marker for noninvasive prenatal diagnosis and screening *(1–4)*. However, these investigations have yet to fully exploit this source of

From: *Methods in Molecular Biology, vol. 444: Prenatal Diagnosis*
Edited by: S. Hahn and L. G. Jackson © Humana Press, Totowa, NJ

genetic material in maternal body fluids other than plasma and serum. Prenatal cytogenetic diagnosis traditionally relies on the analysis of cultured metaphase cells harvested from amniotic fluid (AF) obtained at amniocentesis. Typically, 10–30 ml of AF is removed, and intact cells are cultured for cytogenetic analysis; therefore, they are not available for additional molecular techniques, such as polymerase chain reaction (PCR) amplification or DNA microarray profiling. After removal of the cells, a small amount of the amniotic fluid supernatant is taken to assess levels of cell-free proteins as markers of fetal abnormalities. The remaining supernatant is normally discarded; thus, this fraction can yield cffDNAs for additional genetic testing to potentially augment current standard of prenatal diagnostic care ***(5,6)***.

In 2001, Bianchi et al. ***(5)*** first demonstrated that cffDNAs present in amniotic fluid could be a potential source of genetic material for research and clinical applications. In that preliminary study, they showed that larger quantities (100-200-fold) of cffDNAs per milliliter are present in AF compared with maternal plasma/serum. However, due to difficulties in extracting sufficient amounts of high-quality cffDNAs from AF for effective analysis, only a minor proportion of samples could be further analyzed (e.g., with DNA microarrays, in which a minimum of 100 ng of DNA is necessary) ***(6)***. Our original protocol was based on known protocols for the extraction of cffDNAs from maternal plasma/serum, because specific guidelines for the extraction of this DNA from AF did not exist. Therefore, further investigation was needed to optimize cffDNA extraction from AF supernatant to more fully exploit this promising source of genetic material. Our objective was to develop an approach for the improved isolation of cffDNAs from AF supernatant by using information regarding DNA extraction from urine, because AF is similarly excreted by the fetal urogenital system during the second and third trimesters.

To improve the yield of extracted cffDNAs, the original method ***(6)*** using the Blood and Body Fluid Vacuum Protocol (QIAGEN, Valencia, CA) was changed in the following ways. (1) Increasing vacuum extraction pressure from 400 to 800 mbar to allow for maximal absorption of DNA to mini filters. Although this pressure may exceed that available in some laboratories (e.g., using building vacuum pressure), reduced vacuum leads to lower yield of extracted DNA. (2) Replacing the volume-overloaded Mini Spin columns with Maxi Spin columns (QIAGEN). In the original protocol, 10 ml of amniotic fluid was added to 1 ml of protease, 10 ml of AL lysis buffer (QIAGEN), and 10 ml of 100% ethanol, which exceeded the volumetric capacity of the mini columns. The use of maxi columns allows for larger starting volumes to be processed and therefore a larger quantity of cffDNAs can be obtained compared to mini columns. (3) Substituting AL buffer with AVL buffer (QIAGEN). AL buffer, used in the original protocol, was selected based on prior experience for the isolation of cffDNAs from plasma and serum, because there was no information available on the most suitable buffer for extraction of cffDNAs from AF. However, AVL buffer supplemented with nucleic acid carrier for the extraction of low concentrations of target DNA was selected for the current protocol on the basis of similar qualities between AF and urine, a body fluid in which AVL buffer is recommended for DNA extraction.

From euploid singleton pregnancies ($n = 29$), the median amount of *GAPDH* DNA extracted from 10 ml of AF with the new protocol was 1,700 genomic equivalents (GE)/ml (25th and 75th percentiles, 1,071 and 4,938 GE/ml, respectively) compared with 246 GE/ml (93 and 523.5 GE/ml) by using the original protocol ***(6,7)*** ($p < 0.0001$; *see* **Fig. 1**).

Several advantages have been realized with the protocol developed here. In addition to an improved yield from a greater proportion of samples compared with the original protocol, the current protocol allows for the extraction of cffDNAs from up to 10 samples in <3 h. The replacement of AL buffer

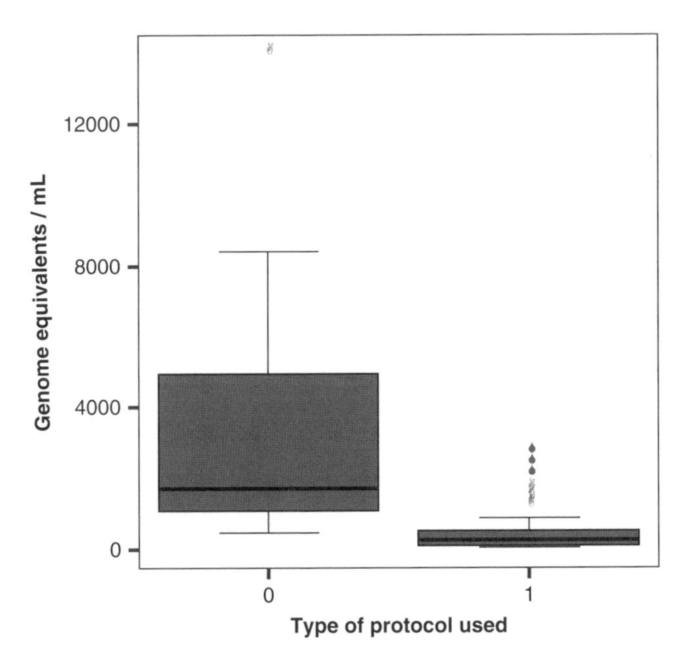

Fig. 1. Comparison of the yield of the extraction mode of cell-free fetal DNA from AF: comparison of the yield of cell-free fetal DNA, as measured by real-time PCR amplification of the *GAPDH* locus, extracted from AF supernatant from euploid singleton pregnancies. 0, new protocol ***(7)***, 1, original protocol ***(6)***. The lines inside the boxes denote medians. The box indicates 25th and 75th percentiles. The whiskers denote the 10th and the 90th percentiles. Symbols indicate data points outside the 10th and 90th percentiles.

with AVL buffer eliminates the need for a heating bath during the lysis step, and fewer overall steps are involved in the protocol (e.g., to reduce potential contamination). However, the cost of cffDNA extraction from a 10-ml AF supernatant sample by using the new protocol is ≈10-fold higher compared with the original protocol ($39 vs $4/sample, respectively), although the advantage of the new protocol with respect to improved DNA yield justifies this higher cost per sample.

For clinical applications, one major advantage of using the AF supernatant is its availability without interfering with current standard of care or compromising fetal health.

2. Materials

2.1. For Each Sample of 10 ml

1. Three 50-ml polypropylene tubes (e.g., BLUE MAX Falcon, BD Biosciences, Franklin Lakes, NJ, cat. no. 352070).
2. One 2-ml microcentrifuge tube (e.g., Fischer Scientific, Pittsburgh, PA, package of 500, cat. no. 02-681-258) for storage of the final eluted sample.
3. One to three 0.5-ml tubes for storage of the final eluted sample.
4. Sterile pipette tips (1,000, 200, and 20 μl).
5. Sterile 10- or 20-ml pipettes.
6. Buffer AVL (155 ml) viral lysis buffer and 4.2 mg of carrier RNA/DNA for 250 reactions (QIAGEN, cat. no. 19073). After arrival, lysis buffer can be stored at room temperature (15–25°C). After addition of RNA/DNA carrier, lysis buffer should be stored at 4°C.
7. QIAamp DNA Blood Maxi kit (10), for 10 DNA maxipreps (QIAGEN cat. no. 51192).
8. Alternatively, QIAamp DNA Blood Maxi kit (50), for 50 DNA maxipreps (QIAGEN, cat. no. 51194).

   a. Maxi columns should be stored at 4°C to maintain constant high yields.
   b. Wash solutions (AW1 and AW2) are supplied in the QIAamp DNA Blood Maxi kit. Storage at room temperature (15–25°C).
   c. AE elution buffer is supplied in the QIAamp DNA Blood Maxi kit. Store at room temperature (15–25°C).
   d. Ethanol 100%, store at room temperature (15–25°C).

2.2. Important Points before Starting

1. All steps of the protocol should be performed at room temperature (15–25°C). During the procedure, work quickly.
2. Wash buffer AW1 and AW2 are supplied. Add volumes of ethanol (100%), as indicated on the bottles.

3. Methods

3.1. Extraction of cffDNAs from 10 ml of AF Supernatant

1. Prepare tubes for extraction: for each sample (10 ml), you need three 50-ml screw cap tubes (polypropylene), one 2-ml microcentrifuge tube, and one to three 0.5-ml tubes. Place tube rack, pipettes, and tubes into the ultraviolet cross-linker for 20 min, leaving caps open.
2. If AF samples are frozen, thaw at 37°C and equilibrate to room temperature (15–25°C) (*see* **Note 1**).
3. Vortex thawed sample briefly.
4. Check buffer AVL for precipitate; if necessary, incubate at 80°C until precipitate is dissolved. Add 1 ml of buffer AVL to one tube of lyophilized carrier RNA/DNA. Dissolve carrier RNA/DNA thoroughly. Transfer to AVL bottle and mix thoroughly before using AVL buffer for the first time (*see* **Note 2**).
5. Pipette 20 ml (or 400% of AF volume) of AVL/carrier RNA/DNA in each of two 50-ml tubes.
6. Add 5 ml of AF into each of the 50-ml tubes. Mix by pulse vortexing for 15 s (*see* **Note 3**).
7. Incubate at room temperature (15–25°C) for 10 min.
8. Briefly centrifuge the two 50-ml tubes at 1,200 rpm to remove residual droplets from the cap.
9. Add 20 ml (or same amount as AVL) of 100% ethanol to each sample. Mix by pulse vortexing for 15 s.
10. Briefly centrifuge the 50-ml tubes at 1,200 rpm (Sorvall RT centrifuge, Sorvall, Newton, CT) to remove residual droplets from the cap.
11. Apply 15 ml (of total 90 ml) from **step 10** to the QIAamp Maxi column, placed in a 50-ml centrifugation tube. Do not moisten the rim of the QIAamp Maxi column. Close the cap and centrifuge at 1,850*g* (3,000 rpm) (Sorvall RT) for 3 min (*see* **Note 4**).
12. Remove the QIAamp Maxi column, discard the filtrate, and place the QIAamp Maxi column back into the 50-ml centrifugation tube. Load 15 ml of the solution from **step 10** onto the QIAamp Maxi column. Close the cap and centrifuge at 1,850*g* (3,000 rpm) (Sorvall RT) for 3 min. Note: wipe off any spillage from the thread of the 50-ml centrifugation tube before reinserting the QIAamp Maxi column. Do not wet the rim of the QIAamp Maxi column. Repeat **step 12** until the whole volume passed through the maxi column.
13. Remove the QIAamp Maxi column, discard the filtrate, and place the QIAamp Maxi column back into the 50-ml centrifugation tube. Note: wipe off any spillage from the thread of the 50-ml centrifugation tube before reinserting the QIAamp Maxi column.
14. Carefully, without moistening the rim, add 5 ml of buffer AW1 to the QIAamp Maxi column. Close the cap and centrifuge at 3,715*g* (4,150 rpm) (Sorvall RT)

for 5 min. Do not discard the flow-through at this stage and continue directly with **step 15**.

15. Carefully, without moistening the rim, add 5 ml of buffer AW2 to the QIAamp Maxi column. Close the cap and centrifuge at 3,715*g* (4,150 rpm) (Sorvall RT) for 25 min (*see* **Note 5**).
16. Discard the 50-ml centrifugation tube containing the filtrate and place the QIAamp Maxi column in a new 50-ml centrifugation tube. Note: wipe off any spillage from the thread of the 50-ml centrifugation tube before reinserting the QIAamp Maxi column.
17. Add 1 ml of buffer AE (15–25°C). Pipet directly to the membrane of the QIAamp Maxi column and close the cap. Incubate at room temperature for 5 minutes and centrifuge at 3,715*g* (4,150 rpm) for 10 min.
18. For maximum yield, pipet 1 ml of fresh buffer AE (15–25°C). Pipet directly to the membrane of the QIAamp Maxi column and close the cap. Incubate at room temperature for 5 min and centrifuge at 3,715*g* (4,150 rpm) (Sorvall RT) for 10 min. Note: less than 1 ml will be eluted from the column, but this has no effect on DNA yield.
19. Transfer the elutant to a 2-ml and additional 500-μl microcentrifuge tubes for further analysis (e.g., PCR, electrophoresis).
20. Analyze samples immediately, store at 4°C for a maximum of 2 weeks, or freeze indefinitely (at –80°C).

4. Notes

1. Samples can be thawed in approximately 15 min, by using a water bath (37°C).
2. AVL buffer that is stored at 4°C after addition of carrier RNA/DNA usually shows small crystalloid precipitates at the bottom of the bottle. Dissolve them by putting the bottle in a warm water bath and vortex the buffer quickly.
3. It is important to mix the sample thoroughly with the lysis buffer to avoid any clotting of the silico-membrane.
4. Use an appropriate centrifuge that allows centrifugation of 50-ml tubes at 3,700–4,000*g* (e.g., Sorval RT). If the solution has not completely passed through the membrane, centrifuge at a slightly higher speed.
5. Please allow the silico-membranes to dry completely. A reduction of the centrifugation time is not recommended.

Acknowledgments

We thank Helene Stroh, Inga Peter, and Janet Cowan with the team of the Tufts-New England Medical Center, Boston, MA, for advice. This work was supported by National Institutes of Health grant R01HD42053 and Swiss National Fund PBBSB-108590.

References

1. Lo, Y. M. D, Corbetta, N., Chamberlain, P. F., et al. (1997) Presence of fetal DNA in maternal plasma and serum. *Lancet* **350,** 485–487.
2. Zhong, X. Y., Laivuori, H., Livingston, J. C., et al. (2001) Elevation of both maternal and fetal extracellular circulating deoxyribonucleic acid concentrations in the plasma of pregnant women with preeclampsia. *Am. J. Obstet. Gynecol.* **184,** 414–419.
3. Wataganara, T., LeShane, E.S., Farina, A., et al. (2003) Maternal serum cell-free fetal DNA levels are increased in cases of trisomy 13, but not trisomy 18. *Hum. Genet.* **112,** 204–208.
4. Farina, A., LeShane, E. S., Lambert-Messerlian, G. M., Canick, J. A., Lee, T., and Neveux, L. M. (2003) Valuation of fetal cell free DNA as a second-trimester marker of Down syndrome pregnancy. *Clin. Chem.* **49,** 239–242.
5. Bianchi, D. W., LeShane, E., and Cowan, J. M. (2001) Large amounts of cell-free fetal DNA are present in amniotic fluid. *Clin. Chem.* **47,** 1867–1869.
6. Larrabee, P. B., Johnson, K. L., Pestova, E., et al. (2004) Microarray analysis of cell-free fetal DNA in amniotic fluid: a prenatal molecular karyotype. *Am. J. Hum. Genet.* **75,** 485–491.
7. Lapaire, O., Stroh, H., Peter, I., et al. (2006) Larger filters and change of lysis buffer significantly improve the quantity of extracted cell-free DNA from amniotic fluid. *Clin. Chem.* **52,** 156–157.

25

Detection of New Screening Markers for Fetal Aneuploidies in Maternal Plasma: *A Proteomic Approach*

Chinnapapagari Satheesh Kumar Reddy, Wolfgang Holzgreve, and Sinuhe Hahn

Summary

Proteomics has brought with it the hope of identifying novel biomarkers for the fetal aneuploidies. This hope is built on the ability of proteomic technologies, such as two-dimensional gel electrophoresis (2-DE) with immobilized pH gradients combined with protein identification by mass spectrometry (MS). The large dynamic range of plasma necessitates the effective removal of abundant plasma proteins to allow analysis of the lower concentration analytes. There are many factors that make this research very challenging, beginning with standardization of sample collection, consistent sample preparation, and continuing through the entire analytical process. Therefore, reproducible sample complexity reduction methods such as depletions or fractionations are an essential first step in biomarker discovery experiments. For qualitative and quantitative evaluation of the proteome, the fluorescent dye stains offer several advantages over traditional staining methods. The sensitivity of the fluorescent dye stains such as SYPRO Ruby is comparable with that of silver staining and also has a broad dynamic range, which allows accurate protein quantification being compatible with MS methods for protein identification. Despite its limitations for proteome analysis 2-DE is currently the workhorse for proteomics. Taking into account the factors such as cost, availability and ease of use, 2-DE electrophoresis is one of the most apposite approaches toward the methodical characterization of proteomes.

Key Words: Chromosomal aneuploidies; biomarker discovery; proteomics; gene expression profiling.

From: *Methods in Molecular Biology, vol. 444: Prenatal Diagnosis*
Edited by: S. Hahn and L. G. Jackson © Humana Press, Totowa, NJ

1. Introduction

Down syndrome (DS) or Trisomy 21 (Ts21) is the most common genetic cause of human mental retardation, affecting approximately 1 in 800 live births, caused by complete or partial triplication of human chromosome 21 *(1)*. The prenatal detection of fetal aneuploidies such as DS is a major obstetrical concern and currently relies on the detection by invasive prenatal diagnostic procedures, such as amniocentesis or chorionic villous sampling, which are associated with a risk of fetal loss or injury. Because no reliable noninvasive alternative seems to be available in the near future, considerable attention is being focused on several screening approaches. Currently, these either rely on ultrasound (nuchal translucency) or serum analytes (pregnancy-associated plasma protein-A, α-fetoprotein, β-human chorionic gondaotropin, or inhibin A). Although great strides have been made in improving the sensitivity of these approaches, which for the combined first trimester test is of the order of 90%, they are still hampered by relatively high false-positive rates (5%). Because this leads to a high number of unnecessary invasive diagnostic procedures, the need has been voiced for more specific screening markers to reduce the current false-positive rate. The many features of DS include neurological, skeletal, cardiovascular, and immunological defects, and they are generally thought to originate from a 1.5-fold increase in the dosage of genes within a critical region of chromosome 21, which is present in triplicate in all cases of DS *(2–4)*. The differential expression of genes located on the extra copy of chromosome 21 has been assumed to be responsible for the phenotypic abnormalities of DS, but this gene dosage hypothesis has not been fully assessed on a genome-wide basis. The expression patterns of these proteins related to phenotypic abnormalities of DS may provide insight into their potential roles in DS.

The importance of plasma as a source of clinically relevant biomarkers/surrogate markers of human disease has increased significantly over the last decade, and modern proteomic methods have evolved and been adapted to meet the demand. Plasma consists mainly of 22 proteins that account for 99% of its protein content. Of interest is the remaining 1%, which is made up of lower abundance circulatory proteins and proteins that are shed or excreted by not only living cells but also by apoptotic and necrotic cells *(5)*. Identification of these proteins might represent potential diagnostic biomarkers of disease states, based upon their change in abundance compared with normal levels. The proteomic analysis of plasma and serum samples represents a formidable challenge due to the presence of a few highly abundant proteins such as albumin and immunoglobulins. Detection of low abundance protein biomarkers, therefore, requires either the specific depletion of high-abundance proteins by using immunoaffinity columns and/or optimized protein fractionation methods based on charge, size, or hydrophobicity. Reproducible sample complexity reduction is an essential first step in biomarker discovery experiments *(6)*.

2. Materials

2.1. Plasma Collection and Storage

1. P100 Blood Collection kit (BD Biosciences, Franklin Lakes, NJ, cat. no. 366456).

2.2. Depletion Of High Abundant Proteins

1. Multiple Affinity Removal Spin Cartridge–High Capacity, 0.45-ml Hu-6HC (Agilent Technologies, Palo Alto, CA, cat. no. 5188–534).

2.3. Postdepletion Cleanup and Concentration

1. Vivaspin 2: 2 ml, 3000 molecular weight cut-off (Sartorius, Goettingen, Germany, cat. no. VS0291).
2. ReadyPrep Two-dimensional (2-D) Cleanup kit (Bio-Rad, Hercules, CA, cat. no. 163–2130).
   a. Prechill wash reagent 2 to –20°C at least for 1 h before use.

2.4. First-Dimension Isoelectric Focusing IEF

1. Immobilized pH gradient (IPG) strips, pH 3–10 (17 cm) (Bio-Rad, cat. no. 163–2000).
2. Bio-Lyte 3-10 Ampholyte (Bio-Rad, cat. no. 163–2094).
3. ReadyPrep proteomic grade water (Bio-Rad, cat. no. 163–2091).
4. Rehydration/sample buffer (Bio-Rad, cat. no. 163–2106).
5. Equilibration buffer I (Bio-Rad, cat. no. 163–2107).
6. Equilibration buffer II (Bio-Rad, cat. no. 163–2108).
7. PROTEAN IEF cell-unit (Bio-Rad, cat. no. 165–4000).

2.5. Second Dimension Sodium Dodecyl Sulfate (SDS)-Polyacrylamide Gel Electrophoresis (PAGE).

2.5.1. Equipment

The second dimension of the 2-D gel electrophoresis (2-DE) was performed in the Anderson's ISODALT system for the simultaneous analysis of 20 samples (instruments produced in the workshop of former Basel Institute for Immunology, Basel, Switzerland).

2.5.2. Horizontal SDS Gels

1. Electrode buffer stock solution (1L of 10 ×): Dissolve 30.3 g of Tris base, 144 g of glycine, 10.0 g of SDS, and 100 mg of sodium azide in deionized water and then fill up to 1,000 ml with deionized water. Filter with Whatman no. 1 filter paper

(Whatman, Maidestone UK, cat. no. 1001125). Electrode buffer stock solution can be kept at room temperature for up to 1 week. Before use, mix 100 ml of buffer stock solution with 900 ml of deionized water to the final working solution.

2.5.3. Vertical SDS Gels

1. Electrode buffer solution (20L): Dissolve 58.0 g of Trizma base, 299.6 g of glycine, and 20.0 g of SDS in 19.9 liters of deionized water.
2. Agarose solution: Suspend 0.5% (w/v) agarose in electrode buffer and melt it in a boiling water bath or in a microwave oven.

2.6. Staining and Quantitation

2.6.1. Silver Staining

1. Fixative: 50% (v/v) methanol, 5% (v/v) acetic acid, and 45% (v/v) deionized water.
2. Wash solution: 50% (v/v) methanol and 50% (v/v) deionized water.
3. Sensitizer: 0.02% (w/v) $Na_2S_2O_3 \cdot 5\ H_2O$ in deionized water.
4. Silver reagent: 0.1% (w/v) silver nitrate in cold (4°C) deionized water.
5. Developer: 0.04% (v/v) formalin and 2% (w/v) sodium carbonate (Na_2CO_3), in 1000 ml of deionized water; add the Formalin-solution (0.4 ml of 37% formaldehyde) just before use.
6. Stop-reagent: 5% (v/v) acetic acid.
7. Gel storage solution: 1% (v/v) acetic acid.

2.6.2. SYPRO Ruby Staining

1. Fixation solution: 10% (v/v) ethanol, 7% (v/v) acetic acid.
2. 330 ml of SYPRO Ruby stain for $16 \times 20\ cm^2$ gels, or 500 ml of SYPRO Ruby stain for $25 \times 20 \times 1\ cm^3$ gels.

2.6.3. Coomassie Blue Staining

1. Fixation solution: 20% trichloroacetic acid (TCA).
2. Washing solution: 40% (v/v) ethanol and 10% (v/v) acetic acid.
3. Dye solution: 0.1% (w/v) Serva Blue R-250 in 40% ethanol, 10% acetic acid.
4. Destaining solution: 40% (v/v) ethanol and 10% (v/v) acetic acid.

2.7. Scanning and Image Analysis

1. Software Progenesis 240 (Nonlinear Dynamics, Newcastle, UK).

2.8. Sample Preparation for Electrospray Ionization (ESI)-Tandem Mass Spectrometry and Matrix-Assisted Laser Desorption Ionization/Time of Flight (MALDI-TOF) Mass Spectrometry (MS)

The following items are required for each peptide digest:

1. Zip-Tip C18 cartridge column (Millipore Corporation, Billerica, MA).
2. Tube A: 100 µl of acetonitrile (high-performance liquid chromatography [HPLC] grade) as wetting solution.
3. Tube B: 1000 µl of water (HPLC grade) as cleaning solution.
4. Tube C: 10–20 µl of sample solution.
   Dissolve the extracted peptides in 10–20 µl of 0.5% (v/v) formic acid. Make sure to wash the lower sides of the PCR tube containing the extracted peptides to achieve maximum dissolution of the peptides.
5. Tube D: 500 µl of 0.5% (v/v) formic acid as desalting/washing solution.
6. Tube E: 100 µl of 0.5% (v/v) formic acid in 1:1 (v/v) water-to-acetonitrile as first extraction solution.
7. Tube F: 100 µl of acetonitrile as secnd extraction solution.
8. Tube G: a clean empty 200 µl of PCR tube.
9. 50 mM NH_4HCO_3, pH 8.8, with 50% acetonitrile.
10. 50 m*M* NH_4HCO_3, pH 8, containing 20 µg/ml trypsin.
11. Matrix-solution: α-cyano-4-hydroxycinnamic acid.
12. Vacuum centrifuge.
13. Mass spectrometer.
14. Data explorer software (e.g. Mascot, http://www.matrixscience.com; MS-Fit: http://prospector.edu).

3. Methods

3.1. Plasma Collection and Storage

1. After venipuncture, collect 8.5 ml of blood in P100 tubes. Invert the tube 8–10 times to mix the protease inhibitors and anticoagulant with the blood sample (*see* **Note 1**).
2. Centrifuge at 2,000–3,000*g* at 4°C for 15 min. This will yield about 2.5–3.0 ml of plasma. Aliquot the plasma in 1-ml volumes.

3.2. Depletion of High Abundant Proteins

Top 6 depletion: MARS (Agilent Technologies, cat. nos. 5188–5241 and 5188–5254).

1. Dilute and filter the sample: Prepare the sample by diluting 14–16 µl of human plasma with buffer A (protease inhibitors may be added to buffer A to prevent protein degradation) to a final volume of 200 µl. For example: if the recommended

cartridge loading capacity on the certificate is 14 μl of plasma, dilute 14 μl of plasma with 186 μL of buffer A to a final volume of 200 μl.

2. Prepare spin cartridge: Remove cartridge cap and plug. Attach Luer-Lock adapter to spin cartridge, draw 4 ml of buffer A into syringe and then attach it to Luer-Lock on spin cartridge. Dispense buffer A through spin cartridge to prepare resin and to remove any trapped air bubbles. With a transfer pipette, remove any excess buffer A from top of the spin cartridge.
3. Apply sample: Remove the Luer-Lock adapter and place spin cartridge in a screw-top collection tube and label it Flow-through 1 or F1. Add 200 μl of diluted serum sample to top of resin bed and centrifuge for 1.5 min at 100*g*. Cap the spin cartridge loosely or leave open during centrifugation so that the sample is able to flow. Collect the flow-through fraction in the collection tube F1. The resin bed and frits should remain moist, not dry after centrifugation.
4. Wash and collect flow-through fraction F1: Add 400 μl of buffer A to the top of the resin bed and centrifuge for 2.5 min at 100*g*. Collect the flow-through fraction into the same F1 collection tube.
5. Wash and collect additional flow-through fraction F2: Place spin cartridge into a new collection tube labeled Flow-through 2 or F2. Add 400 μl of buffer A to the top of the resin bed and centrifuge for 2.5 min at 100*g*. Collect the flow-through fraction into the F2 collection tube.
6. Prepare for elution: Remove the spin cartridge from the F2 collection tube and attach Luer-Lock adapter to the cartridge top.
7. Elute bound fraction: fill a 5-ml plastic Luer-Lock syringe (labeled B) with 2 ml of buffer B and attach to the spin cartridge via the Luer-Lock adapter. Elute bound high-abundant proteins into a new collection tube by slowly pushing buffer B through the spin cartridge (*see* **Note 2**). Save the bound fraction for analysis if desired, or discard it (*see* **Note 3**). Do not push air through the spin cartridge and do not allow the resin bed or frits to run dry.
8. Re-equilibrate: Remove buffer B syringe and attach a 5-ml syringe (labeled A) containing 4 ml of buffer A to the spin cartridge. Re-equilibrate spin cartridge by slowly pushing buffer A through the resin bed. Do not allow the resin bed or frits to run dry by leaving a small aliquot of buffer on the top of the frit. The spin cartridge is ready for the next sample. For storage, leave the resin bed wet with buffer A and leave a layer of buffer A above the top frit. Recap both ends of the spin cartridge tightly.
9. Analyze: Analyze separately or combine flow-through fractions F1 and F2 containing the low-abundant proteins.

3.3. Postdepletion Cleanup and Concentration

After cleanup, the plasma proteins were concentrated using Vivaspin 2 (2 ml, 3,000 molecular weight cut-off). The concentrated proteins were then subjected to a cleanup procedure by using ReadyPrep 2-D Cleanup Kit.

1. Transfer 1–500 μg of protein (*see* **Note 4**) in a final volume of 100 μl into a 1.5-ml microcentrifuge tube.
2. Add 300 μl of precipitating agent 1 to the protein sample (*see* **Note 5**) and mix well by vortexing. Incubate on ice for 15 min.
3. Add 300 μl of precipitating agent 2 to the mixture of protein and precipitating agent 1. Mix well by vortexing.
4. Centrifuge the tube(s) at maximum speed (>12,000*g*) for 5 min to form a tight pellet. Without disturbing the pellet, remove and discard the supernatant by using a pipette.
5. Position the tube in the centrifuge with cap hinge and protein pellet facing outward and centrifuge for 15–30 s to collect any residual liquid at the bottom of the tube. Use a pipette to carefully remove the remaining supernatant.
6. Add 40 μl of wash reagent 1 on top of the pellet. Position the tube in the centrifuge as before and centrifuge at maximum speed (>12,000*g*) for 5 min. With a pipette, remove and discard the wash.
7. Add 25 μl of ReadyPrep proteomic grade water or other ultrapure water on top of the pellet. Vortex the tube 10–20 s. Protein pellets may disperse, but they will not dissolve in the water (*see* **Note 6**).
8. Add 1 ml of wash reagent 2 (prechilled at –20°C for at least 1 h) and 5 μl of wash 2 additive. Vortex the tube for 1 min.
9. Incubate the tube at –20°C for 30 min. Vortex the tube for 30 s every 10 min during the incubation period.
10. After the incubation period, centrifuge the tube at top speed for 5 min to form a tight pellet. Remove and discard the supernatant. Centrifuge the tube briefly (15–30 s) and remove and discard any remaining wash (*see* **Note 7**). The pellet will look white at this stage. Air dry the pellet at room temperature for no more than 5 min (the pellet will look translucent once sufficiently dry) (*see* **Note 8**).
11. Resuspend each pellet by adding an appropriate volume of 2-D rehydration/sample buffer to the pellet. Vortex the tube for at least 30 s. Incubate the tube at room temperature for 3–5 min. Vortex the tube again for ≈1 min or pipette the solution up and down to resuspend completely.
12. Centrifuge the tube at maximum speed for 2–5 min at room temperature to clarify the protein sample. The supernatant can be used directly for IEF in IPG strips or in IEF gels. Store any unused or remaining protein sample in a clean tube at –80°C for later analysis.

3.4. First Dimension IEF

For the first dimension of separation (IEF), 40 μl of the depleted plasma was applied. The first round of the 2-DE was performed using the Bio-Rad PROTEAN IEF cell. This first dimension was based on separation of polypeptides according to charge of the molecules (isoelectric point) by using ampholine carriers (Invitrogen, Carlsbad, CA). The "focusing" process was carried out at

700 V for 20 h. The gel "rod" of the first dimension with the separated polypeptides will then be introduced to the second dimension of 10–20% acrylamide gradient and electric current applied for overnight SDS size separation (constant 140 V).

There are three methods to rehydrate and load the IPG strips depending on the sample type and volume of the load.

Method 1: Rehydration in the rehydration/equilibration tray (prevents cross-contamination of protein samples by using a disposable tray for rehydration). Sample is included in the rehydration solution and the sample is taken up into the IPG strip passively during rehydration (passive rehydration).

Method 2: Rehydration in the PROTEAN IEF focusing tray allows rehydration under current (active rehydration) to improve entry of high-molecular-weight proteins into the IPG strip. Sample is included in the rehydration solution in this method as well. This method also allows the possibility of having IEF commence automatically after rehydration without user intervention.

Method 3: Cup loading (a method in which the sample is applied via sample cups after rehydration, rather than being included in the rehydration solution). This method is less convenient, but it can result in better focusing, particularly when the pH range of the IPG strip is alkaline.

3.4.1. Rehydration in Rehydration/Equilibration Tray

1. Prepare sample in a suitable rehydration buffer containing urea, a nonionic or zwitterionic detergent, carrier ampholytes, and a reducing agent.
2. Pipette the indicated volume (see below) of each sample as a line along the edge of a channel in a new or clean, dry disposable rehydration/equilibration tray. The line of sample should extend along the whole length of the channel except for approximately 1 cm at each end. Care should be taken not to introduce any bubbles, which may interfere with the even distribution of sample in the IPG strip.

	7 cm	11 cm	17 cm	18 cm	24 cm
Rehydration volume (μl)	125	200	300	315	450

3. Load the protein samples into the rehydration/equilibration tray, peel the cover sheet from the IPG strip using forceps and gently place the IPG strips, gel side down, onto the sample. The "+" and the pH range marked on the IPG strip should be legible (*see* **Note 9**).
4. Overlay each of the IPG strips with 2–3 ml of mineral oil to prevent evaporation during the rehydration process (*see* **Note 10**).

5. Cover the rehydration/equilibration tray with the plastic lid provided and leave the tray sitting on a level bench overnight (11–16 h) to rehydrate the IPG strips and load the protein sample.
6. Place a clean, dry PROTEAN IEF focusing tray onto the lab bench. Using forceps, place a paper wick at each end of the channels so that the wire electrode is covered. Pipette 10 μl of deionized (18 MΩ cm) water onto each wick to wet them.
7. Remove the cover from the rehydration/equilibration tray containing the IPG strips. With a forceps, hold the strip vertically for about 7–8 s and blot the tip of the strip on a piece of filter paper to allow the mineral oil to drain. Then, transfer the IPG strip to the corresponding channel in the focusing tray (maintain the gel side down) (*see* **Note 11**).
8. Draining the oil allows removal of unabsorbed protein from the surface of the gel and reduces the incidence of horizontal streaking. Alternatively, the oil can be removed by gentle blotting. Place the IPG strips gel side up on a piece of dry filter paper. Wet a second piece of filter paper and place it gently onto the IPG strips (*see* **Note 12**).
9. Cover each IPG strip with 2–3 ml of fresh mineral oil. Check for any air bubbles trapped beneath the IPG strips, and, if necessary, remove them by lifting the IPG strip from one end and carefully placing them back in the channel. Place the lid onto the focusing tray.
10. Place the focusing tray into the PROTEAN IEF cell and close the cover.
11. Program the PROTEAN IEF cell by using the appropriate protocol with desired rehydration time, volt hours, and voltage gradient. For all IPG strip lengths, use the default cell temperature of 20°C, with a maximum current of 50 μA/IPG strip (*see* **Note 13**).
12. When the electrophoresis run has been completed, remove the IPG strips from the focusing tray and transfer them gel side up into a new clean, dry disposable rehydration/equilibration tray that matches the length of the IPG strip. Hold the IPG strips vertically with forceps and let the mineral oil drain from the IPG strip for ≈5 s before transfer. Maintain the IPG strips in the same order as in the focusing tray (*see* **Note 14**).

3.4.2. Rehydration in PROTEAN IEF Focusing Tray

1. This method is similar to the above-described method (rehydration in the equilibration tray) except for the fact that here the rehydration is done under current (active rehydration) to improve entry of high-molecular-weight proteins into the IPG strip.

3.4.3. Cup Loading

1. In this method, the sample (*see* **Note 15**) is applied via sample cups after rehydration rather than being included in the rehydration solution. This method is adopted to get better focusing, particularly when the pH range of the IPG strip is alkaline.

3.5. Second Dimension SDS-PAGE

The second dimension of the 2-DE was performed in the Anderson's ISODALT system for the simultaneous analysis of 20 samples (instruments produced in the workshop of former Basel Institute for Immunology).

3.5.1. Horizontal SDS-PAGE

1. Fill the buffer tanks of the electrophoresis unit with electrode buffer. Soak 2 sheets of filter paper (size 250 × 200 mm^2) in electrode buffer and put them on the cooling block (15°C). Soak the electrode wicks (size 250 × 100 mm^2) in electrode buffer (*see* **Note 16**).
2. During the equilibration step of the IPG gel strips open the polymerization cassette and pipette a few milliliters of kerosene on the cooling block (15°C) of the electrophoresis unit and put the SDS gel (gel side up) on it (*see* **Note 17**).
3. Place the blotted IPG gel strip(s) gel side down onto the SDS gel surface adjacent to the cathodic wick. No embedding of the IPG gel strip is necessary. If it is desired to co-electrophorese molecular weight (M_r) marker proteins, put a silicone rubber frame onto the SDS gel surface alongside the IPG gel strip and pipette in 5 µl of M_r marker proteins dissolved in SDS-buffer.
4. Put the lid on the electrophoresis unit and start SDS-PAGE with 100 V for 75 min with a limit of 20 mA. When the bromophenol blue tracking dye has completely moved out of the IPG gel strip, interrupt the run, remove the IPG gel strip and move the cathodic electrode wick forward for 4–5 mm so that it now overlaps the former sample application area. Then continue the run at 600 V with a limit of 30 mA until the tracking dye has migrated into the anodic electrode wick. Total running time is ≈6 h (separation distance 180–200 mm). The gel is then fixed in 40% alcohol and 10% acetic acid for at least 1 h and stained with either silver nitrate or Coomassie Blue (*see* **Note 18**).

Running conditions of SDS-pore gradient gels (12-15% T, 3% C, size 250 × 190 × 0.5 mm^3)

Time (min)	Voltage (V)	Current (mA)	Power (W)	Temp (°C)
75	100	20	30	15
Remove the IPG gel strip and move forward the cathodic electrode. Wick so that it overlaps the former IPG strip application area.				
5	600	30	30	15

3.5.2. Vertical SDS-PAGE (8)

First dimension IEF and the equilibration step are performed as described in **Subheading 3.4.** and **Subheading 3.5.1.**, no matter whether the second dimension is run horizontally or vertically. After equilibration, the IPG gel strip is placed on top of the vertical SDS gel.

1. Fill the electrophoresis chamber with electrode buffer and turn on cooling (15°C).
2. Support the SDS gel cassettes (gel size 200 × 250 × 10 mm^3, Laemmli buffer system (*9*)) in a vertical position to facilitate the application of the first dimension IPG strips.
3. Equilibrate the IPG gel strips as described above and immerse them in electrode buffer for a few seconds.
4. Place the IPG gel strip on top of an SDS gel and overlay it with 2 ml of hot agarose solution (75°C). Carefully press the IPG strip with a spatula onto the surface of the SDS gel to achieve complete contact (*see* **Note 19**).
5. Insert the gel cassettes in the electrophoresis apparatus and start electrophoresis. In contrast to the procedure of horizontal SDS-PAGE it is not necessary to remove the IPG gel strips from the surface of the vertical SDS gel once the proteins have migrated out of the IPG gel strip.
6. Run the SDS-PAGE gels according to the following settings: time, 18 h; voltage, 150 V; current, 150 mA; power, 50 W; and temperature, 15°C.
7. Terminate the run when the bromphenol blue tracking dye has migrated off the lower end of the gel.
8. Open the cassettes carefully with a spatula. Use a spatula to remove the agarose overlay from the polyacrylamide gel.
9. Peel the gel off the glass plate carefully, lifting it by the lower edge and place it in a tray containing fixing solution or transfer buffer, respectively. Then, continue with fixing, protein staining or blotting.
10. After termination of the second dimension run (SDS-PAGE), fixing is necessary to immobilize the separated proteins in the gel and to remove any nonprotein components that might interfere with subsequent staining. Depending on gel thickness, the gel is submersed in the fixative for one hour at least, but usually overnight, with gentle shaking. Widely used fixatives are either 20% (w/v) TCA or methanolic (or ethanolic) solutions of acetic acid (e.g., 45:45:10 [v/v/v] methanol-to-distilled water-to acetic acid). A disadvantage of the latter procedure is that low-molecular-weight polypeptides may not be adequately fixed (*see* **Note 20**).

3.6. Staining and Quantitation

3.6.1. Silver Staining Protocol Compatible with MS (*7*)

1. Remove the gel from the electrophoretic cassette.
2. Fix the gels with 50% (v/v) methanol and 5% (v/v) acetic acid for 20 min.
3. Wash with 50% (v/v) methanol for 10 min.
4. Wash with deionized water >2 h (overnight).
5. Sensitize the gels using freshly prepared 0.02% (w/v) $Na_2S_2O_3 \cdot 5H_2O$ for 1 min.
6. Wash with deionized water 2 times 1 min each time.

7. Stain the gels in 0.1% (w/v) silver nitrate 20 min at 4°C.
8. Change tray and wash with deionized water for 1 min.
9. Develop in the developer solution three times 5 min each.
10. Watch the color and change solution when the developer turns yellow.
11. Stop the reaction with 5% (v/v) acetic acid three times for 5 min every time.
12. Store the gels in 1% (v/v) acetic acid (4°C) up to several weeks.

3.6.2. Fluorescent Staining with SYPRO Ruby

1. After removing the gel from the electrophoretic cassette, fix the proteins for 30 min in 10 (v/v) ethanol and 7% (v/v) acetic acid.
2. Place the gel in a high density polypropylene box (*see* **Note 21**). Do NOT use a glass vessel!
3. Add 500 ml of SYPRO Ruby stain solution for a gel with the dimensions 250 × 200 × 10 mm^3.
4. Cover box with a tight-fitting lid and with Aluminium foil to protect the reagent from bright light.
5. Shake the box gently, preferably with a circular action, at least 3 h at room temperature. Pour off the excess stain solution and discard.
6. Place the gel on a Pyrex glass plate.
7. Image the gel using an appropriate fluorescence imager.

3.6.3. Coomassie Brilliant Blue Staining *(10)*

1. Fix proteins before Coomassie Blue staining for >30 min in 20% TCA.
2. Wash briefly (5 min) with 40% (v/v) ethanol / 10% (v/v) acetic acid.
3. Saturate the gel with the dye solution (0.1% (w/v) Serva Blue R-250 in 40% ethanol and 10% acetic acid) for 3 h.
4. Destain the gel (but not completely!) two times 20 min each with 40% ethanol and 10% acetic acid until the background is no longer stained dark blue.
5. Immerse the gel in 500 ml of distilled water containing a few milliliters of acetic acid for up to 48 h to decrease background staining and to increase sensitivity. Detection limit of this stain is better than 1 μg of protein/spot.

3.7. Scanning and Image Analysis

Proteins were visualized by silver and SYPRO Ruby staining. All gels were scanned at 100-μm resolution using the Molecular Imager FX System (Bio-Rad cat. no.170–9400). Image acquisition and analysis of silver stained gels was performed using the instrument equipped with a 480-nm laser and a 600-nm band pass emission filter. Gel images were converted into digital TIF files. Spot detection, pattern evaluation, and normalization were performed using the Progenesis 240 software (Nonlinear Dynamics). Upon spot detection, the entire pattern is inspected for those artefacts that were not removed by the software algorithms. Such spots were eliminated or alternatively merged with other spots if appropriate. At the end of the matching process, the master pattern contains all the spots occurring in each of the images. Quantitative evaluation and comparison of gel patterns is performed either directly with the image analysis data, or upon "normalization." The newly expressed spots, which appeared only in the Ts21 group, and the overexpressed spots, which were >4 times as prominent in the Ts21 group as in the controls, were selected and identified via ESI-MALDI-TOF. Student's *t*-test analysis was performed to compare the means of the relative intensity of each spot between the Ts21 and control subjects. Because in most instances the aim is to identify a set of spots that are reflecting modulation (i.e., either an overexpression or a down-regulation) of the final polypeptide products, the convenient criterion is the inspection of the 'scatter plot." Localization of spots above or below the scoring diagonal lines (delineating the choice of the up- and down-regulation) provides an objective choice for choosing the "candidate spots." Besides that, a careful visual inspection for artefacts, mismatches is a necessary precaution for correct choice. The spots chosen were then further picked for ESI-MALDI analyses.

3.8. Sample Preparation for ESI-MALDI-TOF

3.8.1. In-Gel Trypsin Digestion (11)

To identify the protein spots on the gel pieces, preparative 2-D gels were excised, cut into 1–2-mm^2 pieces, and destained at room temperature by being placed in 50 mM NH_4HCO_3 buffer, pH 8.8, containing 50% acetonitrile (ACN) for 1–2 h.

After washing with 50 ml of ACN, the gel pieces were dehydrated and dried thoroughly in a vacuum centrifuge for a few minutes. The dried gel pieces were rehydrated with 20 ml of 50 mM NH_4HCO_3, pH 8, containing 20 mg/ml trypsin and the proteins in the gel pieces were digested at 37°C overnight. The samples were then dried in a vacuum centrifuge, and 10 ml of double-distilled water was added before MALDI-MS analysis.

3.8.2. Mass Spectrometry

After desalting with Zip-Tip C18 (Millipore Corporation), 1 µl of supernatant was mixed with 1 µl of matrix solution (α-cyano-4-hydroxycinnamic acid) and spotted onto the MALDI target. The dried spots were analyzed in a Voyager DE-STR mass spectrometer (Applied Biosystems, Foster City, CA) and the resultant MALDI-mass spectra were internally calibrated using the Data Explorer software (Applied Biosystems) with trypsin autolysis products. The proteins were finally identified by peptide mass fingerprinting using the software.

1. Attach the Zip-Tip column to a 20-μl micropipette. Adjust the volume setting to 7 μl. Carefully, withdraw acetonitrile (tube A) through the Zip-Tip, and then, while the tip of the Zip-Tip is still under acetonitrile, pipette out the acetonitrile carefully, taking precaution to prevent introducing air bubbles into the Zip-Tip. Repeat this step at least 7–8 times or until no bubbles arise during the pipetting out step. Finally, pipette out the acetonitrile slowly, and while the plunger is still down, immerse the tip of the Zip-Tip into the water (tube B).
2. Slowly withdraw 7 μl of water through the Zip-Tip, and then pipete it out carefully, taking care not to introduce any air into the Zip-Tip (*see* **Note 22**).
3. Pipette out the water, and while the plunger is still down, move the tip of the Zip-Tip into the sample solution (tube C). Carefully, fill the Zip-Tip with the sample solution and slowly push out into the tube (tube C). Repeat at least 10 times to ensure that most of the peptides have been retained on the Zip-Tip.
4. As in **step 2** and **3**, wash the Zip-Tip with 0.5% (v/v) formic acid solution (tube D) at least 10 times to perform the desalting and washing of the peptides.
5. After pipetting out the wash solution and with the plunger is still down, transfer the Zip-Tip to the tube containing the extraction solution (tube E) and slowly fill with the extraction solution. Wait for 20 s to ensure a complete extraction.
6. Pipette out the extracting solution (extract 1) into the empty tube G, and while the plunger is still down, move to the tube F containing the acetonitrile. Slowly withdraw acetonitrile through the Zip-Tip.
7. After waiting for 10 s, pipette out the solution (extract 2) into the tube G containing the extract 1.
8. Store the combined extracts (tube G) in a freezer until required for analysis.

3.9. Bioinformatics (Databases)

Proteins were identified by peptide mass fingerprinting using the software Mascot (http://www.matrixscience.com) and MS-Fit (http://prospector.edu) against National Center for Biotechnology Center and SWISS-PROT databases with the following parameters: human species, one missed cleavage site and a mass tolerance setting of 50 ppm. The criteria used to accept identifications included the extent of sequence coverage, the number of peptides matched (minimum of four), the probability of score (minimum of 70 for the Mowse score) and the mass accuracy.

4. Notes

1. Place the P100 tube in wet ice before centrifuging, centrifuge within 30 min of collection. Freeze aliquots at –80°C until used to avoid freeze-thaw cycles.
2. Do not apply too much pressure on the syringe, press enough for the buffer to form a drop and let the drop trickle through.
3. Do not discard the depletion fraction; instead store them for studying the efficiency of the depletions.

4. Sample quantities >500 μg of protein may reduce the efficiency of the cleanup leading to poor-quality IEF).
5. When adding solution, do not touch protein sample with the pipette tip. The protein may precipitate on the tip causing sample loss.
6. A precipitate may form along the tube wall. In these cases, vortex, pipette, or both the wash solution over the pellet several times to ensure entire pellet is thoroughly washed.
7. Protein pellets will not dissolve in wash reagent 2. If wash reagent 2 is not completely chilled, quantitative recovery may be affected.
8. Do not over dry pellets. Over-dried pellets will be difficult to resuspend.
9. Take care not to get the sample onto the plastic backing of the IPG strips, because this portion of the sample will not be absorbed by the gel material.
10. Add the mineral oil slowly by carefully dripping the oil onto the plastic backing of the IPG strip while moving the pipet along the length of the IPG strip.
11. Remember to observe the correct polarity during the transfer. The "+" marked on the IPG strip should be positioned at the end of the tray marked "+".
12. Carefully pat the paper above the IPG strips to remove the oil. Finally, gently peel back the top layer of filter paper, starting from one end.
13. Refer to the PROTEAN IEF Cell Instruction Manual for details on entering run parameters. Press START to initiate the electrophoresis run.
14. If you are not proceeding directly to the equilibration step, cover the tray containing the IPG strips, wrap it in plastic wrap, and place in a –70°C freezer for storage.
15. Do not apply less than 20 μl of sample solution into the cups. Apply sample near cathode or anode to provide better results. Sample should not exceed a concentration of 10 mg protein/ml to avoid precipitation during sample application. Also, sample solution should be devoid of high salt concentrations. Either desalt or dilute with lysis buffer and apply a larger volume instead. Apply low voltage for slow sample entry.
16. Place them at the edges of the buffer-soaked filter papers and perform a prerun (600 V; 30 mA) for 3 h to remove impurities from the electrode wicks. Then, remove the filter papers and discard them, whereas the purified electrode wicks remain in the electrode buffer tanks and they are used repeatedly.
17. Apply the electrode wicks on the surface of the SDS gel so that they overlap the cathodic and anodic edges of the gel by $\approx$10 mm.
18. Alternatively, wicks can be removed from the plastic backing with the help of a film remover (GE Healthcare, Chalfont St. Giles, UK) and used for blotting.
19. Allow the agarose to solidify for at least 5 min. Repeat this procedure for the remaining IPG strips.
20. When using fluorography, fixing may be carried out in 30% isopropyl alcohol and 10% acetic acid, because methanol can interfere with detection. Several researchers also recommend aqueous solutions of glutaraldehyde for covalently cross-linking proteins to the gel matrix (e.g., for diamine silver staining).

21. Up to three gels may be stained simultaneously in one polypropylene box. Use the filter sets that match the excitation and emission wavelength for SYPRO Ruby. Images also can be obtained by using a simple UV transilluminator. However, good sensitivity and resolution will only be obtained with dedicated imagers. Best is to destain in steps of 10 min with 10% ethanol and 7% acetic acid until you get an optimal image.
22. Repeat this step carefully many times (at least 10 times) to ensure that all acetonitrile has been washed away.

References

1. Korenberg, J. R., Chen, X. N., Schipper, R., et al. (1994) Down syndrome phenotypes: the consequences of chromosomal imbalance. *Proc. Natl. Acad. Sci. USA.* **91,** 4997–5001.
2. Korenberg, J. R., Kawashima, H., Pulst, S. M., Allen, L., Magenis, E., and Epstein, C. J. (1990) Down syndrome: toward a molecular definition of the phenotype. *Am. J. Med. Genet.* **7 (Suppl.),** 91–97.
3. Holtzman, D. M. and Epstein, C. J. (1992) The molecular genetics of Down syndrome. *Mol. Genet. Med.* **2,** 105–120.
4. Antonarakis, S. E., Lyle, R., Dermitzakis, E. T., Reymond, A., and Deutsch, S. (2004) Chromosome 21 and Down syndrome: from genomics to pathophysiology. *Nat. Rev. Genet.* **5,** 725–738.
5. Anderson, N. L., Polanski, M., Pieper, R., et al. (2004) The Human Plasma Proteome. A non-redundant list developed by combination of four separate sources. *Mol. Cell Proteomics* **34,** 311–326.
6. Kuzdzal, S., Lopez, M., Mikulskis, A., et al. (2005) Biomarker discovery and analysis platform: application to Alzheimer's disease. *Biotechniques* **39,** 606–607.
7. Shevchenko, A., Wilm, M., Vorm, O. and Mann, M. (1996) Mass spectrometric sequencing of proteins from silver-stained polyacrylamide gels. *Anal. Chem.* **68,** 850–858.
8. Görg, A., Boguth, G., Obermaier, C., Posch, A. and Weiss, W. (1995) Two-dimensional polyacrylamide gel electrophoresis with immobilized pH gradients in the first dimension (IPG-Dalt): the state of art and the controversy of vertical versus horizontal systems. *Electrophoresis* **16,** 1079–1086.
9. Laemmli, U. K. (1970) Cleavage of structural proteins during the assembly of the head of bacteriophage T4. *Nature* **227,** 680–685.
10. Görg, A. G., Boguth, G., Drews, O., Köpf, A., Lück, C., Reil, G., and Weiss W. (2003) Two-dimensional electrophoresis with immobilized pH gradients for proteome analysis. A Laboratory Manual (http://www.weihenstephan.de/blm/deg).
11. Dupont, A., Tokarski, C., Dekeyzer, O., et al. (2004) Two-dimensional maps and databases of the human macrophage proteome and secretome. *Proteomics* **4,** 1761–1778.

Index

Printed in the United States of America